Bioactivities of Nature Products

Bioactivities of Nature Products

Guest Editors

**Luis Ricardo Hernández
Eugenio Sánchez-Arreola
Edgar R. López-Mena**

 Basel • Beijing • Wuhan • Barcelona • Belgrade • Novi Sad • Cluj • Manchester

Guest Editors

Luis Ricardo Hernández
Chemical and Biological
Sciences
Universidad de las Américas
Puebla
San Andrés Cholula
Mexico

Eugenio Sánchez-Arreola
Chemical and Biological
Sciences
Universidad de las Américas
Puebla
San Andrés Cholula
Mexico

Edgar R. López-Mena
Department of Sciences
Instituto Tecnológico de
Estudios Superiores de
Monterrey
Guadalajara
Mexico

Editorial Office
MDPI AG
Grosspeteranlage 5
4052 Basel, Switzerland

This is a reprint of the Special Issue, published open access by the journal *Plants* (ISSN 2223-7747), freely accessible at: www.mdpi.com/journal/plants/special_issues/E0B39S74K6.

For citation purposes, cite each article independently as indicated on the article page online and using the guide below:

Lastname, A.A.; Lastname, B.B. Article Title. *Journal Name* **Year**, *Volume Number*, Page Range.

ISBN 978-3-7258-3728-1 (Hbk)
ISBN 978-3-7258-3727-4 (PDF)
https://doi.org/10.3390/books978-3-7258-3727-4

© 2025 by the authors. Articles in this book are Open Access and distributed under the Creative Commons Attribution (CC BY) license. The book as a whole is distributed by MDPI under the terms and conditions of the Creative Commons Attribution-NonCommercial-NoDerivs (CC BY-NC-ND) license (https://creativecommons.org/licenses/by-nc-nd/4.0/).

Contents

About the Editors vii

Preface ix

Sofía Isabel Cuevas-Cianca, Cristian Romero-Castillo, José Luis Gálvez-Romero, Eugenio Sánchez-Arreola, Zaida Nelly Juárez and Luis Ricardo Hernández
Latin American Plants against Microorganisms
Reprinted from: *Plants* **2023**, *12*, 3997, https://doi.org/10.3390/plants12233997 1

Saima Jan, Sana Iram, Ommer Bashir, Sheezma Nazir Shah, Mohammad Azhar Kamal and Safikur Rahman et al.
Unleashed Treasures of Solanaceae: Mechanistic Insights into Phytochemicals with Therapeutic Potential for Combatting Human Diseases
Reprinted from: *Plants* **2024**, *13*, 724, https://doi.org/10.3390/plants13050724 32

Abimael López-Pérez, Luicita Lagunez-Rivera, Rodolfo Solano, Aracely Evangelina Chávez-Piña, Gabriela Soledad Barragán-Zarate and Manuel Jiménez-Estrada
Phytochemical Compounds from *Laelia furfuracea* and Their Antioxidant and Anti-Inflammatory Activities
Reprinted from: *Plants* **2025**, *14*, 588, https://doi.org/10.3390/plants14040588 65

Ting-Kang Lin, Jyh-Yih Leu, Yi-Lin Lai, Yu-Chi Chang, Ying-Chien Chung and Hsia-Wei Liu
Application of Microwave-Assisted Water Extraction (MAWE) to Fully Realize Various Physiological Activities of *Melaleuca quinquenervia* Leaf Extract
Reprinted from: *Plants* **2024**, *13*, 3362, https://doi.org/10.3390/plants13233362 79

Manuel Martínez-Lobos, Valentina Silva, Joan Villena, Carlos Jara-Gutiérrez, Waleska E. Vera Quezada and Iván Montenegro et al.
Phytoconstituents, Antioxidant Activity and Cytotoxicity of *Puya chilensis* Mol. Extracts in Colon Cell Lines
Reprinted from: *Plants* **2024**, *13*, 2989, https://doi.org/10.3390/plants13212989 98

Minerva Edith Beltrán-Martínez, Melvin Roberto Tapia-Rodríguez, Jesús Fernando Ayala-Zavala, Agustín Gómez-Álvarez, Ramon Enrique Robles-Zepeda and Heriberto Torres-Moreno et al.
Antimicrobial and Antibiofilm Potential of *Flourensia retinophylla* against *Staphylococcus aureus*
Reprinted from: *Plants* **2024**, *13*, 1671, https://doi.org/10.3390/plants13121671 109

Jorge L. Mejía-Méndez, Ana C. Lorenzo-Leal, Horacio Bach, Edgar R. López-Mena, Diego E. Navarro-López and Luis R. Hernández et al.
Antimicrobial, Cytotoxic, and Anti-Inflammatory Activities of *Tigridia vanhouttei* Extracts
Reprinted from: *Plants* **2023**, *12*, 3136, https://doi.org/10.3390/plants12173136 122

Jorge L. Mejía-Méndez, Horacio Bach, Ana C. Lorenzo-Leal, Diego E. Navarro-López, Edgar R. López-Mena and Luis Ricardo Hernández et al.
Biological Activities and Chemical Profiles of *Kalanchoe fedtschenkoi* Extracts
Reprinted from: *Plants* **2023**, *12*, 1943, https://doi.org/10.3390/plants12101943 135

Nataša Simin, Nemanja Živanović, Biljana Božanić Tanjga, Marija Lesjak, Tijana Narandžić and Mirjana Ljubojević
New Garden Rose (*Rosa* × *hybrida*) Genotypes with Intensely Colored Flowers as Rich Sources of Bioactive Compounds
Reprinted from: *Plants* **2024**, *13*, 424, https://doi.org/10.3390/plants13030424 150

Alejandro Madrid, Evelyn Muñoz, Valentina Silva, Manuel Martínez, Susana Flores and Francisca Valdés et al.
Structure–Activity Relationship of Natural Dihydrochalcones and Chalcones, and Their Respective Oxyalkylated Derivatives as Anti-*Saprolegnia* Agents
Reprinted from: *Plants* **2024**, *13*, 1976, https://doi.org/10.3390/plants13141976 . . . . **169**

Débora Machado de Lima, Anna Lívia Oliveira Santos, Matheus Reis Santos de Melo, Denise Crispim Tavares, Carlos Henrique Gomes Martins and Raquel Maria Ferreira Sousa
Cosmetic Preservative Potential and Chemical Composition of *Lafoensia replicata* Pohl. Leaves
Reprinted from: *Plants* **2024**, *13*, 2011, https://doi.org/10.3390/plants13152011 . . . . **183**

Gabriela Soledad Barragán-Zarate, Beatriz Adriana Pérez-López, Manuel Cuéllar-Martínez, Rodolfo Solano and Luicita Lagunez-Rivera
Chemical Variation of Leaves and Pseudobulbs in *Prosthechea karwinskii* (Orchidaceae) in Oaxaca, Mexico
Reprinted from: *Plants* **2025**, *14*, 608, https://doi.org/10.3390/plants14040608 . . . . **197**

About the Editors

Luis Ricardo Hernández

Dr. Prof. Luis Ricardo Hernández is a full-time professor at the Department of Chemical and Biological Sciences at Universidad de las Américas Puebla (UDLAP). He studied for a Bachelor in Chemistry at the Faculty of Biochemistry, Chemistry, and Pharmacy at Universidad Nacional de Tucumán (UNT), Argentine. At this university, he obtained a doctorate in chemistry. From 1997 to 1999, he made a postdoctoral stay at the "Centro de Investigación y Estudios Avanzados" (CINVESTAV) in Mexico City. Since 1990, he has been devoted to university teaching.

Since 1987, he has worked in the isolation and structural elucidation of secondary metabolites of plants, giving this orientation to his scientific career. He participated in several research projects, sometimes as a collaborator and others as director. In 1999, he was honored as a Researcher by the National Research Council of Argentina, and in 2003, he was recognized as a National Researcher in Mexico, currently as level 2 researcher.

One of the research lines of Dr. Hernández refers to phytoplaguicides, faced with the isolation and structural identification of phytochemicals with potential pesticidal activity, working together with researchers from other universities and farmers in the state of Puebla, Mexico. Moreover, he also researches natural products from plants with antimicrobial, anti-inflammatory, and cytotoxic activities and plant secondary metabolites at trace levels for chemotaxonomic purposes.

Dr. Hernández has about 60 publications in international refereed journals and some book chapters, publications in the field of teaching, and book editions. These publications have reported, among other things, the isolation and spectroscopic characterization of about a hundred new natural compounds, three of which have a new carbon skeleton, which was named "sarcopetalane" because it was isolated from the plant *Croton sarcopetalus*.

Eugenio Sánchez-Arreola

Dr. Eugenio Sánchez-Arreola received his B.Sc. degree in Chemical Pharmaceutical Biology from the Universidad Michoacana de San Nicolás de Hidalgo, Michoacán, Mexico, and his Master's and Ph.D. degrees in chemistry from CINVESTAV, Mexico. As a part-time professor, he joined the School of Chemical Engineering at the National Polytechnical Institute in 1989 and later became a full-time professor in the Department of Chemical and Biological Sciences at Universidad de las Américas Puebla (UDLAP) in 1999.

His research interests include the isolation, characterization, and biological activity of natural products. He has published as an author and co-author over 60 papers in international journals, including *Phytochemistry, Journal of Natural Products, Pharmaceutical Biology, Journal of Ethnopharmacology, Plants, Biomolecules*, and others. He is passionate about spectroscopic and spectrometric techniques to characterize new compounds. He is a level II Sistema Nacional de Investigadores member and won the UDLAP Commitment to Education Award 2013.

Edgar R. López-Mena

Dr. Edgar R. López-Mena received his bachelor's degree in applied mathematics from the Universidad Autonoma de Guadalajara. He obtained his master's and doctoral degrees in physics from the University of Guadalajara. Moreover, he pursued a postdoctoral research position at Nanyang Technological University.

Currently, he is a full-time professor at Instituto Tecnologico y de Estudios Superiores de Monterrey (ITESM) Campus Guadalajara, where his research focuses on developing new rare-earth-based nanostructures for antimicrobial and antioxidant applications, cancer therapy, drug delivery, and solar cells. Additionally, he utilizes machine learning models to predict the biological performance of synthesized nanomaterials based on their physical, chemical, and optical properties. The use of metal-based and polymeric nanostructures to enhance the therapeutic efficacy of plant extracts is also a significant area of interest in his research fields. Considering his extensive experience in materials science, he serves as the national co-leader of the nanotechnology and semiconductor initiative at ITESM. He is a level II Sistema Nacional de Investigadores member due to his significant research, academic, and social activities.

Preface

Bioactive natural products can be found among animals, terrestrial or marine microorganisms, herbs, and plants. Throughout history, bioactive natural compounds have represented an attractive alternative to eradicate or mitigate diseases from distinct etiology. For instance, they can treat infections caused by pathogenic microorganisms, carcinogenic processes, cardiovascular diseases, or nervous system disorders. Conventionally, bioactive molecules can be extracted from natural sources using standard laboratory techniques, separated by chromatography, and identified with spectroscopy techniques. This Special Issue consists of two reviews and ten research articles; the reviews explored current knowledge on the antimicrobial activity of Latin American plants and the constituents of different *Solanum* members and their pharmacological properties, with some emphasis on their mechanism of action. On the other hand, the research articles consisted of documenting the phytochemistry and the biological activities of plant species such as *Laelia furfuracea*, *Melaleuca quinquenervia*, *Puya chilensis*, *Fluorensia retinophylla*, *Tigridia vanhouttei*, *Kalanchoe fedtschenkoi*, and a new garden rose genotype. Also, an article explores the structure–activity relationship of natural and synthetic derivatives of chalcones and dihydrochalcones against *Saprolegnia*. Other works report the potential of *Lafoensia replicata* as a potential cosmetic preservative and the chemical variation of leaves and pseudobulbs of the Mexican endemic orchid *Prosthechea karwinski*. Due to all the above, the *Bioactivities of Natural Products* Special Issue is a fascinating book that updates or reports knowledge on plants' chemistry and biological activities for the first time.

Luis Ricardo Hernández, Eugenio Sánchez-Arreola, and Edgar R. López-Mena
Guest Editors

Review

Latin American Plants against Microorganisms

Sofía Isabel Cuevas-Cianca [1], Cristian Romero-Castillo [2,3], José Luis Gálvez-Romero [4], Eugenio Sánchez-Arreola [1], Zaida Nelly Juárez [3,*] and Luis Ricardo Hernández [1,*]

1. Department of Chemical Biological Sciences, Universidad de las Américas Puebla, Ex Hacienda Sta. Catarina Mártir S/N, San Andrés Cholula 72810, Mexico; sofia.cuevasca@udlap.mx (S.I.C.-C.); eugenio.sanchez@udlap.mx (E.S.-A.)
2. Biotechnology Faculty, Deanship of Biological Sciences, Universidad Popular Autónoma del Estado de Puebla, 21 Sur 1103 Barrio Santiago, Puebla 72410, Mexico; cristian.romero@upaep.edu.mx
3. Chemistry Area, Deanship of Biological Sciences, Universidad Popular Autónoma del Estado de Puebla, 21 Sur 1103 Barrio Santiago, Puebla 72410, Mexico
4. Department of Research ISSSTE Puebla Hospital Regional, Boulevard 14 Sur 4336, Colonia Jardines de San Manuel, Puebla 72570, Mexico; joseluis.galvez@upaep.mx
* Correspondence: zaidanelly.juarez@upaep.mx (Z.N.J.); luisr.hernandez@udlap.mx (L.R.H.)

Citation: Cuevas-Cianca, S.I.; Romero-Castillo, C.; Gálvez-Romero, J.L.; Sánchez-Arreola, E.; Juárez, Z.N.; Hernández, L.R. Latin American Plants against Microorganisms. *Plants* **2023**, *12*, 3997. https://doi.org/10.3390/plants12233997

Academic Editors: Hazem Salaheldin Elshafie and Alison Ung

Received: 5 October 2023
Revised: 14 November 2023
Accepted: 21 November 2023
Published: 28 November 2023

Copyright: © 2023 by the authors. Licensee MDPI, Basel, Switzerland. This article is an open access article distributed under the terms and conditions of the Creative Commons Attribution (CC BY) license (https://creativecommons.org/licenses/by/4.0/).

Abstract: The constant emergence of severe health threats, such as antibacterial resistance or highly transmissible viruses, necessitates the investigation of novel therapeutic approaches for discovering and developing new antimicrobials, which will be critical in combating resistance and ensuring available options. Due to the richness and structural variety of natural compounds, techniques centered on obtaining novel active principles from natural sources have yielded promising results. This review describes natural products and extracts from Latin America with antimicrobial activity against multidrug-resistant strains, as well as classes and subclasses of plant secondary metabolites with antimicrobial activity and the structures of promising compounds for combating drug-resistant pathogenic microbes. The main mechanisms of action of the plant antimicrobial compounds found in medicinal plants are discussed, and extracts of plants with activity against pathogenic fungi and antiviral properties and their possible mechanisms of action are also summarized. For example, the secondary metabolites obtained from *Isatis indigotica* that show activity against SARS-CoV are aloe-emodin, β-sitosterol, hesperetin, indigo, and sinigrin. The structures of the plant antimicrobial compounds found in medicinal plants from Latin America are discussed. Most relevant studies, reviewed in the present work, have focused on evaluating different types of extracts with several classes and subclasses of secondary metabolites with antimicrobial activity. More studies on structure–activity relationships are needed.

Keywords: Latin American plants; microorganisms; bacteria; fungi; virus

1. Introduction

Infectious diseases are a significant source of public health issues. Despite breakthroughs in creating and manufacturing antivirals and antibiotics, bacteria, viruses, and other microorganisms continue to kill millions of people each year.

Antimicrobial resistance is a severe and developing clinical issue that has reduced the therapeutic effectiveness of conventional antibiotics and narrowed the treatment choices for bacterial infections. Antibiotic-resistant bacteria are generally difficult to treat due to reduced membrane penetration, efflux pump overexpression, target site shifting, inactive subpopulations, biofilm growth, and enzymatic destruction. Resistant bacteria are strains resistant to several medicines, resulting in increased infections [1].

Many bacteria may infect and live in their hosts for extended periods. This might be related to host immunosuppression, pathogen immune evasion, and/or inadequate drug clearance. Bacteria that are resistant or tolerant to antibiotics can survive treatment. Persistent bacteria are a transiently antibiotic-tolerant subset of bacterial cells that grow slowly or cease developing but can resume proliferation after exposure to fatal stress. Persistent cell production creates phenotypic variation within a bacterial population, significantly enhancing the odds of effectively responding to environmental change. The existence of resistant cells can lead to the emergence and recurrence of chronic bacterial infections and an increased risk of antibiotic resistance [2].

Emerging viral infections, on the other hand, continue to be a severe concern for worldwide public health. In 1997, it was revealed that a highly virulent avian influenza A (H5N1) virus may be transferred directly from poultry to people, in contrast to previously known human-to-human and livestock-to-human modes of transmission, raising severe fears about a probable influenza pandemic. Several additional avian influenza A virus subtypes (H7N9, H9N2, and H7N3) have also been linked to human sickness, increasing concerns that all influenza A virus subtypes circulating in domestic poultry and cattle in the wild might transmit to people and cause pandemics.

The most recent viral pneumonia epidemic, which began in mid-December 2019 (COVID-19) in Wuhan, China, and has spread swiftly throughout the world, is a stark reminder of our vulnerability to new viral illnesses. Tens of thousands of people are currently infected with SARS-CoV-2 [3].

In the case of fungus, it is believed that roughly 5 million species are extensively ubiquitous in the environment, of which approximately 300 can cause infections in people. However, only 20–25 are commonly seen in the clinic and are the cause of sick patients. Patients with HIV, organ transplant recipients, or those undergoing chemotherapy are examples of such people. The most frequent fungal diseases are *Candida* spp., *Cryptococcus* spp., *Aspergillus* spp., and *Pneumocystis* spp., which cause around 2 million illnesses and 1 million deaths yearly [4].

Because of the above, it is critical to enhance ways of treating infection-related disorders, preventing their spread, and filling the medicine shortage in order to alleviate this public health crisis. In an era of falling antimicrobial efficacy and the fast growth of antibacterial resistance, it is critical to develop novel therapies and tactics based on discovering new active components.

The exploration of active chemicals of natural origin is such potential methodology. Natural goods have served as a source of and the inspiration for many of the pharmaceuticals available today. Although numbers vary depending on the definition of what is deemed a medicine produced from a natural substance, it is reasonable to conclude that, today, natural products are the source of 25% to 50% of the pharmaceuticals on the market. The proportion is much more significant in the case of anti-cancer and anti-infective agents, with over two-thirds of such drugs originating from natural sources. Several recent reviews emphasize the importance of natural products in drug discovery. Many medicines in clinical use are derived from natural products that originated from microbial species, particularly in anti-infectives. However, drugs derived from plants have also made significant contributions. Humanity would undoubtedly be immeasurably poorer without plant-derived natural medicines such as morphine, vinblastine, vincristine, and quinine [5].

We provide a critical review of current research on natural product antibacterial activity and the discovery and classification of secondary metabolites of plants with antimicrobial activity, each with a distinct mechanism of action. The mechanisms of action of natural antifungal agents are also discussed, as are the potential antiviral mechanisms of biocompounds, which include viral replication inhibition through polymerases, proteases, integrases, fusion molecules, and cell membrane adhesion.

The Latin American plants presented in this review were selected from papers published in the last 20 years using databases such as SciFinder®, ScienceDirect®, Scopus®, PubMed®, PLOS, NATURE, and Google Scholar®. For the article search, the keywords "antimicrobial resistance", "antibiotic resistance intrinsic", "antibiotic resistance adaptive", "antibiotic resistance acquired", "antibiotic resistance mechanisms", "antimicrobial activity of medicinal plants multidrug resistant bacteria", "plant extract antimicrobial activity", "plant extract multidrug resistant strains", "plant extract antibiotic resistance", "pathogenic fungi AND bioactive compounds", "plant extracts AND pathogenic fungi", "secondary metabolites AND fungal infections", "drug resistance AND fungi", "secondary metabolites against fungal infections" and "pathogenic fungi AND drug resistance" were used. The search in each database returned the following results: SciFinder® (112 articles), ScienceDirect® (556 articles), Scopus® (1157 articles), PubMed® (2365 articles), PLOS (552 articles), NATURE (409 articles), and Google Scholar® (6354 articles). After a preliminary filter to collect only Latin American plants, 3827 articles were collected; of these, articles discussing non-specific antimicrobial (antibiotic, antifungal, and antiviral) activity were discarded. Only original papers and those published from 2003 to 2023 were considered for data collection.

2. Plant Antimicrobials

2.1. Antimicrobial Resistance

As COVID-19 rages, the antimicrobial resistance (AMR) epidemic continues in the background. AMR causes recurrent microbe (viruses, bacteria, and fungi) infections that lengthen hospital stays and result in preventable deaths. It is estimated that 4.95 million people died due to AMR in 2019 and that by 2050, there will be 10 million annual deaths due to antimicrobial resistance. Two factors primarily cause antimicrobial resistance. The first is the overuse of antimicrobials, which exposes microbes to them regularly, increasing their chances of developing resistance. The second issue is that few new antimicrobial drugs are being developed to replace ineffective ones due to rising drug resistance [6,7].

Compared to non-resistant forms, resistant bacteria are two times more likely to develop into a serious health problem and are three times more likely to lead to death [8,9]. Resistance to first-line antibiotics, such as fluoroquinolones and lactam antibiotics, is responsible for more than two-thirds of AMR-related deaths (carbapenems, cephalosporins, and penicillins). People with low incomes are disproportionately affected by AMR because they have limited access to more expensive second-line antibiotics that may be effective when first-line drugs fail. Physicians should avoid inappropriate antibiotic therapy when, for example, the illness has a viral origin [6,10,11].

There are different mechanisms of resistance to antibiotics (Figure 1). Bacteria produce enzymes that can destroy or alter the structure of the drug, causing the drug to lose its activity during enzymatic inactivation. Drug-inactivating enzymes are classified into three types: hydrolase (primarily lactamase), passivating enzymes (aminoglycoside-inactivating enzyme, chloramphenicol acetyltransferase, and erythromycin esterase), and modifying enzymes (aminoglycoside-modifying enzyme). Similarly, changing the target to which the drug is directed ensures that the antibiotic binds appropriately to the bacteria. This mechanism is primarily seen in Gram-positive bacteria with drug resistance and polymyxin resistance. Changes in outer membrane permeability that result in channel alteration or decreased expression make the bacteria less sensitive. In the drug efflux pump, when the drug is removed from the bacterial cytoplasm, the concentration is much lower than is required for it to exhibit activity, resulting in drug resistance. This process requires energy and works with various antibiotics [7,12–14].

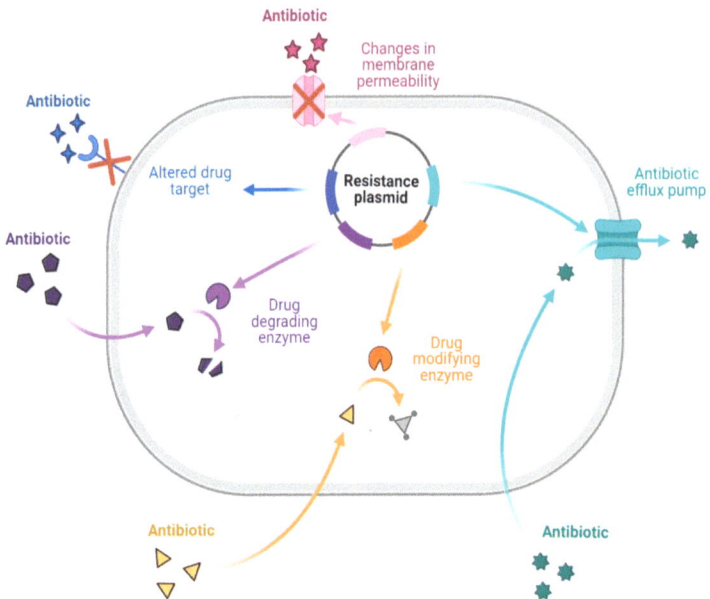

Figure 1. Antibiotic resistance mechanisms [7,12–14]. Created with BioRender.com.

2.2. Natural Products and Plant Extracts with Antimicrobial Activity against MDR Strains

Multidrug resistance (MDR) is a major cause of human suffering because it undermines doctor–patient trust, resulting in massive economic losses. In this world of microbe–man cohabitation, the survival of the human species will be compromised in the absence of health-giving microbes, and there will be no way to avoid the emergence of MDR superbugs. Throughout history, the isolation and identification of biologically active compounds and molecules from nature have resulted in the discovery of new therapeutics, advancing the health and pharmaceutical industries. Phytochemicals are used in the research and development of the pharmaceutical industry as a source of new molecules, leading to the development of novel drugs [15,16].

As shown in Table 1, several classes and subclasses of secondary metabolites (Figure 2) have been isolated from plants with antimicrobial activity, each with a different mechanism of action. This table shows that, depending on the compound class, they share the same kind of mechanism of action.

Regarding essential oils, the essential oil of rosemary (*Rosmarinus officinalis*) was found to have antibacterial activity against three types of MDR acne-causing bacteria: *Staphylococcus aureus*, *Staphylococcus epidermidis*, and *Cutibacterium acnes* [17]. Similarly, volatile oils extracted from cinnamon (*Cinnamomum verum*) and tree basil (*Ocimum gratissimum*) had potent bactericidal activity against MDR *A. baumannii* bacteria [18].

Terminalia bellirica fruits were studied, and it was discovered that the aqueous and methanol extracts had antibacterial activity against all strains of MRSA (Methicillin-resistant *Staphylococcus aureus*), MDR *Acinetobacter* spp., and MDR *P. aeruginosa* [19].

The aqueous, hexane, and ethanol extracts of *Punica granatum* peel demonstrated antibacterial activity against MDR pathogens such as *P. aeruginosa* and *A. baumannii*. Valoneic acid dilactone (aqueous fractions), Hexoside (ethanol fractions), and Coumaric acid (hexane fractions) were discovered to be bioactive compounds [18]. Ethanolic extracts of *Azadirachta indica*, *Allium sativum*, and *Syzygium cumini* were found to have anti-MDR-*Candida* spp. activity. According to a phytochemical analysis of ethanolic plant extracts, all the plants studied contained alkaloids, flavonoids, glycosides, phenols, tannins, and saponins [20].

ESKAPE (*Enterococcus faecium*, *Staphylococcus aureus*, *Klebsiella pneumoniae*, *Acinetobacter baumannii*, *Pseudomonas aeruginosa*, and *Enterobacter species*) MDR pathogens were tested using various extracts. Three ethanolic extracts from *Adiantum capillus-veneris*, *Artemisia absinthium*, and *Martynia annua* were found to inhibit the growth of MDR strains of ESKAPE pathogens [21].

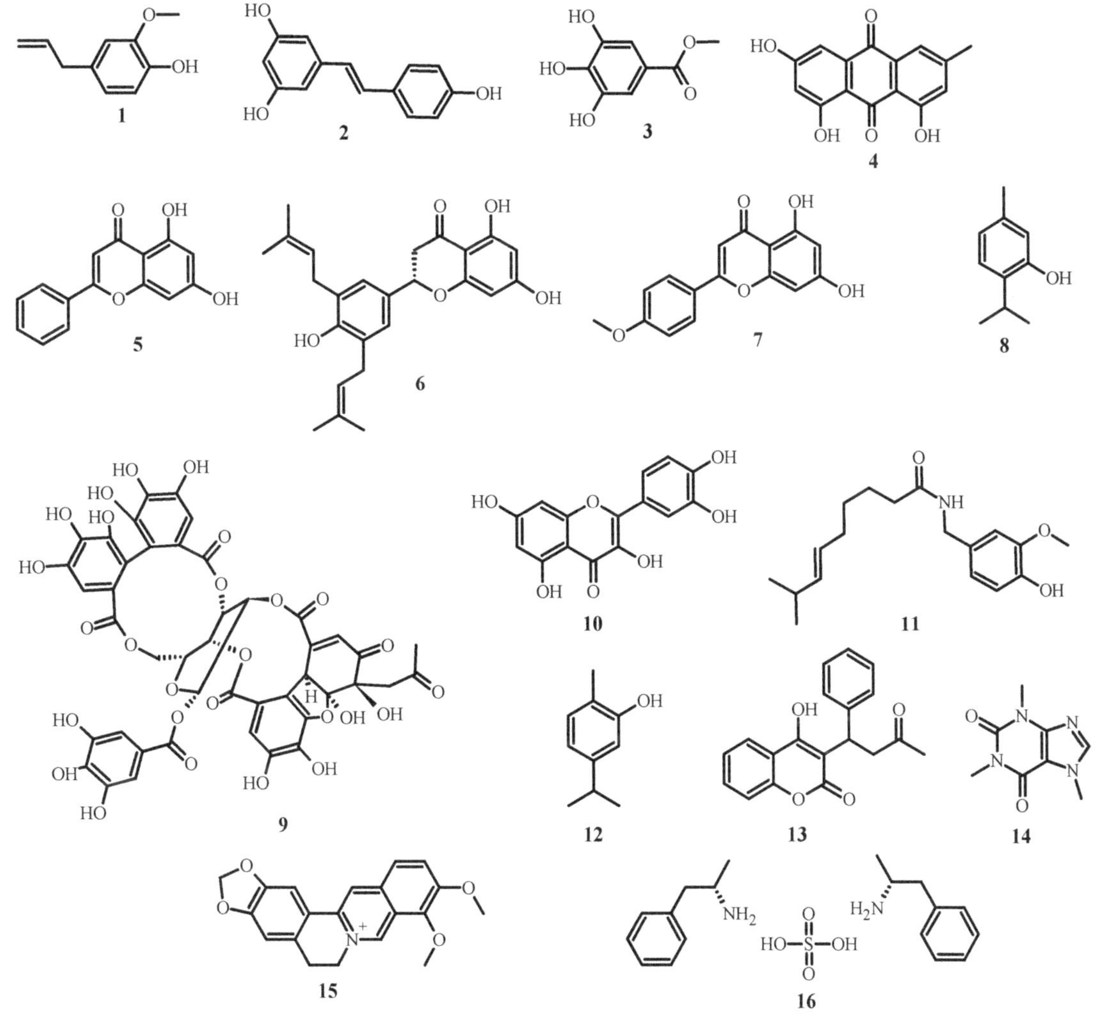

Figure 2. Structures of plant antimicrobial compounds found in medicinal plants from Latin America, from Table 1.

Aside from the plant extracts mentioned above, various plant compounds (Figure 3) with anti-MDR bacteria activity have already been identified. Table 2 lists these compounds, as well as the biological effects they have on specific strains.

Because of the severe problem of MDR properties in microbes, the discovery of alternative drugs from natural products should be one of the primary goals of current research. Understanding the nature of pathogenic microbes, recognizing biofilm formation and architectural scheme, and employing cross-disciplinary techniques are thus critical for discovering new potent and novel drugs.

Table 1. Antimicrobial mechanisms of plant compounds present in Latin American medicinal plants.

Class	Subclass	Examples	Source of the Compound	Mechanism	References
Phenolics	Simple phenols	Eugenol (1)	*Syzygium aromaticum*	Membrane disruption.	[22,23]
		Resveratrol (2)	*Vitis vinifera*	Binds reversibly to ATP synthase.	[22,24]
	Phenolic acids	Methyl gallate (3)	*Euphorbia hyssopifolia*	Inhibits DNA gyrase or ATPase.	[22,25]
	Quinones	Emodin (4)	*Rheum rhabarbarum*	Destroys the integrity of the cell wall and cell membrane.	[22,26]
	Flavonoids	Chrysin (5)	*Passiflora caerulea*	Binds to adhesins.	[27,28]
	Flavones	Abyssinone V (6)	*Erythrina abyssinica*	Complexes with the cell wall, inactivate enzymes and inhibit HIV reverse transcriptase.	[27]
		Acacetin (7)	*Robinia pseudoacacia*	-	[22]
	Flavonols	Quercetin (10)	*Brickellia cavanillesii*	Disrupts bacterial cell walls and cell membranes, disrupt nucleic acid synthesis, inhibit biofilm formation, and reduce expression of virulence factors.	[28,29]
	Tannins	Ellagitannin (9)	*Punica granatum*	Binds to proteins, bind to adhesins, enzyme inhibition, substrate deprivation, complex with the cell wall, membrane disruption, metal ion complexation.	[27]
	Coumarins	Warfarin (13)	*Melilotus officinalis*	Interacts with eukaryotic DNA (antiviral activity).	[27]
Terpenoids		Capsaicin (11)	*Capsicum annuum*	Membrane disruption.	[27]
		Carvacrol (12)	*Xylopia aromatica*	Membrane disruption.	[22,30]
		Thymol (8)		Induces the permeability and depolarization of the cytoplasmic membrane.	[22,31]
Alkaloids		Caffeine (14)	*Coffea arabica*	Inhibits biofilm development.	[22,32]
		Berberine (15)	*Argemone mexicana*	Damages bacterial cells by destroying cellular proteins.	[22,33]
Lectins and polypeptides		Fabatin (16)	*Vicia faba*	Blocks viral fusion or adsorption and forms disulfide bridges.	[27]

Table 2. Different compounds derived from plants with promising activity to combat drug-resistant pathogens (based on [34]).

Name of the Compound	Source of the Compound	Biological Effect on MDR Bacteria	References
9,12,15-Octadecatrienoic acid (**17**)	*Ocimum basilicum*	Used in contesting *E. coli*, *S. aureus*, *K. pneumonia*, *P. aeruginosa*, and *P. mirabilis*.	[34]
Furanone (**18**)	*Vanilla planifolia*	Interferes in the quorum sensing system of *P. aeruginosa*.	[35]
Plumbagin (**19**)	*Plumbago indica*	Has antibacterial properties by binding to the ATP cassette transporter.	[36,37]
Arjunolic acid (**20**)	*Cercidium microphyllum*	Inhibits *E. coli*, *B. subtilis*, and *S. sonnei*.	[38]
1,8-Cineole (**21**)	*Eucalyptus globulus*	Has antibacterial (methicillin-resistant *S. aureus*), antibiofilm, and anti-quorum sensing activities.	[39,40]
Leucoanthocyanidin (**22**)	*Umbellularia californica*	Has a cidal effect against *B. cereus* ATCC14579, *S. pyogens* ATCC10782, and *MRSA* ATCC-BAA-1683.	[41]
Quercetin (**10**)	*Citrus sinensis*	Inhibits the proton motive force (PMF) of *S. aureus* and inhibits *P. aeruginosa* (POA1), *E. coli* O157H7, and *V. harveyi* BB120.	[42]
Warfarin (**13**)	*Dipteryx odorata*	Inhibits *S. viridans*, *S. mutans* and *S. aureus*.	[16]
α-Pinene (**23**)	*Callistemon viminalis*	Suppresses the growth of *B. cereus*, *S. typhi*, *P. aeruginosa*, *B. subtilis*, *E. coli*, and *P. vulgaris*.	[43]
p-Cymen-8-ol (**24**)	*Senecio nutans*	Interferes with the membrane permeability of *V. cholerae*.	[44]
Luteolin (**25**)	*Guazuma ulmifolia*	Has a cidal effect against *M. tuberculosis*.	[45]
Allicin (**26**)	*Allium sativum*	Interferes with the metabolic systems of *H. pylori*, *S. epidermidis*, *B. cepacia*, *P. aeruginosa*, and *S. aureus*.	[46]
Thymol (**8**)	*Lippia sidoides*	Has activity against *L. monocytogen*, *S. typhimurium*, and *E. coli* O157:H7.	[46,47]
Dehydroabietic acid (**27**)	*Pinus elliottii*	Has a cidal effect against *E. faecalis*, *S. haemolyticus*, *S. capitis*, and MDR-*S. epidermidis*.	[48]
Pogostone (**28**)	*Pogostemon cablin*	Is effective against both gram-negative and gram-positive bacteria.	[49]
Apigenin (**29**)	*Mentha pulegium*	Interferes with the growth of *B. cereus*, *E. coli*, and *S. aureus*.	[50]
Isosakuranetin (**30**)	*Hyptis albida*	Inhibits *S. aureus* and *B. subtilis*.	[51]
Guaijaverin (**31**)	*Psidium guajava*	Significantly inhibits the adherence of *S. mutans*.	[52,53]
Zingerone (**32**)	*Zingiber officinale*	Inhibits biofilm formation and attenuation of motility properties in *P. aeruginosa*.	[54–56]

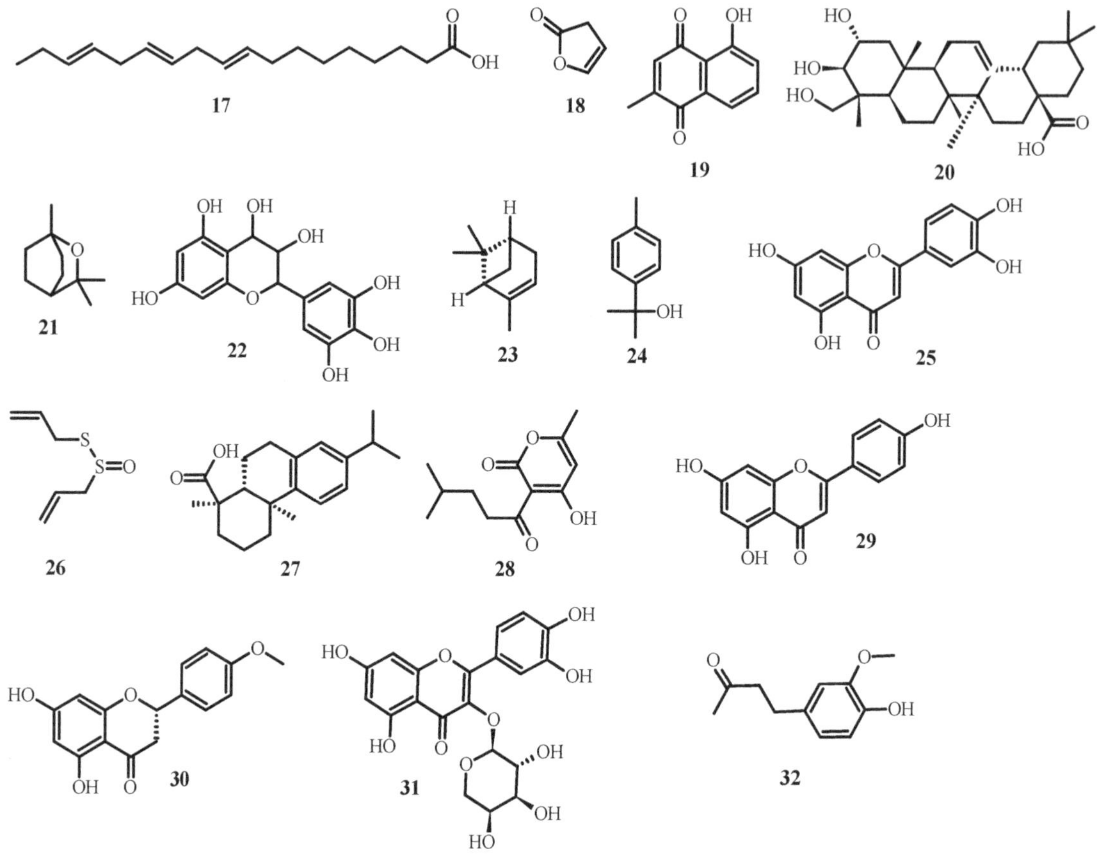

Figure 3. Structures of promising plant-derived compounds merit combating drug-resistant pathogenic microbes, from Table 2.

2.3. Pathogenic Fungi for Human

Fungi are eukaryotic organisms widely distributed across the planet, with more than 700,000 species classified [57]; however, it is estimated that there may be more than 1 million species in existence [58]. Despite these data, the number of fungi that can affect other species is minimal, with less than 0.1% being of medical importance to humans, and less than 50 species being identified as pathogenic fungi. In recent years, fungi adapted to modified ecosystems have significantly impacted human health, as they tend to infect plants and their metabolism, negatively affecting the food web [59,60].

Mycoses are usually superficial, cutaneous, systemic, or opportunistic. A worldwide risk factor is immunosuppression; however, the microbiome imbalance caused by antibiotics must be considered, as it can lead to an even more severe infection [61,62]. It is widely thought that most mycoses are opportunistic. It is extremely important to take into account that mycosis can be considered dangerous due to the entry of several fungi, with cosmopolitan genera such as *Candida*, *Cryptococcus*, and *Aspergillus* being prevalent [63–65], while creating invasive fungal infections (IFIs) that cause high mortality rates worldwide [60,66,67].

2.4. Mechanism of Action and Drug-Resistance of Pathogenic Fungi

Pathogenic fungi create complex signaling cascades that depend on the host and environment [68]. A 2017 review points out the importance of recognizing the pathways involved in fungal pathogenicity and identifying opportunity areas to create better antibiotics [69], even if knowing these factors would make it impossible to create efficient vaccines [70,71]. However, current antifungal drugs have different mechanisms of action (Table 3); the most common mechanisms are directed against the fungal cell wall or membrane, specifically against ergosterol or (1,3)-β-d-glucan biosynthesis, except for pyrimidines and orotomides that target crucial molecules in nucleic acid metabolism [72–75].

Table 3. Mechanisms of action of families of antifungal drugs.

Family of Antifungal Drugs	Mechanism of Action	References
Azoles (fluconazole)	Inhibit fungal cell cytochrome P-450-3-A, disrupting ergosterol synthesis and intoxicating the cell with sterol intermediates.	[72–74]
Polyene (anfotericine B)	Binds to ergosterol and generates pores in the membrane, causing oxidative damage and cell death.	[72,73]
Echinocandins (micafungin)	Inhibit the enzyme 1,3-β-D-glucan synthase, which weakens the cell wall, causing osmotic instability.	[72,73]
Allylamines (terbinafine)	Block the enzyme squalene epoxidase, reducing ergosterol levels and increasing squalene. This increases the permeability of the cell. membrane, causing a decrease in fungal growth.	[72]
Pyrimidines (flucytosine)	Bind to cytosine permease, already in the nucleus, and generate fluorardilic acid, which is incorporated into the RNA, rendering it useless.	[72,74]
Orotomides (olorofim)	Inhibit dihydroorotate dehydrogenase synthesis, preventing the synthesis of DNA and RNA.	[76,77]
Fosmanogepix	Inhibits the enzyme Gwt1, responsible for glycosylphosphatidylinositol synthesis.	[75]

Just as bacteria generate drug resistance, so do fungi; this drug resistance can be described from a clinical point of view, referring to the worsening of an infection despite receiving adequate drug treatment. On the other hand, in the laboratory context, resistance is evaluated through a Minimum Inhibitory Concentration (MIC) assay to determine the growth of the pathogen at different concentrations of antibiotics [68,69,78]. It is necessary to point out the concept of drug tolerance, which is considered as the fungus persistence on the substrate; however, its growth is slow due to multifactorial causes [79,80].

2.5. Latin American Plants with Antifungal Effects

Fungi drug resistance has created a worldwide clinical challenge, and treatment alternatives have been considered, such as including two or more antifungals for one treatment; however, this does not make a significant difference [81]. This is why alternatives should be considered, such as using plant-derived compounds that can act via bypassing common metabolic pathways in fungal pathology. Table 4 summarizes the medicinal plant extracts with antifungal properties.

Table 4. Extracts of Latin American plants with activity against pathogenic fungi.

Species	Extract	Fungi	References
Achyrocline satureioides	Ethanolic	*Fusarium verticillioides*	[82]
Achyrocline tomentosa	Ethanolic	*Fusarium verticillioides*	[82]
Aloysia citriodora	Ethanolic	*Fusarium verticillioides*	[82]
Annona cherimola	Ethanolic	*Fusarium oxysporum*	[83]
Annona muricata L.	Ethanolic	*Candida albicans*	[84]
Aristolochia argentina Griseb.	Ethanolic	*Fusarium verticillioides*	[82,85]
Asclepias curassavica	Hexanic, Methanolic	*Candida albicans*	[86]
Baccharis artemisioides	Ethanolic	*Fusarium verticillioides*	[82]
Baccharis flabellata	Ethanolic	*Fusarium verticillioides*	[82]
Baccharis salicifolia	Ethanolic	*Fusarium verticillioides*	[82]
Bixa orellana	Ethanolic	*Candida albicans*	[87]
Curcuma zedoaria	Acetone, Hexanic	*Candida albicans*	[88,89]
Dalea elegans	Ethanolic	*Fusarium verticillioides*	[82]
Echinacea angustifolia	Ethanolic	*Cryptococcus neoformans*	[90]
Echinacea atrorubens	Ethanolic	*Cryptococcus neoformans*	[91]
Echinacea pallida	Ethanolic	*Candida albicans*	[91]
Echinacea purpurea	Ethanolic	*Saccharomyces cerevisiae*	[90]
Eupatorium buniifolium	Methanolic	*Trichophyton mentagrophytes*	[92]
Euphorbia hyssopifolia	Methanolic	*Aspergillus niger*	[93]
Flourensia oolepis	Ethanolic	*Fusarium verticillioides*	[82]
Gaillardia megapotamica	Ethanolic	*Fusarium verticillioides*	[82]
Galphimia glauca	Hexanic, Methanolic	*Trichophyton mentagrophytes*	[86,94]
Grindelia pulchella	Ethanolic	*Fusarium verticillioides*	[82]
Heterothalamus alienus	Ethanolic	*Fusarium verticillioides*	[82]
Hibiscus sabdariffa	Methanolic	*Candida albicans*	[95]
Kageneckia lanceolata	Ethanolic	*Fusarium verticillioides*	[82]
Larrea cuneifolia	Ethanolic	*Lenzites elegans*	[96]
Larrea divaricata	Ethanolic	*Penicillium notatum*; *Candida* spp.	[96,97]
Lepechinia floribunda	Ethanolic	*Fusarium verticillioides*	[82]
Lippia turbinata	Ethanolic	*Fusarium verticillioides*	[82]
Loeselia mexicana	Ethanolic	*Trichophyton mentagrophytes*	[98]
Lygodium venustum	Ethanolic	*Candida albicans*	[99]
Lysiloma acapulcensis	Hexanic	*Trichophyton mentagrophytes*	[100]
Miconia mexicana	Methanolic	*Candida albicans*	[100]
Microliabum candidum	Ethanolic	*Fusarium verticillioides*	[82]
Minthostachys verticillata	Ethanolic	*Fusarium verticillioides*	[82]
Morrenia brachystephana	Ethanolic	*Fusarium verticillioides*	[82]
Otholobium higuerilla	Ethanolic	*Fusarium verticillioides*	[82]
Passiflora caerulea	Methanolic	*Aspergillus flavus*	[101]
Pimenta dioica	Essential oil	*Fusarium oxysporum*	[102]

Table 4. Cont.

Species	Extract	Fungi	References
Polygonum acuminatum	Dichloromethane	Cryptococcus neoformans	[103]
Salix alba	Methanolic	Aspergillus ornatus	[104]
Salvia cuspidata	Ethanolic	Fusarium verticillioides	[82]
Sebastiania commersoniana	Ethanolic	Candida spp.	[105]
Senecio vira-vira	Ethanolic	Fusarium verticillioides	[82]
Smilax domingensis	Ethanolic	Candida albicans	[106]
Syzygium aromaticum	Essential oil	Candida spp.	[107]
Terminalia triflora	Methanolic	Trichophyton mentagrophytes	[92]
Thalictrum decipiens	Ethanolic	Fusarium verticillioides	[82]
Tithonia diversifolia	Aquous	Fusarium oxysporum	[108]
Trichocline reptans	Ethanolic	Fusarium verticillioides	[82]
Vernonia mollisima	Ethanolic	Fusarium verticillioides	[82]
Vernonia nudiflora	Ethanolic	Fusarium verticillioides	[82]
Vitis vinifera	Aqueous	Candida spp.	[109]
Zanthoxylum coco	Ethanolic	Fusarium verticillioides	[82]
Zinnia peruviana	Ethanolic	Fusarium oxysporum	[96]
Zuccagnia punctata	Ethanolic	Aspergillus niger	[96]
Zuccagnia punctata	Dichloromethane	Candida albicans	[110]

2.6. Medicinal Plant Antiviral Activity against Human-Infecting Viruses

In 2018, over 4400 virus species were classified into 122 families and 7535 [111] subfamilies. Human-infecting viruses include RNA viruses, DNA viruses, retroviruses, bare viruses, and virions, with RNA viruses being the most prevalent. Numerous medicinal plants contain compounds that inhibit the replication of viruses or enhance the immune system. Alkaloids, terpenes, flavonoids, numerous glucosides, and proteins have been recognized as phytochemicals; their metabolites include apigenin (**29**), kaempferol (**34**), and luteolin (**25**), in addition to the triterpenoids oleanolic acid (**35**) and ursolic acid (**36**) [112].

2.6.1. Biological Mechanisms of Antiviral Activity

Plant biocompounds may function similarly to conventional antiviral medications by inhibiting viral replication polymerase, protease, integrase, fusion molecules, and cell membrane binding. For example, the exposure of non-enveloped norovirus to 0.5% of carvacrol (**12**) results in the degradation of its capsid [113]. Some polysaccharides may also deter viruses from attaching to cells, while thiophenes, terpenoids, and polyacetylenes can interact with the membrane of infected cells [114]. Lignans, phenolic compounds, terpenoids, flavonoids, alkaloids, and furocoumarins can all inhibit viral replication. Biocompounds of *Allium sativum* are among the most studied; they have antiviral activity against human, animal, and plant infections. Multiple of these metabolites can strengthen the immune system's response to infections. This biocompound interacts in vivo with thiols such as glutathione and L-cysteine to produce S-allyl-mercapto-glutathione (SAMG) and S-allyl-mercapto-cysteine (SAMC), which can degrade viral protein. In addition, it contains lectins, flavonoids (kaempferol (**34**), quercetin (**10**), and myricetin (**37**)), polysaccharides (fructan), steroids, saponins, fatty acids (lauric (**38**) and linoleic acid (**39**)), diverse enzymes, vitamins (A, B1, and C), allixin (**40**), minerals (Ca, Cu, Fe, K, Mg, Zn, and Se), and amino acids [115].

Table 5 summarizes the medicinal plant extracts and their possible mechanisms of action (Figure 4), while Table 6 discusses the medicinal plant biocompounds with antiviral properties (Figures 5–7).

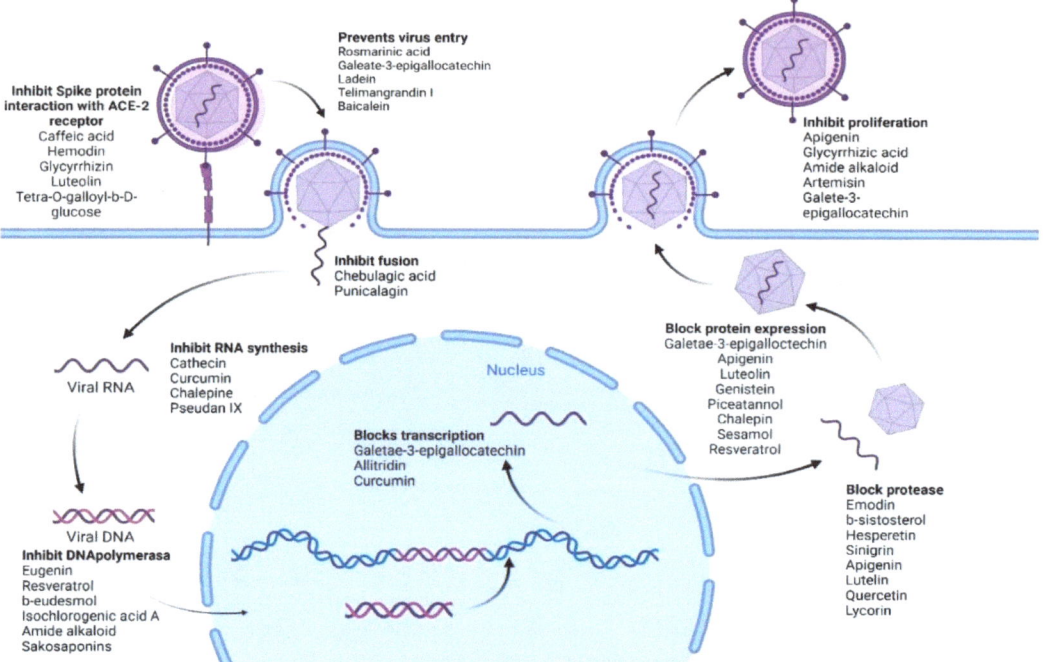

Figure 4. Various biocompounds' potential antiviral mechanisms of action. Created with BioRender.com.

2.6.2. Antiviral-Active Extracts for Respiratory Infections

The leading cause of morbidity in humans is viral respiratory tract infections, with rhinovirus, influenza, respiratory syncytial virus (RSV), and human coronavirus having the most significant impact.

Several extracts of medicinal plants exhibit antiviral activity in vitro; the ethanolic, ethyl acetate, and hexane extracts of *Echinacea pallida* var. *angustifolia* root inhibit rhinovirus replication [116]. The *Echinacea purpurea* ethanolic extract inhibits the invasion of HcoV-299E (coronavirus) into cells [117]. The ethanolic extract of *Sambucus formosana* Nakai seeds inhibits the binding of HCoV-NL63 (coronavirus) [118]. Aqueous extracts of *Plantago asiatica* and *Clerodendrum trichotomun* inhibit RSV (respiratory syncytial virus) replication [119].

2.6.3. Extracts and Biocompounds with Activity against Human Herpes Viruses

Several medicinal plant extracts have in vitro anti-herpes simplex activity: the hexane, dichloromethane, and methanolic extracts of *Clinacanthus mutans* and *C. siamensis* inhibit the formation of HS-1 and HS-2 viral plaques [120]. Caffeic acid and chlorogenic acid are inhibitors of HS replication [121]. *Polygonum minus* methanolic extract inhibits HS adhesion [122]. *Aloe vera* glycerol extract prevents HS-2 replication [123]. *Lysimachia mauritiana* ethanolic extract inhibits varicella-zoster virus replication [89].

Alkaloids, glycosides, taxol derivatives, terpenes, flavonoids, ellagitannin, catechin, phenolic acids, triterpenoids, monoterpenoids, and steroids have been identified as active against Herpes simplex types 1 and 2 [124].

Carvacrol (**12**), extracted from the essential oil of Mexican oregano (*Lippia graveolens*), demonstrates antiviral activity against RNA and DNA viruses (primarily herpes viruses) [125].

Coumarins imperatorin (**41**) and phellopterin (**42**), isolated from *Angelica archeangelica* L., exhibit antiviral activity against herpes simplex virus type 1 and, most likely, Coxsackievirus B3 [126].

Eugenin (**43**) is a biocompound extracted from *Geum japonicum* and *Syzygium aromaticum*. Eugenin (**43**) inhibits the DNA polymerase of the Herpes simplex virus, which appears to be its mechanism of action. Also, it inhibits Herpes simplex virus activity in both Vero cells and mice [127].

The monoterpene aldehydes citral a (**45**), citral b (**46**), and citronellal (**44**) are the biocompounds found to have anti-herpes virus activity in *Melissa officinalis* essential oil [128]. Moreover, rosmarinic acid (**47**) from the hydroalcoholic leaf extract of *M. officinalis* demonstrates anti-herpes simplex type 2 activity [129]. The potential mechanism of action is to prevent virus entry into cells [130].

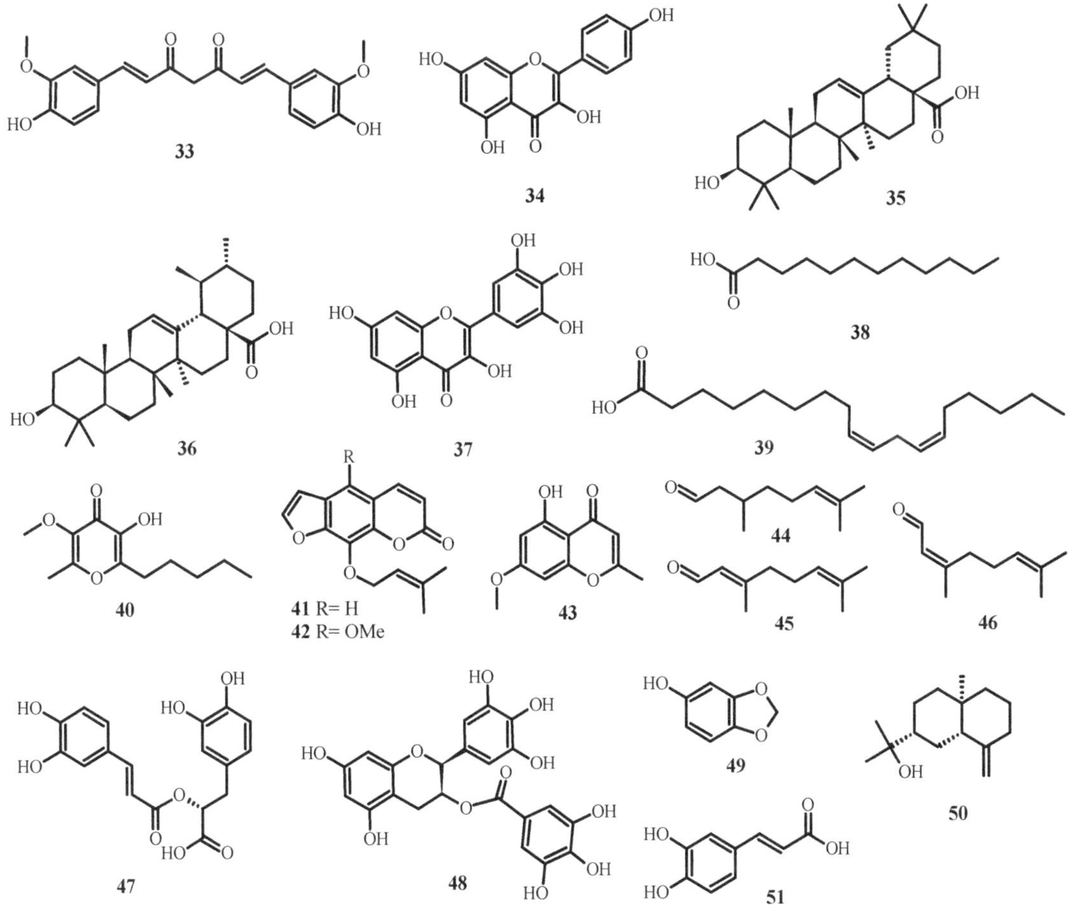

Figure 5. Antiviral biological compounds (Table 6).

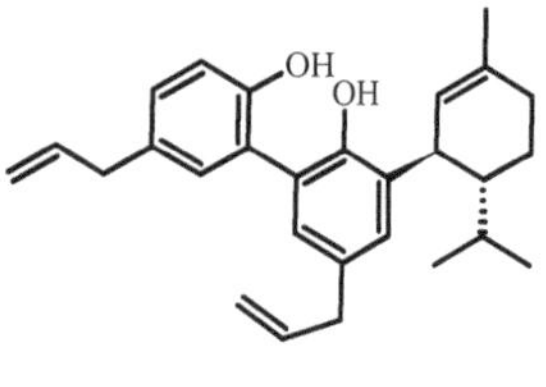

Figure 6. Antiviral biological compounds (Table 6).

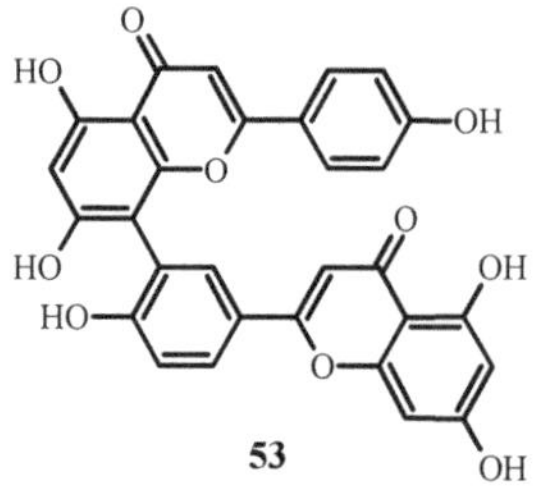

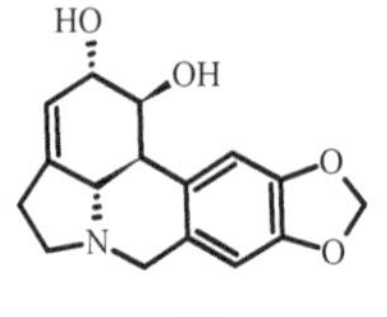

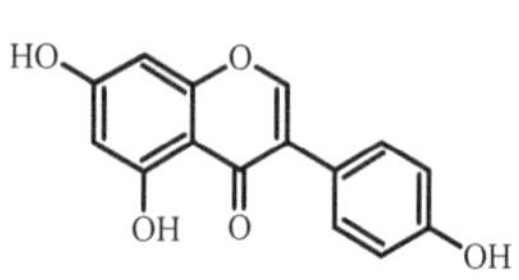

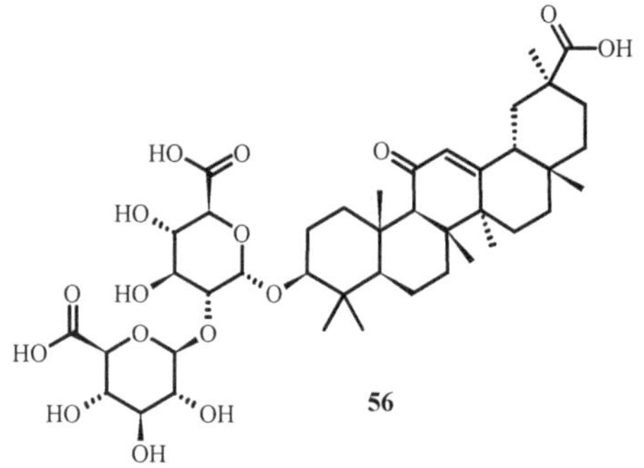

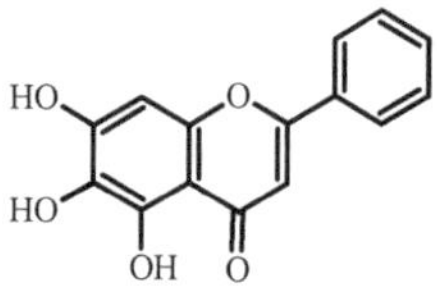

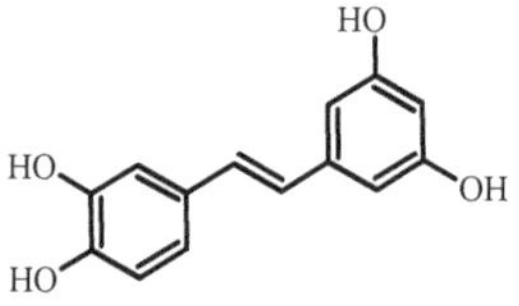

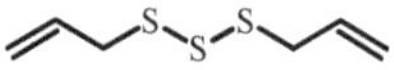

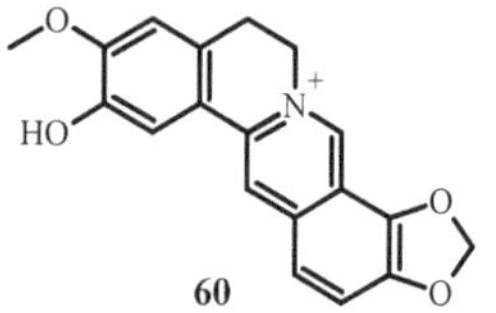

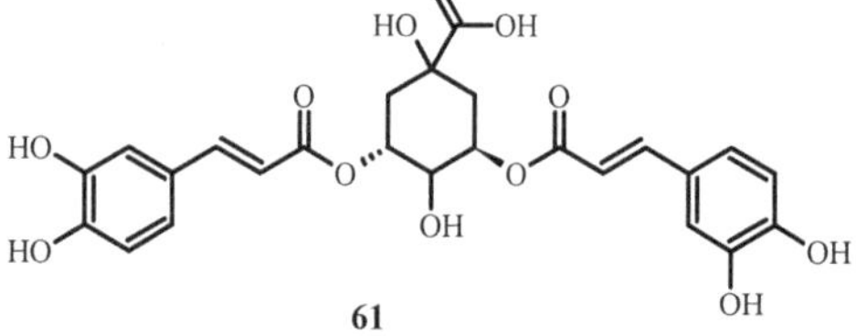

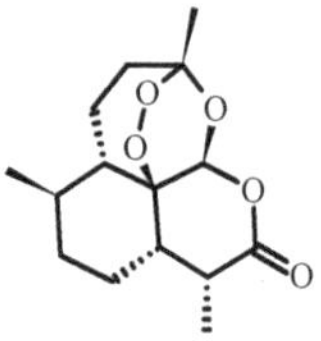

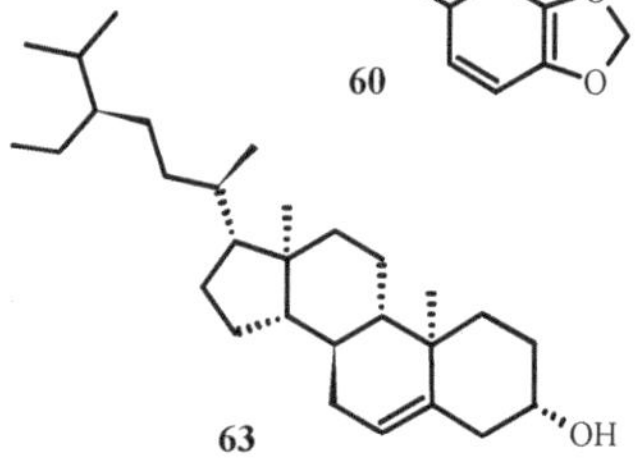

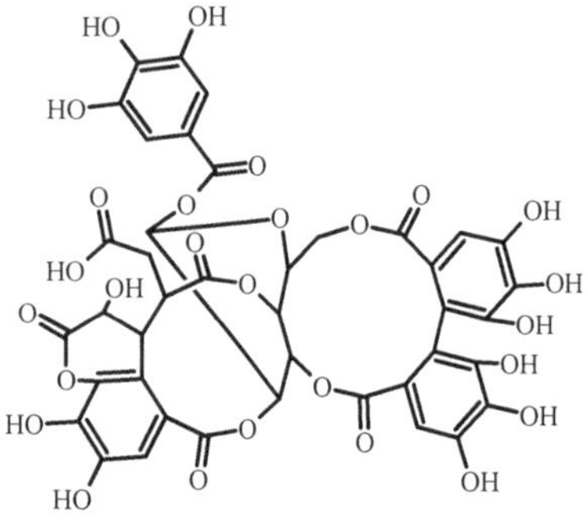

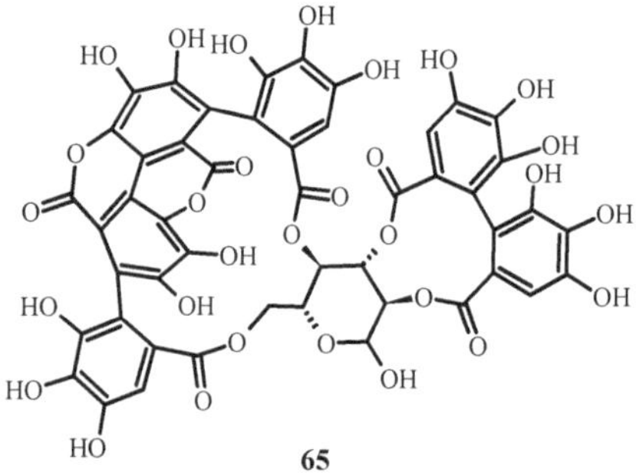

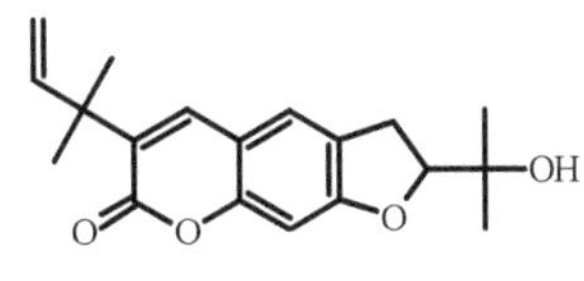
66

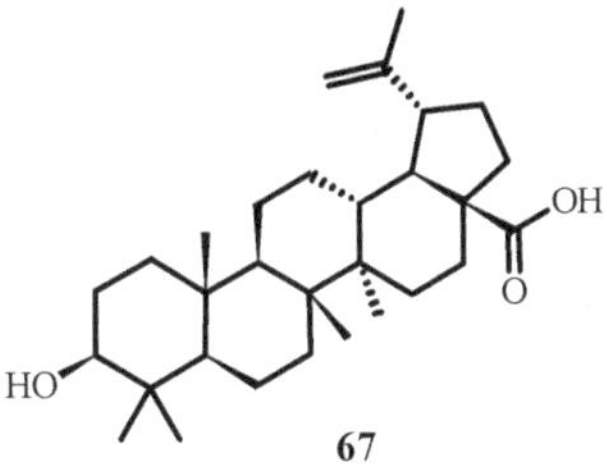
67

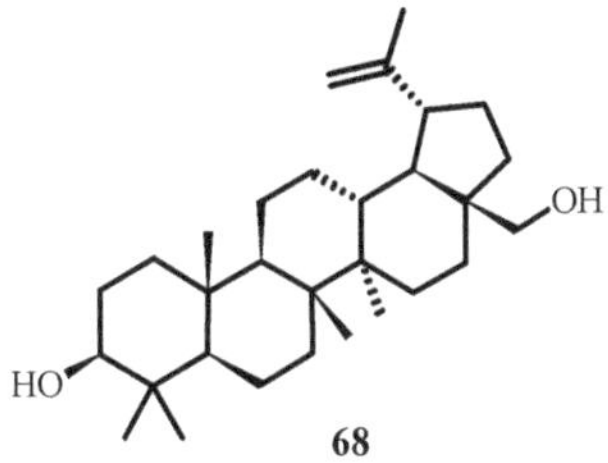
68

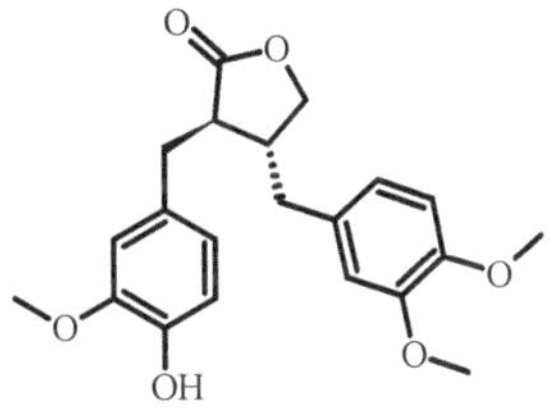
69

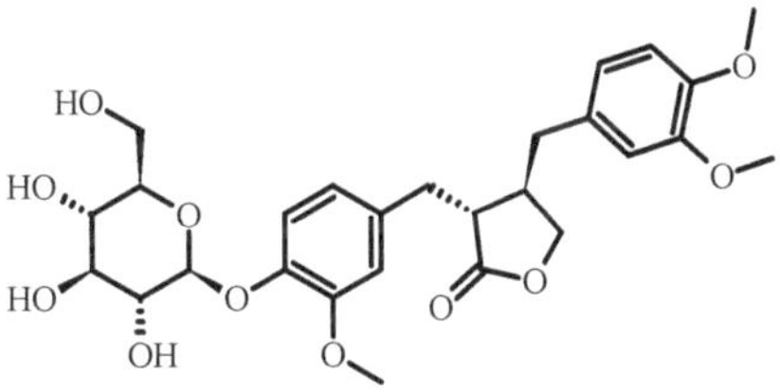
70

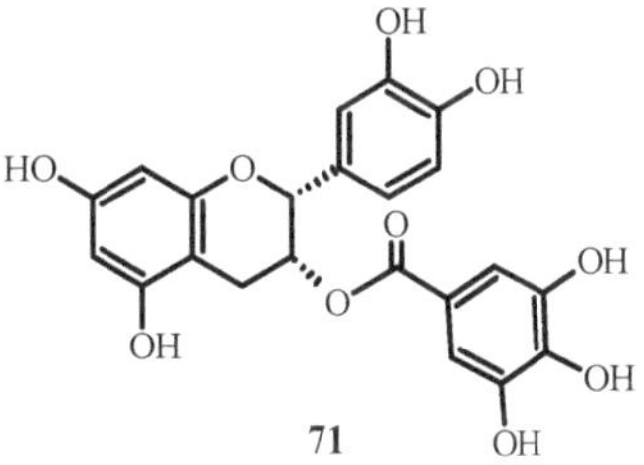
71

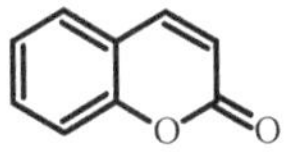
72

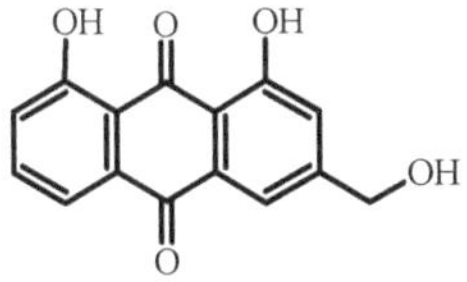
73

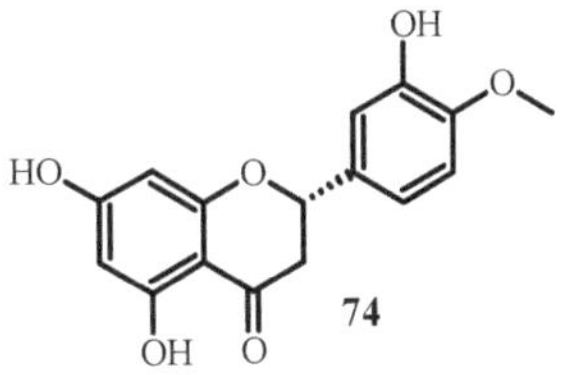
74

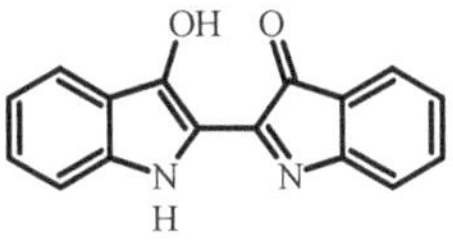
75

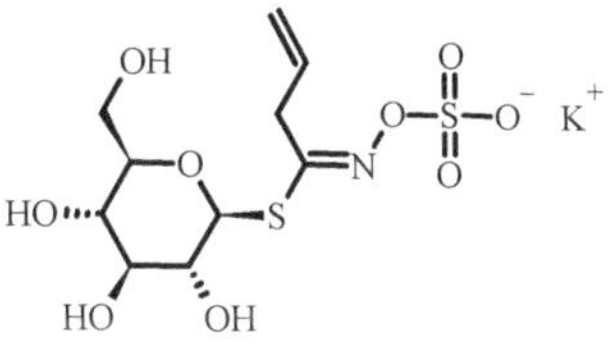
76

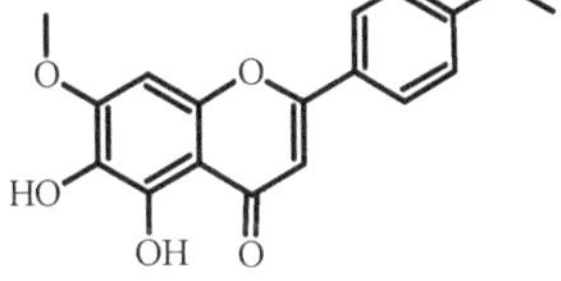
77

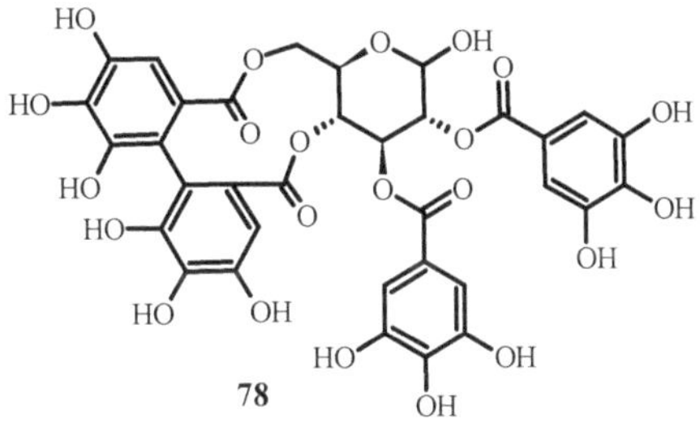
78

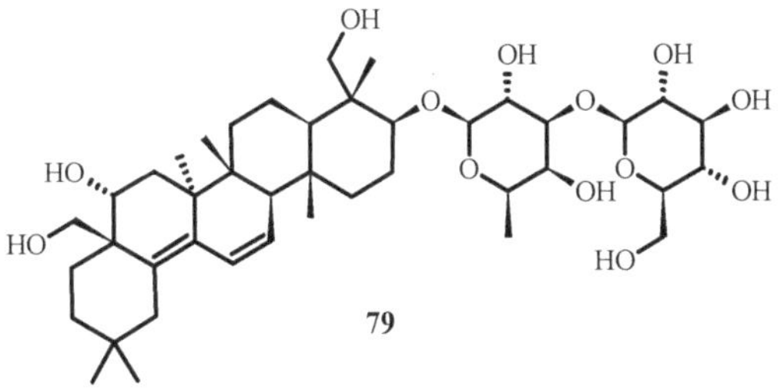
79

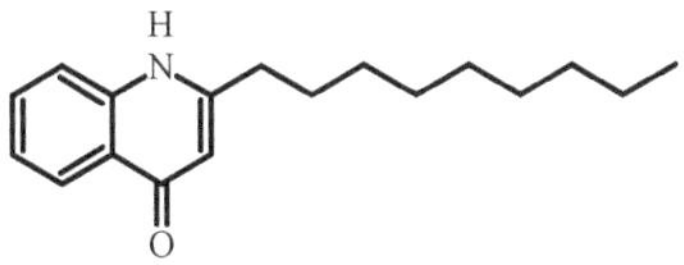
80

Figure 7. Antiviral biological compounds (Table 6).

2.6.4. Activity against Epstein-Barr Virus

Epstein–Barr (EBV) is a herpes virus that affects 90 percent of the world's population and is linked to numerous immunological and neoplastic diseases.

Epigallocatechin-3-gallate (**48**), a catechin derived from *Camellia sinensis*, inhibits the spontaneous lytic infection of infected cells and blocks their transcription and protein expression via the ERK1/2 (extracellular-regulated kinase 12) and PI3-K/Akt (phosphatidylinositol-3-kinase) pathways [131].

The compounds sesamol (**49**) and resveratrol (**2**), along with sesame and sunflower essential oils, inhibit the early antigen activation in vitro of the Epstein–Barr virus [132].

Konoshima et al. [133] found that monoterpenylmagnolol (**52**) and β-eudesmol (**50**), extracted from *Magnolia officinalis*, inhibit replication (EBV) in Raji cells.

Berberine (**15**) is an alkaloid derived from several medicinal plants (*Cortidis rhizome*, *Coptis chinensis*, and *Barnerini vulgaris*) that inhibits cell proliferation and induces apoptosis in Epstein–Barr virus-infected cells via the inhibition of p-STAT3 and the overexpression of EBNA1 [134].

Curcumin (**33**) is highly effective at reducing TPA-, butyrate-, and TGF-b-induced levels of BZLF1 mRNA and TPA-induced luciferase mRNA, indicating that it inhibits three main EBV pathways [135].

Apigenin (**29**) inhibits the expression of the EBV lytic proteins Zta, Rta, EAD, and DNase in B and epithelial cells. In addition, it decreases the number of EBV-reactivating cells detectable via immunofluorescence analysis. Additionally, apigenin (**29**) has been found to significantly reduce EBV virus production [136].

Glycyrrhizic acid (**56**) (18-GL or GL) possesses a wide range of antiviral activities, pharmacological effects, and sites of action. In vitro, GL (**56**) inhibits Epstein–Barr virus (EBV) infection by interfering with an early stage in the EBV replication cycle (possibly attachment or penetration) [137].

The flavonoid luteolin (**25**) inhibits EBV reactivation significantly. In EBV-positive epithelial and B cell lines, 25 inhibits the expression of EBV-lytic gene-encoded proteins. In addition, it decreases the number of EBV-reactivating cells detected via immunofluorescence and virion production. Moreover, 25 decreases the activities of the promoters of the immediate–early genes Zta (Zp) and Rta (Rp). It inhibits the activity of Sp1-luc, indicating that the disruption of Sp1 binding is involved in the mechanism of inhibition [138].

2.6.5. Anti-Cytomegalovirus Activities

Human cytomegalovirus (hCMV) is a pervasive herpesvirus that causes a latent infection that persists throughout the host's lifetime and can be reactivated when immunity is compromised.

Genistein (**55**) and baicalein (**57**) are antiviral flavonoids against HCMV. The primary mode of action of genistein's antiviral activity against HCMV is to inhibit the function of immediate–early proteins. Baicalein's antiviral activity against HCMV works primarily by inhibiting the kinase activity of EGFR to prevent viral entry [139].

Supplementation with piceatannol (**58**) inhibits the lytic changes caused by hCMV infection. In addition, piceatannol dose-dependently inhibits the expression of hCMV immediate–early (IE) and early (E) proteins and the replication of hCMV DNA [140].

Resveratrol (**2**) inhibits human cytomegalovirus DNA replication to undetectable levels during the second (late) phase of virus-induced phosphatidylinositol-3-kinase signaling and transcription factor activation [141].

Allitridin (**59**), a compound extracted from *A. sativum*, reduces the amount of viral DNA in cytomegalovirus-infected cells by inhibiting the transcription of the IE gene [142].

2.6.6. Anti-HIV Activity of Extracts and Biocompounds

Among the extracts that inhibit in vitro HIV activity or replication is an aqueous extract of *Salvia miltiorrhiza* that inhibits HIV-1 integrase [143]. *Rhaphiolepsis indica* methanolic extract inhibits its replication [144]. *Acacia arabica*'s n-butanol fraction inhibits the activity of viral proteases and Tat [145]. The *Phyllanthus amarus* ethanolic and aqueous extracts inhibit its replication [146]. The *Olea europaea* aqueous extract inhibits cell–cell infection [147]. *Hyssopus officinalis* L. aqueous extract inhibits its replication [148]. The reverse transcriptase is inhibited by the methanolic extract of *Terminalia sericea* [149], the n-hexane fraction of *Phyllanthus emblica* and *Cassia occidentalis*, and the pine cone extract of *Pinus yunnanensis* [150]; floral extracts of *Calendula officinalis* inhibit HIV-1 reverse transcriptase activity [151]. *Cassine xylocarpa*'s lupane-type pentacyclic triterpenoid also possesses anti-HIV activity [152].

2.6.7. Antiviral Activity of Extracts and Biocompounds against Hepatitis B and C Viruses

The secondary metabolites and extracts listed below have demonstrated in vitro activity against HBV (Hepatitis virus B): isochlorogenic acid A (**61**), obtained from *Laggera alata*, inhibits replication and decreases the stability of its core protein [153]. Amide alkaloids from *Piper longum* [154] and dehydrocheilanthifoline (**60**), isolated from *Corydalis saxifolia*, inhibit its replication [155]. Saikosaponins (*Bupleurum* species) inhibit the replication and expression of its surface antigen [156]; the ethanolic extract of *Polygonum cuspidatum* inhibits its surface antigen expression [157]; and curcumin (**33**) (*Curcuma longa*) decreases the expression of the PGC-1a coactivator, required for its transcription [158]. Glycyrrhizinic acid (**56**) (*Glycyrrhiza glabra*), artemisinin (**62**) (*Artemisia annua*), and LPRP-Et-97,543 compound (*Liriope platyphylla*) all inhibit its viral production [159]. On the other hand, epigallocatechin-3-gallate (**48**) (*Camellia sinensis*) inhibits its viral replication [160].

Concerning the hepatitis C virus, flavonolignans (*Silybum marianum*) possess antiviral and antioxidant properties [161], while curcumin (**33**) (*Curcuma longa*) inhibits its viral replication via the Akt-SREBP-1 pathway [162]. Epigallocatechin-3-gallate (**48**) [163] and ladanein (**77**) [164] inhibit viral entry; griffithsin inhibits viral cell–cell transmission [165], tellimagrandin I (**78**) (*Rosa rugosa*) inhibits viral invasion [159], chebulagic acid (**64**) and punicalagin (**65**) (*Terminalia chebula*) inhibit the viral particles necessary for their fusion and cell–cell transmission [166], saikosaponin B2 (**79**) (*Bupleurum kaoi*) prevents viral binding, and chalepine (**66**) and pseudan IX (**80**) (*Ruta angustifolia*) decrease viral protein synthesis and RNA replication [159,167].

Betulinic acid (**67**) and betulin (**68**), derived from *Betula alba* L., exhibit anti-hepatitis C virus activity. Shikov et al. (2011) [168] suggested that **68** can induce TNF-α expression and thereby enhance the Th1-type immune cell response in patients with chronic hepatitis C virus.

2.6.8. Anti-Influenza Activity of Extracts and Biomolecules

Moradi (2019) [169] reported that ethanolic and polyphenolic extracts of *Punica granatum* inhibit influenza replication and virions. *Geranium sanguineum* polyphenolic, methanolic, and ethanolic extracts have antiviral properties [170]. Glycyrrhizin (**56**) from *Glycyrrhiza glabra* induces the apoptosis of H5N1-infected cells [171]; polyphenols from *Chenomeles sinensis* inhibit the binding of its hemagglutinins [172]; and the *Sambucus nigra* fruit inhibits viral entry and modulates cytokine release. It has been shown in other studies to inhibit hemagglutins and the replication of influenza viruses: A/Shangdong 9/93 (H3N2), A/Beijing 32/92 (H3N2), A/Texas 36/91 (H1N1), A/Singapore 6/86 (H1N1), type B/Panama 45/90, B/Yamagata 16/88, and B/Ann Arbor [173].

The *Phyllanthus embolica* aqueous extract inhibits hemagglutinin and viruses in infected cells [174]. Catechin derived from *Camellia sinensis* inhibits both RNA synthesis and neuraminidase activity [175].

Arctigenin (**69**) and arcitiin (**70**), extracted from the fruits of *Arctium lappa* L., exhibit potent anti-influenza A virus activity in vitro [176].

Echinacea extract is active against influenza A/B viruses (H3N2, H1N1, H5N1, H7N7, and S-OIV), Respiratory Syncytial Virus, and Herpes Simplex [177]. On the other hand, it also induces the production of IL-6 and IL-8 (CXCL8) and other cytokines with antiviral properties [178]. In a clinical trial, it was demonstrated to be as effective as oseltamivir in reducing influenza symptoms if administered at the onset of the disease [179].

On the other hand, the monoterpene aldehydes citral a (**45**) and citral b (**46**), from *Melissa officinalis*, exhibit synergistic activity with oseltamivir against the H9N2 influenza virus [180].

Wyde et al. [181] found that polyphenolic polymers derived from the Euphorbiaceae shrub are active in vitro against parainfluenza virus type 3, Respiratory Syncytial Virus, and influenza viruses.

2.6.9. Extracts In Vitro Possess Anti-Papillomavirus Activity

Their growth is inhibited by polyphenon E (**71**) (poly E) and epigallocatechin gallate (**48**) from *Camellia sinensis* [182]. Artemisinin (**62**) (*Artemisia absintium*) inhibits the expression of HPV-39, induces apoptosis, and reduces the proliferation of infected cells in ME-180 cells [183]. Curcumin (**33**) (*Curcuma longa* L.) has been utilized to boost immunity against HPV. *Hamamelis virginiana* tannins inhibit HPV-16, *Ficus religiosa* aqueous extract induces the apoptosis of HPV-16 and 18 infected cervical cells, and the *Phyllanthus emblica* fruit inhibits HPV-16 and 18 carcinogenic gene expression. The chloroplast leaf extract of *Bryophyllum pinnatum* inhibits HPV-18 transcription in cervical cancer cells, whereas the soluble extract of *Pinellia pedatisecta* inhibits HPV-E6 expression in multiple cell lines [184].

2.6.10. In Vitro Activity of Extracts against Dengue and Chikungunya Viruses

Coumarin (**72**) and the ether extract of *Alternanthera philoxeroides* [185], as well as aqueous and chloroform extracts of *Carioca papaya* [186], inhibit dengue virus. *Sambucus nigra* methanolic extract protects against dengue serotype 2 [187].

Vernonia amygdalina ethyl acetate extract reduces the Chikungunya viral burden [188]. Chikungunya helicases and proteases are inhibited by aqueous extracts of *Picrorhiza kurrooa*, *Ocimum tenuiflorum*, and *Terminalia chebula* [189].

2.6.11. Antiviral Activity of In Vitro Extracted Compounds against SARS-CoV

The following organisms were evaluated for their anti-SARS-CoV-1 activity: *Lycoris radiate, Artemisia annua, Pyrrosia lingua, Lindera aggregata, Isatis indigotica* (inhibition of 3CL protease) [190], *Rheum officinale* Bail, *Polygonum multiforum* Thunb. (inhibit ACE2 protein interaction with spike protein) [191], *Gentiana scabana, Dioscorea batatas, Casssia tora, Taxillus chinensis, Cibotium barometz* (inhibit 3CL protease) [192], and ethanolic extracts of *Anthemis hyalina, Nigella sativa,* and *Citrus sinensis* (increase IL-8, modify TRPA, TRPM, and TRPV gene expression) [193]. Some purified secondary metabolites that show activity against SARS-CoV are as follows: aloe-emodin (**73**), β-sitosterol (**63**), hesperetin (**74**), indigo (**75**), and sinigrin (**76**) (obtained from *Isatis indigotica*) [194], amentoflavone (**53**), apigenin (**29**), luteolin (**25**), quercetin (**10**) (obtained from *Torreya nucifera*), which inhibit 3CL protease [195], and lycorine (**54**) (obtained from *Lycoris radiata*) [190].

In the case of SARS-CoV-2 (which causes COVID-19), the following secondary metabolites (Figure 8) may be advantageous [196]:

1. To inhibit the binding of the spike protein to the ACE-2 receptor: caffeic acid (**51**), emodin (**82**), glycyrrhizin (**56**), luteolin (**25**), and tetra-O-galloyl-β-D-glucose (**81**).
2. To prevent virus transcription: cepharanthin (**83**), fangquinoline (**84**), forystoside A (**85**), tetrandin (**87**), coumaroyltyramine (**86**), cryptoansionone (**88**), kaempferol (**34**), moupinamide (**89**), N-*cis*-feruloyltyramine (**90**), quercetin (**10**), tanshinone IIa (**91**), and tryptanthrine (**92**).
3. To inhibit viral translation: tryptanthrine (**92**).
4. To inhibit the cellular discharge of virions: emodin (**82**) and kaempferol (**34**).

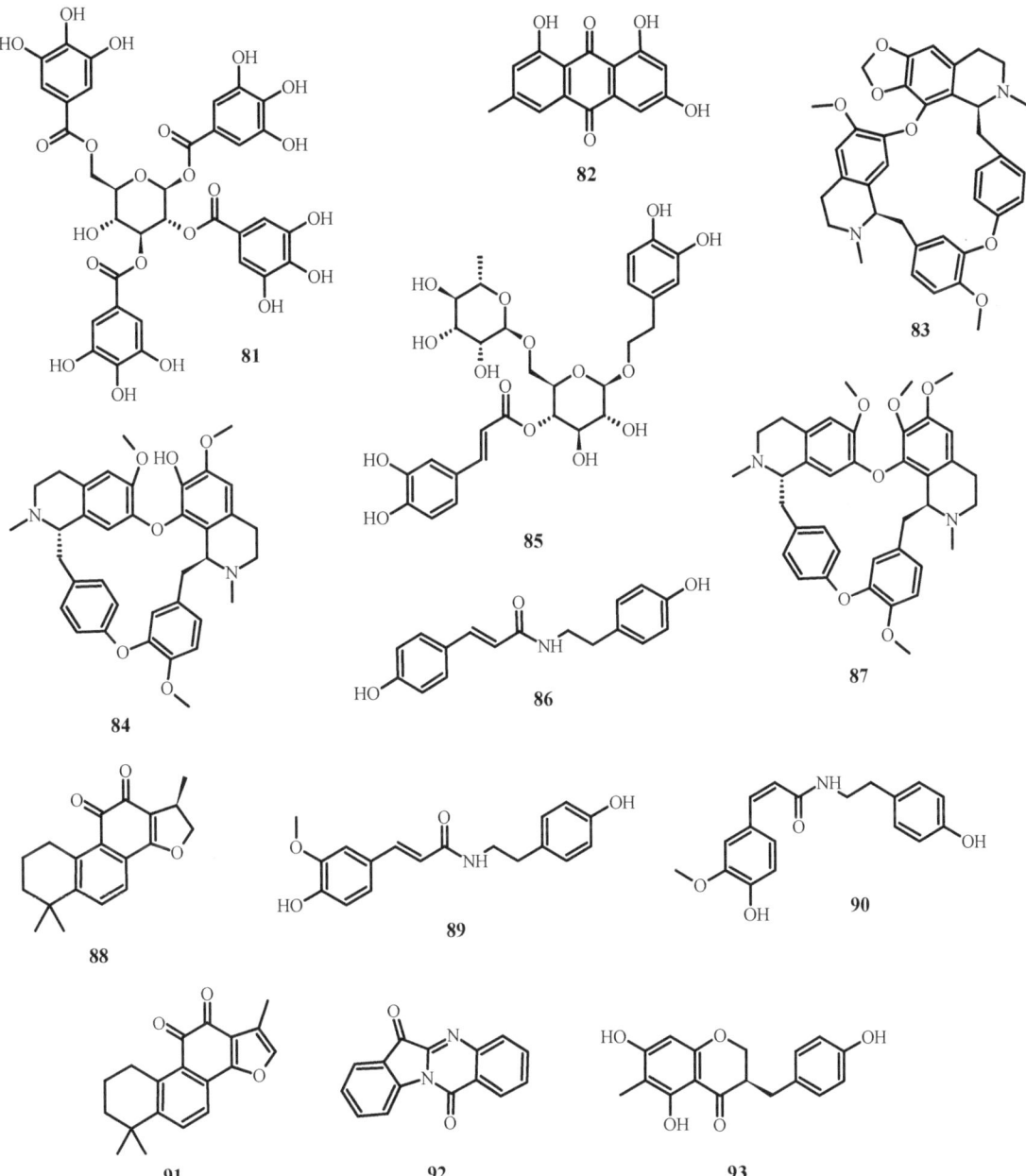

Figure 8. Natural products reported to have anti-SARS-CoV-2 activity.

Table 5. Antiviral extracts derived from plants.

Plant	Extract	Virus	Possible Antiviral Mechanism	References
Echinacea pallida var. *angustifolia*	Hexane	Rhinovirus	Impedes replication.	[116]
Echinacea purpurea	Ethanolic	Coronavirus HcoV-299E	Prevents the invasion of cells.	[117]
Sambucus formosana Nakai	Ethanolic	HCoV-NL63 (coronavirus)	Prevents bonding.	[118]
Plantago asiatica	Aqueous extract	Respiratory syncytial virus	Replication inhibition.	[119]
Clerodendrum trichotomun	Aqueous extract	Respiratory syncytial virus	Replication inhibition.	[119]
Clinacanthus mutans *Clinacanthus siamensis*	Hexane, dichloromethane, and methanolic	Herpes simplex-1 and 2	Inhibit viral plaques.	[120]
Polygonum minus	Methanolic	Herpes simplex-1 and 2	Inhibits adhesion.	[122]
Aloe vera	Glycerol	Herpes simplex 2	Impedes replication.	[123]
Lysimachia mauritania	Ethanolic extract	Varicella-zoster	Impedes replication.	[89]
Sesamum indicum *Helianthus annuus*	Sesame essential oil and Sunflower essential oil	Epstein-Barr Virus	Inhibit precocious antigen activation.	[132]
Salvia miltiorrhiza	Aqueous extract	HIV-1	Interferes with integrase activation.	[143]
Rhaphiolepsis indica	Methanolic extract	HIV-1	Impedes replication.	[144]
Acacia arabica	N-butanol fraction	HIV-1	Inhibits viral proteases and Tat activity.	[145]
Phyllanthus amarus Schum.	Ethanolic and aqueous extract	HIV-1	Impedes replication.	[146]
Olea europaea	Aqueous extract	HIV-1	Prevents infections between cells.	[147]
Hyssopus officinalis L.	Aqueous extract	HIV-1	Inhibits replication.	[148]
Polygonum cuspidatum	Ethanolic extract	Hepatitis virus B	Inhibits surface antigen expression.	[157]
Punica granatum	Ethanolic and polyphenolic extracts	Influenza virus	Inhibits influenza replication and virions.	[169]
Geranium sanguineum	Polyphenolic, methanolic, and ethanolic	Influenza virus	No study.	[170]
Chenomeles sinensis	Polyphenols	Influenza virus	Inhibits the attachment of its hemagglutinins.	[172]
Sambucus nigra	Aqueous extract	Influenza virus	Modulates cytokine release and inhibits viral entrance.	[173]
Phyllanthus emblica	Aqueous extract	Influenza virus	Prevents hemagglutinins and viruses from infecting infected cells.	[174]
Echinacea purpurea	Aqueous extract	Influenza A/B viruses H3N2, H1N1, H5N1, H7N7, and S-OIV	Induces IL-6 and IL-8 production.	[177]
Euphorbiacea shrub	Polyphenolic polymers	Influenza	No study.	[181]
Ficus religiosa	Aqueous extract	Papillomavirus	HPV-16 apoptosis is induced.	[184]
Bryophyllum pinnatum	Chloroplast extract	Papillomavirus	Suppresses HPV-18 transcription.	[184]
Pinellia pedatisecta	Soluble extract	Papillomavirus	Inhibits the HPV-E6 expression in multiple cell lines.	[184]
Carioca papaya	Aqueous and chloroplast extract	Chikungunya	Stops the dengue virus.	[186]
Sambucus nigra	Methanolic extract	Dengue serotype 2	Defends against infection.	[187]
Vernonia amygdalina	Ethyl acetate extract	Chikungunya	Minimizes the viral burden.	[188]
Picrorhiza kurrooa *Ocimum tenuiflorum* *Terminalia chebula*	Aqueous extracts	Chikungunya	Block helicases and proteases.	[189]
Lycoris radiate, Artemisia annua, Pyrrosia lingua, Lindera aggregata, and *Isatis indigotica*	Different extracts	SARS-CoV-1	Obstruct 3CL protease.	[190]
Rheum officinale Bail, *Polygonum multiforum* Thunb	Different extracts	SARS-CoV-1	Inhibit the interaction between ACE2 and spike proteins.	[191]
Gentiana scabana, Dioscorea batatas, Casssia tora, Taxillus chinensis, and *Cibotium barometz*	Different extracts	SARS-CoV-1	Prevent 3CL protease.	[192]
Anthemis hyalina, Nigella sativa, and *Citrus sinensis*	Ethanolic extracts	SARS-CoV-1	Increase IL-8 and modulate gene expression of TRPA, TRPM, and TRPV.	[193]

Table 6. Antiviral biological compounds.

Secondary Metabolite Class	Biocompound (Species)	Virus	Potential Antiviral Mechanism	Reference
Menthane monoterpenoids	Carvacrol (12) (*Lippia graveolens*)	Herpes viruses	No study.	[125]
Furocoumarin	Imperatorin (41) and phellopterin (42) (*Angelica archangelica*)	Herpes simplex virus type 1 Coxsackievirus B3	No study.	[126]
Chromone	Eugenin (43) (*Geum japonicum, Syzygium aromaticum*)	Herpes simplex virus	Prevents DNA polymerase.	[127]
Cinnamic acid derivative	Rosmarinic acid (47) (*M. officinalis*)	Herpes simplex type 2	Prevents virus entry into cells.	[129]
Flavan-3-ol	Epigallocatechin-3-gallate (48) (*Camellia sinensis*)	Epstein–Barr Virus	Blocks transcription and protein expression via ERK1/2 (extracellular-regulated-kinase 12) and PI3-K/Akt (phosphatidylinositol-3-kinase) pathways.	[131]
Phenol, Monomeric stilbene	Sesamol (49), resveratrol (2) (*Sesamum indicum*)	Epstein–Barr Virus	Inhibit early antigen activation.	[132]
Isoquinoline alkaloid	Berberine (15) (*Barnerini vulgaris*)	Epstein–Barr Virus	Inhibits cell proliferation and induces apoptosis in Epstein–Barr virus-infected cells by inhibiting p-STAT3.	[134]
Linear diarylheptanoid	Curcumin (33) (*Curcuma longa*)	Epstein–Barr Virus	Inhibits TPA-, butyrate-, and TGF-b induced levels of BZLF1 mRNA	[135]
Flavone	Apigenin (29) (purchased from Sigma-Aldrich Co., St. Louis, MO, USA)	Epstein–Barr Virus	Inhibits lytic proteins Zta, Rta, EAD, and DNase in B and epithelial cells and reduces the production of EBV viruses.	[136]
Oleanane triterpenoid	Glycyrrhizic acid (56) (*Glycyrrhiza radix*)	Epstein–Barr Virus	Interferes with the initial phase of EBV replication.	[137]
Flavone	Luteolin (25) (purchased from Sigma-Aldrich Co.)	Epstein–Barr Virus	Inhibits the expression of proteins encoded by the EBV lytic gene.	[138]
Isoflavone	Genistein (55) (purchased from Sigma-Aldrich)	Cytomegalovirus	Inhibits immediate-early (ie) protein function.	[139]
Flavone	Baicalein (57) (purchased from Sigma-Aldrich)	Cytomegalovirus	Inhibits EGFR's kinase activity to prevent viral entry.	[139]
Monomeric stilbene	Piceatannol (58) (purchased from Sigma-Aldrich)	Cytomegalovirus	Inhibits the lytic modifications and expression of hCMV early (E) and immediate—early (IE) proteins.	[140]
Monomeric stilbene	Resveratrol (2) (purchased from Sigma-Aldrich)	Cytomegalovirus	Reduces DNA replication.	[141]
Sulfide	Allitridin (59) (*A. sativum*)	Cytomegalovirus	Inhibits the IE genes' transcription.	[142]
Neolignan	Monoterpenylmagnolol (52) and β-eudesmol (50) (*Magnolia officinalis*)	Epstein–Barr Virus	Impede replication.	[133]
Cinnamic acid derivative	Isochlorogenic acid A (61) (*Laggera alata*)	Hepatitis virus B	Impedes replication.	[153]
Alkaloid	Amide alkaloids (*Piper longum*)	Hepatitis virus B	Inhibit replication and surface antigen expression.	[154]
Saponin	Saikosaponins (*Bupleurum* species)	Hepatitis virus B	Inhibit replication and surface antigen expression.	[156]
Protoberberine alkaloid	Dehydrocheilanthifoline (60) (*Corydalis saxifolia*)	Hepatitis virus B	Prevents reproduction.	[155]
Linear diarylheptanoid	Curcumin (33) (*Curcuma longa*)	Hepatitis virus B	Decreases Transcription.	[158]

Table 6. *Cont.*

Secondary Metabolite Class	Biocompound (Species)	Virus	Potential Antiviral Mechanism	Reference
Oleanane triterpenoid	Glycyrrhizinic acid (56) (*Glycyrrhiza glabra*)	Hepatitis virus B	Prevents viral reproduction.	[159,197]
Sesquiterpene lactone	Artemisinin (62) (*Artemisia annua*)	Hepatitis virus B	Prevents viral reproduction.	[159,197]
Isoflavonoid	LPRP-Et-97543 (93) (*Liriope platyphylla*)	Hepatitis virus B	Prevents viral reproduction.	[159,197]
Flavan-3-ol	Epigallocatechin-3-gallate (48) (*Camellia sinensis*)	Hepatitis virus B	Prevents viral reproduction.	[160]
Lignan	Flavonolignans (*Silybum marianum*)	Hepatitis C virus	No study.	[161]
Linear diarylheptanoid	Curcumin (33) (*Curcuma longa*)	Hepatitis C virus	Inhibits viral replication by blocking Akt-SREBP-1.	[162]
Flavan-3-ol	Epigallocatechin-3-gallate (48) (*Camellia sinensis*)	Hepatitis C virus	Inhibits viral introduction.	[163]
Flavone	Ladanein (77) (*Marrubium peregrinum*)	Hepatitis C virus	Inhibits viral introduction.	[164]
Peptide	Recombinant Griffithsin (*Nicotiana benthamiana*)	Hepatitis C virus	Inhibits viral cell–cell transmission.	[165]
Gallotannin	Tellimagrandin I (78) (*Rosae rugosae*)	Hepatitis C virus	Prevents viral penetration.	[159]
Benzopyran tannin and phenol	Chebulagic acid (64) and punicalagin (65) (*Terminalia chebula* Retz)	Hepatitis C virus	Inhibit fusion and cell–cell transmission.	[166]
Oleanane triterpenoid	Saikosaponin B2 (79) (*Bupleurum kaoi*)	Hepatitis C virus	Prevents viral attachment.	[159]
Furocoumarin, Quinoline alkaloid	Chalepine (66), pseudan IX (80) (*Ruta angustifolia*)	Hepatitis C virus	Reduce viral protein synthesis and viral RNA replication.	[159]
Lupane triterpenoids	Betulinic acid (67) and betulin (68) (*Betula alba* L)	Hepatitis C virus	Induce expression of TNF-α.	[168]
Oleanane triterpenoid	Glycyrrhizin (56) (*Glycyrrhiza glabra*)	Influenza virus	Initiates cell death in H5N1-infected cells.	[171]
Catechin	Catechins (*Camellia sinensis*)	Influenza virus	Inhibit both RNA synthesis and neuraminidase activity.	[175]
Dibenzylbutyrolactone lignans	Arctigenin (69) and arcitiin (70) (*Arctium lappa*)	Influenza virus	Anti-influenza A virus in vitro activity.	[176]
Monoterpenaldehydes	Citral a (45) and citral b (46) (*Melissa officinalis*)	H9N2 influenza virus	Have synergistic activity with oseltamivir.	[180]
Flavan-3-ols	Polyphenon E (poly E) (71) and epigallocatechin gallate (48) (*Camellia sinensis*)	Papillomavirus	Impede growth.	[182]
Sesquiterpene lactone	Artemisinin (62) (*Artemisia absintium*)	Papillomavirus	In ME-180 cells, this compound inhibits the expression of HPV-39, induces apoptosis, and reduces the proliferation of infected cells.	[183]
Tannin	Tannins (*Hamamelis virginiana*)	Papillomavirus	Inhibit HPV-16	[184]
Benzopyrone	Coumarin (33) (*Alternanthera philoxeroides*)	Chikungunya	Stops the dengue virus.	[185]
Anthraquinone, Stigmastane steroid, Flavanone, Anthranilic acid alkaloid, Glucosinolate	Emodin (82), β-sistosterol (63), hesperetin (74), indigo (75), and sinigrin (76) (*Isatis indigotica*)	SARS-CoV-1	Block the 3CL protease.	[194]

Table 6. *Cont.*

Secondary Metabolite Class	Biocompound (Species)	Virus	Potential Antiviral Mechanism	Reference
Flavones, Flavonol	Amentoflavone (53), apigenin (29), luteolin (25), quercetin (10) (*Torreya nucifera*)	SARS-CoV-1	Block the 3CL protease.	[195]
Indolizidine alkaloid	Lycorine (54) (*Lycoris radiata*)	SARS-CoV-1	Block 3CL protease.	[190]
Cinammic acid derivative, Anthraquinone, Oleanane triterpenoid, Flavonoid, Gallotannin	Caffeic acid (51), emodin (82), glycyrrhizin (56), luteolin (25), and tetra-O-galloyl-β-D-glucose (81)	SARS-CoV-2	Inhibit the spike protein's interaction with the ACE-2 receptor.	[196]

2.6.12. Molecules with Antiviral Activity Identified In Silico

Computational models enable us to simulate the interaction between the biocompound and the virus's target molecule [198]. Quercetin-7-O-glucoside inhibits influenza virus RNA polymerase. Quercetagetin, a flavonoid with activity against HVC through the inhibition of RNA bound to NS5B non-structural polymerase [199], naringenin, and quercetin, could inhibit hepatitis C virus proteases [200], and β-amyrin could inhibit hepatitis D virus proteases [201].

Luteolin could block SARS-CoV-2 entrance into cells [202]; isothymol and curcumin can block angiotensin-converting enzyme receptor (ACE2) activity [203]; gingerol binds to the spike protein; and quercetin with proteases [204], enterodiol, taxifolin, eriodictyol, leucopelargonidin, morin, and myricetin were found to exhibit remarkable binding affinities against the major protease (Mpro) and potato-like protease (PLpro) [205].

3. Conclusions

The Latin American plant species studied in the last 20 years have shown various secondary metabolites and families of natural products that could be used to fight against antimicrobial resistance. Of particular interest, due to the events experienced by humanity in recent years, are antivirals. Many studies still need to be carried out to determine the structure–activity relationship of different compounds. However, it is assumed that natural products belonging to the same family will act similarly, but this still needs to be corroborated. The great wealth that Latin America presents regarding plant species variety can be used to benefit global health.

Author Contributions: Conceptualization, L.R.H. and Z.N.J.; methodology, S.I.C.-C., C.R.-C. and J.L.G.-R.; validation, Z.N.J., L.R.H. and E.S.-A.; resources, E.S.-A.; data curation, L.R.H.; writing—original draft preparation, Z.N.J.; writing—review and editing, L.R.H.; supervision, E.S.-A.; project administration, L.R.H. All authors have read and agreed to the published version of the manuscript.

Funding: This research received no external funding and the APC was funded by the annual institutional budget of L.R.H.

Conflicts of Interest: The authors declare no conflict of interest.

References

1. Zhong, C.; Zhang, F.; Yao, J.; Zhu, Y.; Zhu, N.; Zhang, J.; Ouyang, X.; Zhang, T.; Li, B.; Xie, J.; et al. New Antimicrobial Peptides with Repeating Unit against Multidrug-Resistant Bacteria. *ACS Infect. Dis.* **2021**, *7*, 1619–1637. [CrossRef]
2. Fisher, R.A.; Gollan, B.; Helaine, S. Persistent Bacterial Infections and Persister Cells. *Nat. Rev. Microbiol.* **2017**, *15*, 453–464. [CrossRef] [PubMed]
3. Luo, G.G.; Gao, S.J. Global Health Concerns Stirred by Emerging Viral Infections. *J. Med. Virol.* **2020**, *92*, 399–400. [CrossRef] [PubMed]
4. Ji, C.; Liu, N.; Tu, J.; Li, Z.; Han, G.; Li, J.; Sheng, C. Drug Repurposing of Haloperidol: Discovery of New Benzocyclane Derivatives as Potent Antifungal Agents against Cryptococcosis and Candidiasis. *ACS Infect. Dis.* **2020**, *6*, 768–786. [CrossRef]

5. Kingston, D.G.I. Modern Natural Products Drug Discovery and Its Relevance to Biodiversity Conservation. *J. Nat. Prod.* **2011**, *74*, 496–511. [CrossRef]
6. O'Neill, J. *Tackling Drug-Resistance Infections Globally: Final Report and Recommendations*; Government of the United Kingdom: London, UK, 2016.
7. Zhu, Y.; Huang, W.E.; Yang, Q. Clinical Perspective of Antimicrobial Resistance in Bacteria. *Infect. Drug Resist.* **2022**, *15*, 735–746. [CrossRef] [PubMed]
8. Dadgostar, P. Antimicrobial Resistance: Implications and Costs. *Infect. Drug Resist.* **2019**, *12*, 3903–3910. [CrossRef] [PubMed]
9. Rodríguez-Baño, J.; Rossolini, G.M.; Schultsz, C.; Tacconelli, E.; Murthy, S.; Ohmagari, N.; Holmes, A.; Bachmann, T.; Goossens, H.; Canton, R.; et al. Antimicrobial Resistance Research in a Post-Pandemic World: Insights on Antimicrobial Resistance Research in the COVID-19 Pandemic. *J. Glob. Antimicrob. Resist.* **2021**, *25*, 5–7. [CrossRef] [PubMed]
10. Laxminarayan, R. The Overlooked Pandemic of Antimicrobial Resistance. *Lancet* **2022**, *399*, 606–607. [CrossRef]
11. Rizvi, S.G.; Ahammad, S.Z. COVID-19 and Antimicrobial Resistance: A Cross-Study. *Sci. Total Environ.* **2022**, *807*, 150873. [CrossRef]
12. Christaki, E.; Marcou, M.; Tofarides, A. Antimicrobial Resistance in Bacteria: Mechanisms, Evolution, and Persistence. *J. Mol. Evol.* **2020**, *88*, 26–40. [CrossRef] [PubMed]
13. Reygaert, W.C. An Overview of the Antimicrobial Resistance Mechanisms of Bacteria. *AIMS Microbiol.* **2018**, *4*, 482–501. [CrossRef] [PubMed]
14. Murugaiyan, J.; Kumar, P.A.; Rao, G.S.; Iskandar, K.; Hawser, S.; Hays, J.P.; Mohsen, Y.; Adukkadukkam, S.; Awuah, W.A.; Jose, R.A.M.; et al. Progress in Alternative Strategies to Combat Antimicrobial Resistance: Focus on Antibiotics. *Antibiotics* **2022**, *11*, 200. [CrossRef] [PubMed]
15. Anand, U.; Jacobo-Herrera, N.; Altemimi, A.; Lakhssassi, N. A Comprehensive Review on Medicinal Plants as Antimicrobial Therapeutics: Potential Avenues of Biocompatible Drug Discovery. *Metabolites* **2019**, *9*, 258. [CrossRef]
16. Srivastava, J.; Chandra, H.; Nautiyal, A.R.; Kalra, S.J.S. Antimicrobial Resistance (AMR) and Plant-Derived Antimicrobials (PDAms) as an Alternative Drug Line to Control Infections. *3 Biotech* **2014**, *4*, 451–460. [CrossRef]
17. Esmael, A.; Hassan, M.G.; Amer, M.M.; Abdelrahman, S.; Hamed, A.M.; Abd-raboh, H.A.; Foda, M.F. Antimicrobial Activity of Certain Natural-Based Plant Oils against the Antibiotic-Resistant Acne Bacteria. *Saudi J. Biol. Sci.* **2020**, *27*, 448–455. [CrossRef]
18. Intorasoot, A.; Chornchoem, P.; Sookkhee, S.; Intorasoot, S. Bactericidal Activity of Herbal Volatile Oil Extracts against Multidrug Resistant *Acinetobacter Baumannii*. *J. Intercult. Ethnopharmacol.* **2017**, *6*, 1. [CrossRef]
19. Dharmaratne, M.P.J.; Manoraj, A.; Thevanesam, V.; Ekanayake, A.; Kumar, N.S.; Liyanapathirana, V.; Abeyratne, E.; Bandara, B.M.R. *Terminalia Bellirica* Fruit Extracts: *In-Vitro* Antibacterial Activity against Selected Multidrug-Resistant Bacteria, Radical Scavenging Activity and Cytotoxicity Study on BHK-21 Cells. *BMC Complement. Altern. Med.* **2018**, *18*, 325. [CrossRef]
20. Khan, S.; Imran, M.; Imran, M.; Pindari, N. Antimicrobial Activity of Various Ethanolic Plant Extracts against Pathogenic Multi-Drug Resistant *Candida* spp. *Bioinformation* **2017**, *13*, 67–72. [CrossRef]
21. Khan, M.F.; Tang, H.; Lyles, J.T.; Pineau, R.; Mashwani, Z.-R.; Quave, C.L. Antibacterial Properties of Medicinal Plants From Pakistan against Multidrug-Resistant ESKAPE Pathogens. *Front. Pharmacol.* **2018**, *9*, 815. [CrossRef]
22. Alibi, S.; Crespo, D.; Navas, J. Plant-Derivatives Small Molecules with Antibacterial Activity. *Antibiotics* **2021**, *10*, 231. [CrossRef]
23. Jeyakumar, G.E.; Lawrence, R. Mechanisms of Bactericidal Action of Eugenol against *Escherichia coli*. *J. Herb. Med.* **2021**, *26*, 100406. [CrossRef]
24. Vestergaard, M.; Ingmer, H. Antibacterial and Antifungal Properties of Resveratrol. *Int. J. Antimicrob. Agents* **2019**, *53*, 716–723. [CrossRef] [PubMed]
25. Liang, H.; Huang, Q.; Zou, L.; Wei, P.; Lu, J.; Zhang, Y. Methyl Gallate: Review of Pharmacological Activity. *Pharmacol. Res.* **2023**, *194*, 106849. [CrossRef] [PubMed]
26. Dong, X.; Fu, J.; Yin, X.; Cao, S.; Li, X.; Lin, L.; Huyiligeqi; Ni, J. Emodin: A Review of Its Pharmacology, Toxicity and Pharmacokinetics. *Phytother. Res.* **2016**, *30*, 1207–1218. [CrossRef] [PubMed]
27. Ferdes, M. Antimicrobial Compounds from Plants. In *Fighting Antimicrobial Resistance*; Budimir, A., Ed.; IAPC Publishing: Zagreb, Croatia, 2018; pp. 243–271. ISBN 978-953-56942-6-7.
28. Nguyen, T.L.A.; Bhattacharya, D. Antimicrobial Activity of Quercetin: An Approach to Its Mechanistic Principle. *Molecules* **2022**, *27*, 2494. [CrossRef]
29. Farhadi, F.; Khameneh, B.; Iranshahi, M.; Iranshahy, M. Antibacterial Activity of Flavonoids and Their Structure–Activity Relationship: An Update Review. *Phytother. Res.* **2019**, *33*, 13–40. [CrossRef]
30. Asadi, S.; Nayeri-Fasaei, B.; Zahraei-Salehi, T.; Yahya-Rayat, R.; Shams, N.; Sharifi, A. Antibacterial and Anti-Biofilm Properties of Carvacrol Alone and in Combination with Cefixime against *Escherichia coli*. *BMC Microbiol.* **2023**, *23*, 55. [CrossRef]
31. Tian, L.; Wang, X.; Liu, R.; Zhang, D.; Wang, X.; Sun, R.; Guo, W.; Yang, S.; Li, H.; Gong, G. Antibacterial Mechanism of Thymol against *Enterobacter sakazakii*. *Food Control* **2021**, *123*, 107716. [CrossRef]
32. Chakraborty, P.; Dastidar, D.G.; Paul, P.; Dutta, S.; Basu, D.; Sharma, S.R.; Basu, S.; Sarker, R.K.; Sen, A.; Sarkar, A.; et al. Inhibition of Biofilm Formation of *Pseudomonas aeruginosa* by Caffeine: A Potential Approach for Sustainable Management of Biofilm. *Arch. Microbiol.* **2020**, *202*, 623–635. [CrossRef]
33. Peng, L.; Kang, S.; Yin, Z.; Jia, R.; Song, X.; Li, L.; Li, Z.; Zou, Y.; Liang, X.; Li, L.; et al. Antibacterial Activity and Mechanism of Berberine against *Streptococcus agalactiae*. *Int. J. Clin. Exp. Pathol.* **2015**, *8*, 5217–5223.

34. Mohanad, J.K.; Azhar, A.S.; Imad, H.H. Evaluation of Anti-Bacterial Activity and Bioactive Chemical Analysis of *Ocimum basilicum* Using Fourier Transform Infrared (FT-IR) and Gas Chromatography-Mass Spectrometry (GC-MS) Techniques. *J. Pharmacogn. Phytother.* **2016**, *8*, 127–146. [CrossRef]
35. Choo, J.H.; Rukayadi, Y.; Hwang, J.-K. Inhibition of Bacterial Quorum Sensing by Vanilla Extract. *Lett. Appl. Microbiol.* **2006**, *42*, 637–641. [CrossRef]
36. Ohene-Agyei, T.; Mowla, R.; Rahman, T.; Venter, H. Phytochemicals Increase the Antibacterial Activity of Antibiotics by Acting on a Drug Efflux Pump. *MicrobiologyOpen* **2014**, *3*, 885–896. [CrossRef] [PubMed]
37. Castro, F.A.V.; Mariani, D.; Panek, A.D.; Eleutherio, E.C.A.; Pereira, M.D. Cytotoxicity Mechanism of Two Naphthoquinones (Menadione and Plumbagin) in *Saccharomyces cerevisiae*. *PLoS ONE* **2008**, *3*, e3999. [CrossRef] [PubMed]
38. Ghosh, J.; Sil, P.C. Arjunolic Acid: A New Multifunctional Therapeutic Promise of Alternative Medicine. *Biochimie* **2013**, *95*, 1098–1109. [CrossRef] [PubMed]
39. Merghni, A.; Noumi, E.; Hadded, O.; Dridi, N.; Panwar, H.; Ceylan, O.; Mastouri, M.; Snoussi, M. Assessment of the Antibiofilm and Antiquorum Sensing Activities of *Eucalyptus globulus* Essential Oil and Its Main Component 1,8-Cineole against Methicillin-Resistant *Staphylococcus aureus* Strains. *Microb. Pathog.* **2018**, *118*, 74–80. [CrossRef]
40. McLean, R.J.C.; Pierson, L.S.; Fuqua, C. A Simple Screening Protocol for the Identification of Quorum Signal Antagonists. *J. Microbiol. Methods* **2004**, *58*, 351–360. [CrossRef]
41. Carranza, M.G.; Sevigny, M.B.; Banerjee, D.; Fox-Cubley, L. Antibacterial Activity of Native California Medicinal Plant Extracts Isolated from *Rhamnus Californica* and *Umbellularia Californica*. *Ann. Clin. Microbiol. Antimicrob.* **2015**, *14*, 29. [CrossRef]
42. Bouyahya, A.; Dakka, N.; Et-Touys, A.; Abrini, J.; Bakri, Y. Medicinal Plant Products Targeting Quorum Sensing for Combating Bacterial Infections. *Asian Pac. J. Trop. Med.* **2017**, *10*, 729–743. [CrossRef]
43. Salem, M.Z.; Ali, H.M.; El-Shanhorey, N.A.; Abdel-Megeed, A. Evaluation of Extracts and Essential Oil from *Callistemon viminalis* Leaves: Antibacterial and Antioxidant Activities, Total Phenolic and Flavonoid Contents. *Asian Pac. J. Trop. Med.* **2013**, *6*, 785–791. [CrossRef]
44. Paredes, A.; Leyton, Y.; Riquelme, C.; Morales, G. A Plant from the Altiplano of Northern Chile *Senecio nutans*, Inhibits the *Vibrio cholerae* Pathogen. *SpringerPlus* **2016**, *5*, 1788. [CrossRef] [PubMed]
45. Alvin, A.; Miller, K.I.; Neilan, B.A. Exploring the Potential of Endophytes from Medicinal Plants as Sources of Antimycobacterial Compounds. *Microbiol. Res.* **2014**, *169*, 483–495. [CrossRef] [PubMed]
46. Barbieri, R.; Coppo, E.; Marchese, A.; Daglia, M.; Sobarzo-Sánchez, E.; Nabavi, S.F.; Nabavi, S.M. Phytochemicals for Human Disease: An Update on Plant-Derived Compounds Antibacterial Activity. *Microbiol. Res.* **2017**, *196*, 44–68. [CrossRef]
47. Botelho, M.A.; Nogueira, N.A.P.; Bastos, G.M.; Fonseca, S.G.C.; Lemos, T.L.G.; Matos, F.J.A.; Montenegro, D.; Heukelbach, J.; Rao, V.S.; Brito, G.A.C. Antimicrobial Activity of the Essential Oil from *Lippia sidoides*, Carvacrol and Thymol against Oral Pathogens. *Braz. J. Med. Biol. Res.* **2007**, *40*, 349–356. [CrossRef]
48. Subramani, R.; Narayanasamy, M.; Feussner, K.-D. Plant-Derived Antimicrobials to Fight against Multi-Drug-Resistant Human Pathogens. *3 Biotech* **2017**, *7*, 172. [CrossRef]
49. Swamy, M.; Sinniah, U. A Comprehensive Review on the Phytochemical Constituents and Pharmacological Activities of *Pogostemon cablin* Benth.: An Aromatic Medicinal Plant of Industrial Importance. *Molecules* **2015**, *20*, 8521–8547. [CrossRef]
50. Proestos, C.; Chorianopoulos, N.; Nychas, G.-J.E.; Komaitis, M. RP-HPLC Analysis of the Phenolic Compounds of Plant Extracts. Investigation of Their Antioxidant Capacity and Antimicrobial Activity. *J. Agric. Food Chem.* **2005**, *53*, 1190–1195. [CrossRef] [PubMed]
51. Rojas, A.; Hernandez, L.; Pereda-Miranda, R.; Mata, R. Screening for Antimicrobial Activity of Crude Drug Extracts and Pure Natural Products from Mexican Medicinal Plants. *J. Ethnopharmacol.* **1992**, *35*, 275–283. [CrossRef]
52. Song, X.; Xia, Y.-X.; He, Z.-D.; Zhang, H.-J. A Review of Natural Products with Anti-Biofilm Activity. *Curr. Org. Chem.* **2018**, *22*, 789–817. [CrossRef]
53. Prabu, G.R.; Gnanamani, A.; Sadulla, S. Guaijaverin—A Plant Flavonoid as Potential Antiplaque Agent against *Streptococcus mutans*. *J. Appl. Microbiol.* **2006**, *101*, 487–495. [CrossRef] [PubMed]
54. Ali, B.H.; Blunden, G.; Tanira, M.O.; Nemmar, A. Some Phytochemical, Pharmacological and Toxicological Properties of Ginger (*Zingiber officinale* Roscoe): A Review of Recent Research. *Food Chem. Toxicol.* **2008**, *46*, 409–420. [CrossRef]
55. Kim, H.-S.; Park, H.-D. Ginger Extract Inhibits Biofilm Formation by *Pseudomonas aeruginosa* PA14. *PLoS ONE* **2013**, *8*, e76106. [CrossRef]
56. Kumar, L.; Chhibber, S.; Harjai, K. Zingerone Inhibit Biofilm Formation and Improve Antibiofilm Efficacy of Ciprofloxacin against *Pseudomonas aeruginosa* PAO1. *Fitoterapia* **2013**, *90*, 73–78. [CrossRef] [PubMed]
57. GBIF. Sistema Global de Información Sobre Biodiversidad. Available online: https://www.gbif.org/es/ (accessed on 25 July 2023).
58. Cheek, M.; Nic Lughadha, E.; Kirk, P.; Lindon, H.; Carretero, J.; Looney, B.; Douglas, B.; Haelewaters, D.; Gaya, E.; Llewellyn, T.; et al. New Scientific Discoveries: Plants and Fungi. *Plants People Planet* **2020**, *2*, 371–388. [CrossRef]
59. Fisher, M.C.; Henk, D.A.; Briggs, C.J.; Brownstein, J.S.; Madoff, L.C.; McCraw, S.L.; Gurr, S.J. Emerging Fungal Threats to Animal, Plant and Ecosystem Health. *Nature* **2012**, *484*, 186–194. [CrossRef]
60. Rokas, A. Evolution of the Human Pathogenic Lifestyle in Fungi. *Nat. Microbiol.* **2022**, *7*, 607–619. [CrossRef]
61. Restrepo-Rivera, L.M.; Cardona-Castro, N. Micobioma: Diversidad Fúngica En El Cuerpo Humano. *CES Med.* **2021**, *35*, 113–125. [CrossRef]

62. Seelbinder, B.; Chen, J.; Brunke, S.; Vazquez-Uribe, R.; Santhaman, R.; Meyer, A.C.; De Oliveira Lino, F.S.; Chan, K.F.; Loos, D.; Imamovic, L.; et al. Antibiotics Create a Shift from Mutualism to Competition in Human Gut Communities with a Longer-Lasting Impact on Fungi than Bacteria. *Microbiome* **2020**, *8*, 133. [CrossRef]
63. Bongomin, F.; Gago, S.; Oladele, R.O.; Denning, D.W. Global and Multi-National Prevalence of Fungal Diseases—Estimate Precision. *J. Fungi* **2017**, *3*, 57. [CrossRef]
64. Guarro, J. Taxonomía y Biología de Los Hongos Causantes de Infección En Humanos. *Enferm. Infecc. Microbiol. Clin.* **2012**, *30*, 33–39. [CrossRef] [PubMed]
65. Köhler, J.R.; Casadevall, A.; Perfect, J. The Spectrum of Fungi That Infects Humans. *Cold Spring Harb. Perspect. Med.* **2015**, *5*, a019273. [CrossRef] [PubMed]
66. Brown, G.D.; Denning, D.W.; Gow, N.A.R.; Levitz, S.M.; Netea, M.G.; White, T.C. Hidden Killers: Human Fungal Infections. *Sci. Transl. Med.* **2012**, *4*, 165rv13. [CrossRef] [PubMed]
67. Pemán, J.; Salavert, M. Epidemiología General de La Enfermedad Fúngica Invasora. *Enfermedades Infecc. Microbiol. Clínica* **2012**, *30*, 90–98. [CrossRef]
68. Shapiro, R.S.; Robbins, N.; Cowen, L.E. Regulatory Circuitry Governing Fungal Development, Drug Resistance, and Disease. *Microbiol. Mol. Biol. Rev.* **2011**, *75*, 213–267. [CrossRef]
69. Perfect, J.R. The Antifungal Pipeline: A Reality Check. *Nat. Rev. Drug Discov.* **2017**, *16*, 603–616. [CrossRef]
70. Oliveira, L.V.N.; Wang, R.; Specht, C.A.; Levitz, S.M. Vaccines for Human Fungal Diseases: Close but Still a Long Way to Go. *Npj Vaccines* **2021**, *6*, 33. [CrossRef]
71. Romani, L. Immunity to Fungal Infections. *Nat. Rev. Immunol.* **2011**, *11*, 275–288. [CrossRef]
72. Carmo, A.; Rocha, M.; Pereirinha, P.; Tomé, R.; Costa, E. Antifungals: From Pharmacokinetics to Clinical Practice. *Antibiotics* **2023**, *12*, 884. [CrossRef]
73. Quiles-Melero, I.; García-Rodríguez, J. Systemic Antifungal Drugs. *Rev. Iberoam. Micol.* **2021**, *38*, 42–46. [CrossRef]
74. Ruiz-Camps, I.; Cuenca-Estrella, M. Antifungals for Systemic Use. *Enferm. Infecc. Microbiol. Clin.* **2009**, *27*, 353–362. [CrossRef] [PubMed]
75. Shaw, K.J.; Ibrahim, A.S. Fosmanogepix: A Review of the First-in-Class Broad Spectrum Agent for the Treatment of Invasive Fungal Infections. *J. Fungi* **2020**, *6*, 239. [CrossRef] [PubMed]
76. Gonzalez-Lara, M.F.; Sifuentes-Osornio, J.; Ostrosky-Zeichner, L. Drugs in Clinical Development for Fungal Infections. *Drugs* **2017**, *77*, 1505–1518. [CrossRef] [PubMed]
77. Hope, W.W.; McEntee, L.; Livermore, J.; Whalley, S.; Johnson, A.; Farrington, N.; Kolamunnage-Dona, R.; Schwartz, J.; Kennedy, A.; Law, D.; et al. Pharmacodynamics of the Orotomides against *Aspergillus fumigatus*: New Opportunities for Treatment of Multidrug-Resistant Fungal Disease. *mBio* **2017**, *8*, e01157-17. [CrossRef] [PubMed]
78. Fairlamb, A.H.; Gow, N.A.R.; Matthews, K.R.; Waters, A.P. Drug Resistance in Eukaryotic Microorganisms. *Nat. Microbiol.* **2016**, *1*, 16092. [CrossRef] [PubMed]
79. Berman, J.; Krysan, D.J. Drug Resistance and Tolerance in Fungi. *Nat. Rev. Microbiol.* **2020**, *18*, 319–331. [CrossRef] [PubMed]
80. Fisher, M.C.; Alastruey-Izquierdo, A.; Berman, J.; Bicanic, T.; Bignell, E.M.; Bowyer, P.; Bromley, M.; Brüggemann, R.; Garber, G.; Cornely, O.A.; et al. Tackling the Emerging Threat of Antifungal Resistance to Human Health. *Nat. Rev. Microbiol.* **2022**, *20*, 557–571. [CrossRef]
81. Shrestha, S.K.; Fosso, M.Y.; Garneau-Tsodikova, S. A Combination Approach to Treating Fungal Infections. *Sci. Rep.* **2015**, *5*, 17070. [CrossRef]
82. Carpinella, M.C.; Andrione, D.G.; Ruiz, G.; Palacios, S.M. Screening for Acetylcholinesterase Inhibitory Activity in Plant Extracts from Argentina. *Phytother. Res.* **2010**, *24*, 259–263. [CrossRef]
83. Méndez-Chávez, M.; Ledesma-Escobar, C.A.; Hidalgo-Morales, M.; Rodríguez-Jimenes, G.D.C.; Robles-Olvera, V.J. Antifungal Activity Screening of Fractions from *Annona cherimola* Mill. Leaf Extract against *Fusarium oxysporum*. *Arch. Microbiol.* **2022**, *204*, 330. [CrossRef]
84. Campos, L.M.; Silva, T.P.; De Oliveira Lemos, A.S.; Mendonça Diniz, I.O.; Palazzi, C.; Novaes Da Rocha, V.; De Freitas Araújo, M.G.; Melo, R.C.N.; Fabri, R.L. Antibiofilm Potential of *Annona muricata* L. Ethanolic Extract against Multi-Drug Resistant Candida Albicans. *J. Ethnopharmacol.* **2023**, *315*, 116682. [CrossRef] [PubMed]
85. Diaz Napal, G.N.; Buffa, L.M.; Nolli, L.C.; Defagó, M.T.; Valladares, G.R.; Carpinella, M.C.; Ruiz, G.; Palacios, S.M. Screening of Native Plants from Central Argentina against the Leaf-Cutting Ant *Acromyrmex lundi* (Guérin) and Its Symbiotic Fungus. *Ind. Crops Prod.* **2015**, *76*, 275–280. [CrossRef]
86. Navarro García, V.M.; Gonzalez, A.; Fuentes, M.; Aviles, M.; Rios, M.Y.; Zepeda, G.; Rojas, M.G. Antifungal Activities of Nine Traditional Mexican Medicinal Plants. *J. Ethnopharmacol.* **2003**, *87*, 85–88. [CrossRef]
87. Poma-Castillo, L.; Espinoza-Poma, M. Antifungal Activity of Ethanol-Extracted Bixa Orellana (L) (Achiote) on Candida Albicans, at Six Different Concentrations. *J. Contemp. Dent. Pract.* **2019**, *20*, 1159–1163. [CrossRef]
88. Wilson, B.; Abraham, G.; Manju, V.S.; Mathew, M.; Vimala, B.; Sundaresan, S.; Nambisan, B. Antimicrobial Activity of Curcuma Zedoaria and Curcuma Malabarica Tubers. *J. Ethnopharmacol.* **2005**, *99*, 147–151. [CrossRef] [PubMed]
89. Chen, X.; Wang, C.; Xu, L.; Chen, X.; Wang, W.; Yang, G.; Tan, R.X.; Li, E.; Jin, Y. A Laboratory Evaluation of Medicinal Herbs Used in China for the Treatment of Hand, Foot, and Mouth Disease. *Evid. Based Complement. Alternat. Med.* **2013**, *2013*, 504563. [CrossRef]

90. Mir-Rashed, N.; Cruz, I.; Jessulat, M.; Dumontier, M.; Chesnais, C.; Juliana, N.; Amiguet, V.T.; Golshani, A.; Arnason, J.T.; Smith, M.L. Disruption of Fungal Cell Wall by Antifungal Echinacea Extracts. *Med. Mycol.* **2010**, *48*, 949–958. [CrossRef]
91. Merali, S.; Binns, S.; Paulin-Levasseur, M.; Ficker, C.; Smith, M.; Baum, B.; Brovelli, E.; Arnason, J.T. Antifungal and Anti-Inflammatory Activity of the Genus Echinacea. *Pharm. Biol.* **2003**, *41*, 412–420. [CrossRef]
92. Muschietti, L.; Derita, M.; Sülsen, V.; De Dios Muñoz, J.; Ferraro, G.; Zacchino, S.; Martino, V. In Vitro Antifungal Assay of Traditional Argentine Medicinal Plants. *J. Ethnopharmacol.* **2005**, *102*, 233–238. [CrossRef]
93. Fred-Jaiyesimi, A.A.; Abo, K.A. Phytochemical and Antimicrobial Analysis of the Crude Extract, Petroleum Ether and Chloroform Fractions of Euphorbia Heterophylla Linn Whole Plant. *Pharmacogn. J.* **2010**, *2*, 1–4. [CrossRef]
94. Sharma, A.; Angulo-Bejarano, P.; Madariaga-Navarrete, A.; Oza, G.; Iqbal, H.; Cardoso-Taketa, A.; Luisa Villarreal, M. Multidisciplinary Investigations on Galphimia Glauca: A Mexican Medicinal Plant with Pharmacological Potential. *Molecules* **2018**, *23*, 2985. [CrossRef]
95. Alshami, I.; Alharbi, A.E. *Hibiscus sabdariffa* Extract Inhibits in Vitro Biofilm Formation Capacity of *Candida albicans* Isolated from Recurrent Urinary Tract Infections. *Asian Pac. J. Trop. Biomed.* **2014**, *4*, 104–108. [CrossRef] [PubMed]
96. Quiroga, E.N.; Sampietro, A.R.; Vattuone, M.A. Screening Antifungal Activities of Selected Medicinal Plants. *J. Ethnopharmacol.* **2001**, *74*, 89–96. [CrossRef] [PubMed]
97. Moreno, M.A.; Córdoba, S.; Zampini, I.C.; Mercado, M.I.; Ponessa, G.; Sayago, J.E.; Ramos, L.L.P.; Schmeda-Hirschmann, G.; Isla, M.I. Argentinean *Larrea* Dry Extracts with Potential Use in Vaginal Candidiasis. *Nat. Prod. Commun.* **2018**, *13*, 1934578X1801300. [CrossRef]
98. Navarro-García, V.M.; Rojas, G.; Avilés, M.; Fuentes, M.; Zepeda, G. In Vitro Antifungal Activity of Coumarin Extracted from *Loeselia Mexicana* Brand: Antifungal Coumarins from *Loeselia mexicana*. *Mycoses* **2011**, *54*, e569–e571. [CrossRef]
99. Morais-Braga, M.F.B.; Souza, T.M.; Santos, K.K.A.; Andrade, J.C.; Guedes, G.M.M.; Tintino, S.R.; Sobral-Souza, C.E.; Costa, J.G.M.; Menezes, I.R.A.; Saraiva, A.A.F.; et al. Antimicrobial and Modulatory Activity of Ethanol Extract of the Leaves from *Lygodium venustum* SW. *Am. Fern J.* **2012**, *102*, 154–160. [CrossRef]
100. Navarro García, V.M.; Rojas, G.; Gerardo Zepeda, L.; Aviles, M.; Fuentes, M.; Herrera, A.; Jiménez, E. Antifungal and Antibacterial Activity of Four Selected Mexican Medicinal Plants. *Pharm. Biol.* **2006**, *44*, 297–300. [CrossRef]
101. AL-Rubaey, N.K.F.; Abbas, F.M.; Hameed, I.H. Antibacterial and Anti-Fungal Activity of Methanolic Extract of *Passiflora caerulea*. *Indian J. Public Health Res. Dev.* **2019**, *10*, 930. [CrossRef]
102. Zabka, M.; Pavela, R.; Slezakova, L. Antifungal Effect of Pimenta Dioica Essential Oil against Dangerous Pathogenic and Toxinogenic Fungi. *Ind. Crops Prod.* **2009**, *30*, 250–253. [CrossRef]
103. Derita, M.G.; Leiva, M.L.; Zacchino, S.A. Influence of Plant Part, Season of Collection and Content of the Main Active Constituent, on the Antifungal Properties of Polygonum Acuminatum Kunth. *J. Ethnopharmacol.* **2009**, *124*, 377–383. [CrossRef]
104. Javed, B.; Farooq, F.; Ibrahim, M.; Abbas, H.A.B.; Jawwad, H.; Zehra, S.S.; Ahmad, H.M.; Sarwer, A.; Malik, K.; Nawaz, K. Antibacterial and Antifungal Activity of Methanolic Extracts of *Salix alba* L. against Various Disease Causing Pathogens. *Braz. J. Biol.* **2023**, *83*, e243332. [CrossRef] [PubMed]
105. Hnatyszyn, O.; Juárez, S.; Ouviña, A.; Martino, V.; Zacchino, S.; Ferraro, G. Phytochemical Analysis and Antifungal Evaluation of *Sebastiania commersoniana*. Extracts. *Pharm. Biol.* **2007**, *45*, 404–406. [CrossRef]
106. Cáceres, A.; Cruz, S.M.; Martínez, V.; Gaitán, I.; Santizo, A.; Gattuso, S.; Gattuso, M. Ethnobotanical, Pharmacognostical, Pharmacological and Phytochemical Studies on Smilax Domingensis in Guatemala. *Rev. Bras. Farmacogn.* **2012**, *22*, 239–248. [CrossRef]
107. Pinto, E.; Vale-Silva, L.; Cavaleiro, C.; Salgueiro, L. Antifungal Activity of the Clove Essential Oil from Syzygium Aromaticum on Candida, Aspergillus and Dermatophyte Species. *J. Med. Microbiol.* **2009**, *58*, 1454–1462. [CrossRef]
108. Awere, C.A.; Githae, E.W.; Gichumbi, J.M. Phytochemical Analysis and Antifungal Activity of *Tithonia diversifolia* and *Kigelia africana* Extracts against Fusarium Oxysporum in Tomato. *Afr. J. Agric. Res.* **2021**, *17*, 726–732. [CrossRef]
109. Simonetti, G.; Brasili, E.; Pasqua, G. Antifungal Activity of Phenolic and Polyphenolic Compounds from Different Matrices of *Vitis vinifera* L. against Human Pathogens. *Molecules* **2020**, *25*, 3748. [CrossRef]
110. Svetaz, L.; Agüero, M.; Alvarez, S.; Luna, L.; Feresin, G.; Derita, M.; Tapia, A.; Zacchino, S. Antifungal Activity of *Zuccagnia Punctata* Cav.: Evidence for the Mechanism of Action. *Planta Med.* **2007**, *73*, 1074–1080. [CrossRef] [PubMed]
111. Lefkowitz, E.J.; Dempsey, D.M.; Hendrickson, R.C.; Orton, R.J.; Siddell, S.G.; Smith, D.B. Virus Taxonomy: The Database of the International Committee on Taxonomy of Viruses (ICTV). *Nucleic Acids Res.* **2018**, *46*, D708–D717. [CrossRef]
112. Ben-Shabat, S.; Yarmolinsky, L.; Porat, D.; Dahan, A. Antiviral Effect of Phytochemicals from Medicinal Plants: Applications and Drug Delivery Strategies. *Drug Deliv. Transl. Res.* **2020**, *10*, 354–367. [CrossRef]
113. Cliver, D.O. Capsid and Infectivity in Virus Detection. *Food Environ. Virol.* **2009**, *1*, 123–128. [CrossRef]
114. Jassim, S.A.; Naji, M.A. Novel Antiviral Agents: A Medicinal Plant Perspective. *J. Appl. Microbiol.* **2003**, *95*, 412–427. [CrossRef]
115. Sharma, N. Efficacy of Garlic and Onion against Virus. *Int. J. Res. Pharm. Sci.* **2019**, *10*, 3578–3586. [CrossRef]
116. Hudson, J.; Vimalanathan, S. Echinacea—A Source of Potent Antivirals for Respiratory Virus Infections. *Pharmaceuticals* **2011**, *4*, 1019–1031. [CrossRef]
117. Signer, J.; Jonsdottir, H.R.; Albrich, W.C.; Strasser, M.; Züst, R.; Ryter, S.; Ackermann-Gäumann, R.; Lenz, N.; Siegrist, D.; Suter, A. In Vitro Virucidal Activity of Echinaforce®, an *Echinacea purpurea* Preparation, against Coronaviruses, Including Common Cold Coronavirus 229E and SARS-CoV-2. *Virol. J.* **2020**, *17*, 136. [CrossRef]

118. Weng, J.-R.; Lin, C.-S.; Lai, H.-C.; Lin, Y.-P.; Wang, C.-Y.; Tsai, Y.-C.; Wu, K.-C.; Huang, S.-H.; Lin, C.-W. Antiviral Activity of *Sambucus Formosana Nakai* Ethanol Extract and Related Phenolic Acid Constituents against Human Coronavirus NL63. *Virus Res.* **2019**, *273*, 197767. [CrossRef]
119. Chathuranga, K.; Kim, M.S.; Lee, H.-C.; Kim, T.-H.; Kim, J.-H.; Gayan Chathuranga, W.A.; Ekanayaka, P.; Wijerathne, H.; Cho, W.-K.; Kim, H.I. Anti-Respiratory Syncytial Virus Activity of *Plantago asiatica* and *Clerodendrum trichotomum* Extracts in Vitro and in Vivo. *Viruses* **2019**, *11*, 604. [CrossRef]
120. Kunsorn, P.; Ruangrungsi, N.; Lipipun, V.; Khanboon, A.; Rungsihirunrat, K. The Identities and Anti-Herpes Simplex Virus Activity of *Clinacanthus nutans* and *Clinacanthus siamensis*. *Asian Pac. J. Trop. Biomed.* **2013**, *3*, 284–290. [CrossRef]
121. Chiang, L.C.; Chiang, W.; Chang, M.Y.; Ng, L.T.; Lin, C.C. Antiviral Activity of *Plantago major* Extracts and Related Compounds in Vitro. *Antivir. Res.* **2002**, *55*, 53–62. [CrossRef]
122. Wahab, N.Z.A.; Bunawan, H.; Ibrahim, N. Cytotoxicity and Antiviral Activity of Methanol Extract from *Polygonum minus*. In Proceedings of the AIP Conference Proceedings, Selangor, Malaysia, 15–16 April 2015; AIP Publishing: Melville, NY, USA, 2015; Volume 1678.
123. Zandi, K.; Zadeh, M.A.; Sartavi, K.; Rastian, Z. Antiviral Activity of *Aloe vera* against Herpes Simplex Virus Type 2: An in Vitro Study. *Afr. J. Biotechnol.* **2007**, *6*, 1770–1773. [CrossRef]
124. van de Sand, L.; Bormann, M.; Schmitz, Y.; Heilingloh, C.S.; Witzke, O.; Krawczyk, A. Antiviral Active Compounds Derived from Natural Sources Against Herpes Simplex Viruses. *Viruses* **2021**, *13*, 1386. [CrossRef]
125. Pilau, M.R.; Alves, S.H.; Weiblen, R.; Arenhart, S.; Cueto, A.P.; Lovato, L.T. Antiviral Activity of the *Lippia graveolens* (Mexican Oregano) Essential Oil and Its Main Compound Carvacrol against Human and Animal Viruses. *Braz. J. Microbiol.* **2011**, *42*, 1616–1624. [CrossRef]
126. Rajtar, B.; Skalicka-Woźniak, K.; Świątek, Ł.; Stec, A.; Boguszewska, A.; Polz-Dacewicz, M. Antiviral Effect of Compounds Derived from *Angelica archangelica* L. on Herpes Simplex Virus-1 and Coxsackievirus B3 Infections. *Food Chem. Toxicol.* **2017**, *109*, 1026–1031. [CrossRef]
127. Kurokawa, M.; Hozumi, T.; Basnet, P.; Nakano, M.; Kadota, S.; Namba, T.; Kawana, T.; Shiraki, K. Purification and Characterization of Eugeniin as an Anti-Herpesvirus Compound from Geum Japonicum Andsyzygium Aromaticum. *J. Pharmacol. Exp. Ther.* **1998**, *284*, 728–735.
128. Schnitzler, P.; Schuhmacher, A.; Astani, A.; Reichling, J. *Melissa officinalis* Oil Affects Infectivity of Enveloped Herpesviruses. *Phytomedicine* **2008**, *15*, 734–740. [CrossRef]
129. Mazzanti, G.; Battinelli, L.; Pompeo, C.; Serrilli, A.M.; Rossi, R.; Sauzullo, I.; Mengoni, F.; Vullo, V. Inhibitory Activity of *Melissa officinalis* L. Extract on Herpes Simplex Virus Type 2 Replication. *Nat. Prod. Res.* **2008**, *22*, 1433–1440. [CrossRef]
130. Astani, A.; Heidary Navid, M.; Schnitzler, P. Attachment and Penetration of Acyclovir-Resistant Herpes Simplex Virus Are Inhibited by *Melissa officinalis* Extract. *Phytother. Res.* **2014**, *28*, 1547–1552. [CrossRef]
131. Liu, S.; Li, H.; Chen, L.; Yang, L.; Li, L.; Tao, Y.; Li, W.; Li, Z.; Liu, H.; Tang, M. (-)-Epigallocatechin-3-Gallate Inhibition of Epstein–Barr Virus Spontaneous Lytic Infection Involves ERK1/2 and PI3-K/Akt Signaling in EBV-Positive Cells. *Carcinogenesis* **2013**, *34*, 627–637. [CrossRef]
132. Kapadia, G.J.; Azuine, M.A.; Tokuda, H.; Takasaki, M.; Mukainaka, T.; Konoshima, T.; Nishino, H. Chemopreventive Effect of Resveratrol, Sesamol, Sesame Oil and Sunflower Oil in the Epstein–Barr Virus Early Antigen Activation Assay and the Mouse Skin Two-Stage Carcinogenesis. *Pharmacol. Res.* **2002**, *45*, 499–505. [CrossRef]
133. Konoshima, T.; Takasaki, M.; Kozuka, M.; Tokuda, H.; Nishino, H.; Iwashima, A.; Haruna, M.; Ito, K.; Tanabe, M. Inhibitory Effects on Epstein-Barr Virus Activation and Anti-Tumor Promoting Activities of Neolignans from *Magnolia officinalis*. *Planta Med.* **1990**, *56*, 653. [CrossRef]
134. Wang, C.; Wang, H.; Zhang, Y.; Guo, W.; Long, C.; Wang, J.; Liu, L.; Sun, X. Berberine Inhibits the Proliferation of Human Nasopharyngeal Carcinoma Cells via an Epstein-Barr Virus Nuclear Antigen 1-Dependent Mechanism. *Oncol. Rep.* **2017**, *37*, 2109–2120. [CrossRef]
135. Hergenhahn, M.; Soto, U.; Weninger, A.; Polack, A.; Hsu, C.-H.; Cheng, A.-L.; Rösl, F. The Chemopreventive Compound Curcumin Is an Efficient Inhibitor of Epstein-Barr Virus BZLF1 Transcription in Raji DR-LUC Cells. *Mol. Carcinog.* **2002**, *33*, 137–145. [CrossRef]
136. Wu, C.-C.; Fang, C.-Y.; Cheng, Y.-J.; Hsu, H.-Y.; Chou, S.-P.; Huang, S.-Y.; Tsai, C.-H.; Chen, J.-Y. Inhibition of Epstein-Barr Virus Reactivation by the Flavonoid Apigenin. *J. Biomed. Sci.* **2017**, *24*, 2. [CrossRef]
137. Lin, J.-C.; Cherng, J.-M.; Hung, M.-S.; Baltina, L.A.; Baltina, L.; Kondratenko, R. Inhibitory Effects of Some Derivatives of Glycyrrhizic Acid against Epstein-Barr Virus Infection: Structure–Activity Relationships. *Antivir. Res.* **2008**, *79*, 6–11. [CrossRef]
138. Wu, C.-C.; Fang, C.-Y.; Hsu, H.-Y.; Chen, Y.-J.; Chou, S.-P.; Huang, S.-Y.; Cheng, Y.-J.; Lin, S.-F.; Chang, Y.; Tsai, C.-H. Luteolin Inhibits Epstein-Barr Virus Lytic Reactivation by Repressing the Promoter Activities of Immediate-Early Genes. *Antivir. Res.* **2016**, *132*, 99–110. [CrossRef]
139. Evers, D.L.; Chao, C.-F.; Wang, X.; Zhang, Z.; Huong, S.-M.; Huang, E.-S. Human Cytomegalovirus-Inhibitory Flavonoids: Studies on Antiviral Activity and Mechanism of Action. *Antivir. Res.* **2005**, *68*, 124–134. [CrossRef]
140. Wang, S.-Y.; Zhang, J.; Xu, X.-G.; Su, H.-L.; Xing, W.-M.; Zhang, Z.-S.; Jin, W.-H.; Dai, J.-H.; Wang, Y.-Z.; He, X.-Y. Inhibitory Effects of Piceatannol on Human Cytomegalovirus (hCMV) in Vitro. *J. Microbiol.* **2020**, *58*, 716–723. [CrossRef]

141. Evers, D.L.; Wang, X.; Huong, S.-M.; Huang, D.Y.; Huang, E.-S. 3,4′,5-Trihydroxy-Trans-Stilbene (Resveratrol) Inhibits Human Cytomegalovirus Replication and Virus-Induced Cellular Signaling. *Antivir. Res.* **2004**, *63*, 85–95. [CrossRef]
142. Zhen, H.; Fang, F.; Ye, D.; Shu, S.; Zhou, Y.; Dong, Y.; Nie, X.; Li, G. Experimental Study on the Action of Allitridin against Human Cytomegalovirus in Vitro: Inhibitory Effects on Immediate-Early Genes. *Antivir. Res.* **2006**, *72*, 68–74. [CrossRef]
143. Abd-Elazem, I.S.; Chen, H.S.; Bates, R.B.; Huang, R.C.C. Isolation of Two Highly Potent and Non-Toxic Inhibitors of Human Immunodeficiency Virus Type 1 (HIV-1) Integrase from *Salvia miltiorrhiza*. *Antivir. Res.* **2002**, *55*, 91–106. [CrossRef]
144. Locher, C.P.; Witvrouw, M.; De Béthune, M.P.; Burch, M.T.; Mower, H.F.; Davis, H.; Lasure, A.; Pauwels, R.; De Clercq, E.; Vlietinck, A.J. Antiviral Activity of Hawaiian Medicinal Plants against Human Immunodeficiency Virus Type-1 (HIV-1). *Phytomed. Int. J. Phytother. Phytopharm.* **1996**, *2*, 259–264. [CrossRef]
145. Nutan; Modi, M.; Dezzutti, C.S.; Kulshreshtha, S.; Rawat, A.K.S.; Srivastava, S.K.; Malhotra, S.; Verma, A.; Ranga, U.; Gupta, S.K. Extracts from *Acacia catechu* Suppress HIV-1 Replication by Inhibiting the Activities of the Viral Protease and Tat. *Virol. J.* **2013**, *10*, 309. [CrossRef]
146. Notka, F.; Meier, G.; Wagner, R. Concerted Inhibitory Activities of *Phyllanthus Amarus* on HIV Replication in Vitro and Ex Vivo. *Antivir. Res.* **2004**, *64*, 93–102. [CrossRef]
147. Lee-Huang, S.; Zhang, L.; Huang, P.L.; Chang, Y.-T.; Huang, P.L. Anti-HIV Activity of Olive Leaf Extract (OLE) and Modulation of Host Cell Gene Expression by HIV-1 Infection and OLE Treatment. *Biochem. Biophys. Res. Commun.* **2003**, *307*, 1029–1037. [CrossRef]
148. Mukhtar, M.; Arshad, M.; Ahmad, M.; Pomerantz, R.J.; Wigdahl, B.; Parveen, Z. Antiviral Potentials of Medicinal Plants. *Virus Res.* **2008**, *131*, 111–120. [CrossRef]
149. Bessong, P.O.; Obi, C.L.; Andréola, M.-L.; Rojas, L.B.; Pouységu, L.; Igumbor, E.; Meyer, J.J.M.; Quideau, S.; Litvak, S. Evaluation of Selected South African Medicinal Plants for Inhibitory Properties against Human Immunodeficiency Virus Type 1 Reverse Transcriptase and Integrase. *J. Ethnopharmacol.* **2005**, *99*, 83–91. [CrossRef]
150. Zhang, X.; Yang, L.-M.; Liu, G.-M.; Liu, Y.-J.; Zheng, C.-B.; Lv, Y.-J.; Li, H.-Z.; Zheng, Y.-T. Potent Anti-HIV Activities and Mechanisms of Action of a Pine Cone Extract from *Pinus yunnanensis*. *Molecules* **2012**, *17*, 6916–6929. [CrossRef]
151. Kalvatchev, Z.; Walder, R.; Garzaro, D. Anti-HIV Activity of Extracts from *Calendula officinalis* Flowers. *Biomed. Pharmacother.* **1997**, *51*, 176–180. [CrossRef]
152. Callies, O.; Bedoya, L.M.; Beltrán, M.; Munoz, A.; Calderón, P.O.; Osorio, A.A.; Jiménez, I.A.; Alcami, J.; Bazzocchi, I.L. Isolation, Structural Modification, and HIV Inhibition of Pentacyclic Lupane-Type Triterpenoids from *Cassine xylocarpa* and *Maytenus cuzcoina*. *J. Nat. Prod.* **2015**, *78*, 1045–1055. [CrossRef]
153. Hao, B.-J.; Wu, Y.-H.; Wang, J.-G.; Hu, S.-Q.; Keil, D.J.; Hu, H.-J.; Lou, J.-D.; Zhao, Y. Hepatoprotective and Antiviral Properties of Isochlorogenic Acid A from *Laggera alata* against Hepatitis B Virus Infection. *J. Ethnopharmacol.* **2012**, *144*, 190–194. [CrossRef]
154. Jiang, Z.-Y.; Liu, W.-F.; Zhang, X.-M.; Luo, J.; Ma, Y.-B.; Chen, J.-J. Anti-HBV Active Constituents from *Piper longum*. *Bioorg. Med. Chem. Lett.* **2013**, *23*, 2123–2127. [CrossRef]
155. Zeng, F.-L.; Xiang, Y.-F.; Liang, Z.-R.; Wang, X.; Huang, D.-E.; Zhu, S.-N.; Li, M.-M.; Yang, D.-P.; Wang, D.-M.; Wang, Y.-F. Anti-Hepatitis B Virus Effects of Dehydrocheilanthifoline from *Corydalis saxicola*. *Am. J. Chin. Med.* **2013**, *41*, 119–130. [CrossRef]
156. Chiang, L.-C.; Ng, L.T.; Liu, L.-T.; Shieh, D.-E.; Lin, C.-C. Cytotoxicity and Anti-Hepatitis B Virus Activities of Saikosaponins from *Bupleurum* Species. *Planta Med.* **2003**, *69*, 705–709. [CrossRef]
157. Chiang, L.-C.; Ng, L.-T.; Cheng, P.-W.; Chiang, W.; Lin, C.-C. Antiviral Activities of Extracts and Selected Pure Constituents of *Ocimum basilicum*. *Clin. Exp. Pharmacol. Physiol.* **2005**, *32*, 811–816. [CrossRef]
158. Rechtman, M.M.; Har-Noy, O.; Bar-Yishay, I.; Fishman, S.; Adamovich, Y.; Shaul, Y.; Halpern, Z.; Shlomai, A. Curcumin Inhibits Hepatitis B Virus via Down-Regulation of the Metabolic Coactivator PGC-1alpha. *FEBS Lett.* **2010**, *584*, 2485–2490. [CrossRef]
159. Ali, S.I.; Sheikh, W.M.; Rather, M.A.; Venkatesalu, V.; Muzamil Bashir, S.; Nabi, S.U. Medicinal Plants: Treasure for Antiviral Drug Discovery. *Phytother. Res. PTR* **2021**, *35*, 3447–3483. [CrossRef]
160. Karamese, M.; Aydogdu, S.; Karamese, S.A.; Altoparlak, U.; Gundogdu, C. Preventive Effects of a Major Component of Green Tea, Epigallocathechin-3-Gallate, on Hepatitis-B Virus DNA Replication. *Asian Pac. J. Cancer Prev.* **2015**, *16*, 4199–4202. [CrossRef]
161. Polyak, S.J.; Morishima, C.; Lohmann, V.; Pal, S.; Lee, D.Y.W.; Liu, Y.; Graf, T.N.; Oberlies, N.H. Identification of Hepatoprotective Flavonolignans from Silymarin. *Proc. Natl. Acad. Sci. USA* **2010**, *107*, 5995–5999. [CrossRef]
162. Kim, K.; Kim, K.H.; Kim, H.Y.; Cho, H.K.; Sakamoto, N.; Cheong, J. Curcumin Inhibits Hepatitis C Virus Replication via Suppressing the Akt-SREBP-1 Pathway. *FEBS Lett.* **2010**, *584*, 707–712. [CrossRef]
163. Calland, N.; Albecka, A.; Belouzard, S.; Wychowski, C.; Duverlie, G.; Descamps, V.; Hober, D.; Dubuisson, J.; Rouillé, Y.; Séron, K. (-)-Epigallocatechin-3-Gallate Is a New Inhibitor of Hepatitis C Virus Entry. *Hepatology* **2012**, *55*, 720–729. [CrossRef]
164. Haid, S.; Novodomská, A.; Gentzsch, J.; Grethe, C.; Geuenich, S.; Bankwitz, D.; Chhatwal, P.; Jannack, B.; Hennebelle, T.; Bailleul, F.; et al. A Plant-Derived Flavonoid Inhibits Entry of All HCV Genotypes into Human Hepatocytes. *Gastroenterology* **2012**, *143*, 213–222.e5. [CrossRef]
165. Meuleman, P.; Albecka, A.; Belouzard, S.; Vercauteren, K.; Verhoye, L.; Wychowski, C.; Leroux-Roels, G.; Palmer, K.E.; Dubuisson, J. Griffithsin Has Antiviral Activity against Hepatitis C Virus. *Antimicrob. Agents Chemother.* **2011**, *55*, 5159–5167. [CrossRef] [PubMed]

166. Lin, L.-T.; Chen, T.-Y.; Lin, S.-C.; Chung, C.-Y.; Lin, T.-C.; Wang, G.-H.; Anderson, R.; Lin, C.-C.; Richardson, C.D. Broad-Spectrum Antiviral Activity of Chebulagic Acid and Punicalagin against Viruses That Use Glycosaminoglycans for Entry. *BMC Microbiol.* **2013**, *13*, 187. [CrossRef]
167. Wahyuni, T.S.; Widyawaruyanti, A.; Lusida, M.I.; Fuad, A.; Soetjipto; Fuchino, H.; Kawahara, N.; Hayashi, Y.; Aoki, C.; Hotta, H. Inhibition of Hepatitis C Virus Replication by Chalepin and Pseudane IX Isolated from *Ruta angustifolia* Leaves. *Fitoterapia* **2014**, *99*, 276–283. [CrossRef] [PubMed]
168. Shikov, A.N.; Djachuk, G.I.; Sergeev, D.V.; Pozharitskaya, O.N.; Esaulenko, E.V.; Kosman, V.M.; Makarov, V.G. Birch Bark Extract as Therapy for Chronic Hepatitis C–a Pilot Study. *Phytomedicine* **2011**, *18*, 807–810. [CrossRef] [PubMed]
169. Moradi, M.-T.; Karimi, A.; Shahrani, M.; Hashemi, L.; Ghaffari-Goosheh, M.-S. Anti-Influenza Virus Activity and Phenolic Content of Pomegranate (*Punica granatum* L.) Peel Extract and Fractions. *Avicenna J. Med. Biotechnol.* **2019**, *11*, 285–291.
170. Pantev, A.; Ivancheva, S.; Staneva, L.; Serkedjieva, J. Biologically Active Constituents of a Polyphenol Extract from *Geranium sanguineum* L. with Anti-Influenza Activity. *Z. Naturforschung C J. Biosci.* **2006**, *61*, 508–516. [CrossRef] [PubMed]
171. Michaelis, M.; Geiler, J.; Naczk, P.; Sithisarn, P.; Ogbomo, H.; Altenbrandt, B.; Leutz, A.; Doerr, H.W.; Cinatl, J. Glycyrrhizin Inhibits Highly Pathogenic H5N1 Influenza A Virus-Induced pro-Inflammatory Cytokine and Chemokine Expression in Human Macrophages. *Med. Microbiol. Immunol.* **2010**, *199*, 291–297. [CrossRef] [PubMed]
172. Sawai, R.; Kuroda, K.; Shibata, T.; Gomyou, R.; Osawa, K.; Shimizu, K. Anti-Influenza Virus Activity of *Chaenomeles sinensis*. *J. Ethnopharmacol.* **2008**, *118*, 108–112. [CrossRef]
173. Zakay-Rones, Z.; Varsano, N.; Zlotnik, M.; Manor, O.; Regev, L.; Schlesinger, M.; Mumcuoglu, M. Inhibition of Several Strains of Influenza Virus in Vitro and Reduction of Symptoms by an Elderberry Extract (*Sambucus nigra* L.) during an Outbreak of Influenza B Panama. *J. Altern. Complement. Med.* **1995**, *1*, 361–369. [CrossRef]
174. Liu, G.; Xiong, S.; Xiang, Y.-F.; Guo, C.-W.; Ge, F.; Yang, C.-R.; Zhang, Y.-J.; Wang, Y.-F.; Kitazato, K. Antiviral Activity and Possible Mechanisms of Action of Pentagalloylglucose (PGG) against Influenza A Virus. *Arch. Virol.* **2011**, *156*, 1359–1369. [CrossRef]
175. Song, J.-M.; Lee, K.-H.; Seong, B.-L. Antiviral Effect of Catechins in Green Tea on Influenza Virus. *Antivir. Res.* **2005**, *68*, 66–74. [CrossRef]
176. Hayashi, K.; Narutaki, K.; Nagaoka, Y.; Hayashi, T.; Uesato, S. Therapeutic Effect of Arctiin and Arctigenin in Immunocompetent and Immunocompromised Mice Infected with Influenza A Virus. *Biol. Pharm. Bull.* **2010**, *33*, 1199–1205. [CrossRef]
177. Schapowal, A. Efficacy and Safety of Echinaforce® in Respiratory Tract Infections. *Wien. Med. Wochenschr. 1946* **2013**, *163*, 102–105. [CrossRef]
178. Dall'Acqua, S.; Perissutti, B.; Grabnar, I.; Farra, R.; Comar, M.; Agostinis, C.; Caristi, G.; Golob, S.; Voinovich, D. Pharmacokinetics and Immunomodulatory Effect of Lipophilic Echinacea Extract Formulated in Softgel Capsules. *Eur. J. Pharm. Biopharm.* **2015**, *97*, 8–14. [CrossRef] [PubMed]
179. Rauš, K.; Pleschka, S.; Klein, P.; Schoop, R.; Fisher, P. Effect of an Echinacea-Based Hot Drink versus Oseltamivir in Influenza Treatment: A Randomized, Double-Blind, Double-Dummy, Multicenter, Noninferiority Clinical Trial. *Curr. Ther. Res.* **2015**, *77*, 66–72. [CrossRef] [PubMed]
180. Pourghanbari, G.; Nili, H.; Moattari, A.; Mohammadi, A.; Iraji, A. Antiviral Activity of the Oseltamivir and *Melissa officinalis* L. Essential Oil against Avian Influenza A Virus (H9N2). *Virus Dis.* **2016**, *27*, 170–178. [CrossRef] [PubMed]
181. Wyde, P.R.; Meyerson, L.R.; Gilbert, B.E. In Vitro Evaluation of the Antiviral Activity of SP-303, an Euphorbiaceae Shrub Extract, against a Panel of Respiratory Viruses. *Drug Dev. Res.* **1993**, *28*, 467–472. [CrossRef]
182. Zou, C.; Liu, H.; Feugang, J.M.; Hao, Z.; Chow, H.-H.S.; Garcia, F. Green Tea Compound in Chemoprevention of Cervical Cancer. *Int. J. Gynecol. Cancer Off. J. Int. Gynecol. Cancer Soc.* **2010**, *20*, 617–624. [CrossRef]
183. Mondal, A.; Chatterji, U. Artemisinin Represses Telomerase Subunits and Induces Apoptosis in HPV-39 Infected Human Cervical Cancer Cells. *J. Cell. Biochem.* **2015**, *116*, 1968–1981. [CrossRef]
184. Yarnell, E. Herbs against Human Papillomavirus. *Altern. Complement. Ther.* **2015**, *21*, 71–76. [CrossRef]
185. Jiang, W.-L.; Luo, X.-L.; Kuang, S.-J. Effects of *Alternanthera philoxeroides* Griseb against dengue virus in vitro. *1 Jun Yi Xue Xue Bao Acad. J. First Med. Coll. PLA* **2005**, *25*, 454–456.
186. Salim, F.; Abu, N.A.; Yaakob, H.; Kadir, L.; Zainol, N.; Taher, Z. Interaction of *Carica papaya* L. Leaves Optimum Extract on Virus Dengue Infected Cells. *Sci. Int.* **2018**, *30*, 437–441.
187. Castillo-Maldonado, I.; Moreno-Altamirano, M.M.B.; Serrano-Gallardo, L.B. Anti-Dengue Serotype-2 Activity Effect of *Sambucus nigra* Leaves-and Flowers-Derived Compounds. *Virol. Res. Rev.* **2017**, *1*, 1–5. [CrossRef]
188. Chan, Y.S.; Khoo, K.S.; Sit, N.W.W. Investigation of Twenty Selected Medicinal Plants from Malaysia for Anti-Chikungunya Virus Activity. *Int. Microbiol. Off. J. Span. Soc. Microbiol.* **2016**, *19*, 175–182. [CrossRef]
189. Raghavendhar, S.; Tripati, P.K.; Ray, P.; Patel, A.K. Evaluation of Medicinal Herbs for Anti-CHIKV Activity. *Virology* **2019**, *533*, 45–49. [CrossRef] [PubMed]
190. Li, S.-Y.; Chen, C.; Zhang, H.-Q.; Guo, H.-Y.; Wang, H.; Wang, L.; Zhang, X.; Hua, S.-N.; Yu, J.; Xiao, P.-G.; et al. Identification of Natural Compounds with Antiviral Activities against SARS-Associated Coronavirus. *Antivir. Res.* **2005**, *67*, 18–23. [CrossRef]
191. Ho, T.-Y.; Wu, S.-L.; Chen, J.-C.; Li, C.-C.; Hsiang, C.-Y. Emodin Blocks the SARS Coronavirus Spike Protein and Angiotensin-Converting Enzyme 2 Interaction. *Antivir. Res.* **2007**, *74*, 92–101. [CrossRef] [PubMed]

192. Wen, C.-C.; Shyur, L.-F.; Jan, J.-T.; Liang, P.-H.; Kuo, C.-J.; Arulselvan, P.; Wu, J.-B.; Kuo, S.-C.; Yang, N.-S. Traditional Chinese Medicine Herbal Extracts of *Cibotium barometz, Gentiana scabra, Dioscorea batatas, Cassia tora*, and *Taxillus chinensis* Inhibit SARS-CoV Replication. *J. Tradit. Complement. Med.* **2011**, *1*, 41–50. [CrossRef] [PubMed]
193. Ulasli, M.; Gurses, S.A.; Bayraktar, R.; Yumrutas, O.; Oztuzcu, S.; Igci, M.; Igci, Y.Z.; Cakmak, E.A.; Arslan, A. The Effects of *Nigella sativa* (Ns), *Anthemis hyalina* (Ah) and *Citrus sinensis* (Cs) Extracts on the Replication of Coronavirus and the Expression of TRP Genes Family. *Mol. Biol. Rep.* **2014**, *41*, 1703–1711. [CrossRef]
194. Lin, C.-W.; Tsai, F.-J.; Tsai, C.-H.; Lai, C.-C.; Wan, L.; Ho, T.-Y.; Hsieh, C.-C.; Chao, P.-D.L. Anti-SARS Coronavirus 3C-like Protease Effects of Isatis Indigotica Root and Plant-Derived Phenolic Compounds. *Antivir. Res.* **2005**, *68*, 36–42. [CrossRef]
195. Ryu, Y.B.; Jeong, H.J.; Kim, J.H.; Kim, Y.M.; Park, J.-Y.; Kim, D.; Nguyen, T.T.H.; Park, S.-J.; Chang, J.S.; Park, K.H.; et al. Biflavonoids from Torreya Nucifera Displaying SARS-CoV 3CL(pro) Inhibition. *Bioorg. Med. Chem.* **2010**, *18*, 7940–7947. [CrossRef]
196. Remali, J.; Aizat, W.M. A Review on Plant Bioactive Compounds and Their Modes of Action Against Coronavirus Infection. *Front. Pharmacol.* **2020**, *11*, 589044. [CrossRef]
197. Huang, T.-J.; Tsai, Y.-C.; Chiang, S.-Y.; Wang, G.-J.; Kuo, Y.-C.; Chang, Y.-C.; Wu, Y.-Y.; Wu, Y.-C. Anti-Viral Effect of a Compound Isolated from *Liriope platyphylla* against Hepatitis B Virus in Vitro. *Virus Res.* **2014**, *192*, 16–24. [CrossRef]
198. Gansukh, E.; Kazibwe, Z.; Pandurangan, M.; Judy, G.; Kim, D.H. Probing the Impact of Quercetin-7-O-Glucoside on Influenza Virus Replication Influence. *Phytomedicine* **2016**, *23*, 958–967. [CrossRef]
199. Akher, F.B.; Farrokhzadeh, A.; Ramharack, P.; Shunmugam, L.; Van Heerden, F.R.; Soliman, M.E.S. Discovery of Novel Natural Flavonoids as Potent Antiviral Candidates against Hepatitis C Virus NS5B Polymerase. *Med. Hypotheses* **2019**, *132*, 109359. [CrossRef]
200. Sajitha Lulu, S.; Thabitha, A.; Vino, S.; Mohana Priya, A.; Rout, M. Naringenin and Quercetin—Potential Anti-HCV Agents for NS2 Protease Targets. *Nat. Prod. Res.* **2016**, *30*, 464–468. [CrossRef] [PubMed]
201. Moharana, M.; Pattanayak, S.K.; Khan, F. Molecular Recognition of Bio-Active Triterpenoids from *Swertia chirayita* towards Hepatitis Delta Antigen: A Mechanism through Docking, Dynamics Simulation, Gibbs Free Energy Landscape. *J. Biomol. Struct. Dyn.* **2023**, 1–14. [CrossRef] [PubMed]
202. Shadrack, D.M.; Deogratias, G.; Kiruri, L.W.; Onoka, I.; Vianney, J.-M.; Swai, H.; Nyandoro, S.S. Luteolin: A Blocker of SARS-CoV-2 Cell Entry Based on Relaxed Complex Scheme, Molecular Dynamics Simulation, and Metadynamics. *J. Mol. Model.* **2021**, *27*, 221. [CrossRef] [PubMed]
203. Abdelli, I.; Hassani, F.; Bekkel Brikci, S.; Ghalem, S. In Silico Study the Inhibition of Angiotensin Converting Enzyme 2 Receptor of COVID-19 by *Ammoides verticillata* Components Harvested from Western Algeria. *J. Biomol. Struct. Dyn.* **2021**, *39*, 3263–3276. [CrossRef] [PubMed]
204. Kumar Verma, A.; Kumar, V.; Singh, S.; Goswami, B.C.; Camps, I.; Sekar, A.; Yoon, S.; Lee, K.W. Repurposing Potential of Ayurvedic Medicinal Plants Derived Active Principles against SARS-CoV-2 Associated Target Proteins Revealed by Molecular Docking, Molecular Dynamics and MM-PBSA Studies. *Biomed. Pharmacother.* **2021**, *137*, 111356. [CrossRef] [PubMed]
205. Rudrapal, M.; Issahaku, A.R.; Agoni, C.; Bendale, A.R.; Nagar, A.; Soliman, M.E.S.; Lokwani, D. In Silico Screening of Phytopolyphenolics for the Identification of Bioactive Compounds as Novel Protease Inhibitors Effective against SARS-CoV-2. *J. Biomol. Struct. Dyn.* **2022**, *40*, 10437–10453. [CrossRef] [PubMed]

Disclaimer/Publisher's Note: The statements, opinions and data contained in all publications are solely those of the individual author(s) and contributor(s) and not of MDPI and/or the editor(s). MDPI and/or the editor(s) disclaim responsibility for any injury to people or property resulting from any ideas, methods, instructions or products referred to in the content.

Review

Unleashed Treasures of Solanaceae: Mechanistic Insights into Phytochemicals with Therapeutic Potential for Combatting Human Diseases

Saima Jan [1,†], Sana Iram [2,†], Ommer Bashir [3], Sheezma Nazir Shah [1], Mohammad Azhar Kamal [4], Safikur Rahman [5], Jihoe Kim [2,*] and Arif Tasleem Jan [1,*]

1. School of Biosciences and Biotechnology, Baba Ghulam Shah Badshah University, Rajouri 185234, Jammu and Kashmir, India; saimajan.scholar@bgsbu.ac.in (S.J.); shahsheezma@gmail.com (S.N.S.)
2. Department of Medical Biotechnology, Yeungnam University, Gyeongsan 712-749, Republic of Korea; sanairam157@yu.ac.kr
3. Department of School Education, Srinagar 190001, Jammu and Kashmir, India; ommerbashir@gmail.com
4. Department of Pharmaceutics, College of Pharmacy, Prince Sattam Bin AbdulAziz University, Alkharj 11942, Saudi Arabia; ma.kamal@psau.edu.sa
5. Department of Botany, Munshi Singh College, BR Ambedkar Bihar University, Muzaffarpur 845401, Bihar, India; shafique2@gmail.com
* Correspondence: kimjihoe@ynu.ac.kr (J.K.); atasleem@bgsbu.ac.in (A.T.J.)
† These authors contributed equally to this work.

Abstract: Plants that possess a diverse range of bioactive compounds are essential for maintaining human health and survival. The diversity of bioactive compounds with distinct therapeutic potential contributes to their role in health systems, in addition to their function as a source of nutrients. Studies on the genetic makeup and composition of bioactive compounds have revealed them to be rich in steroidal alkaloids, saponins, terpenes, flavonoids, and phenolics. The Solanaceae family, having a rich abundance of bioactive compounds with varying degrees of pharmacological activities, holds significant promise in the management of different diseases. Investigation into Solanum species has revealed them to exhibit a wide range of pharmacological properties, including antioxidant, hepatoprotective, cardioprotective, nephroprotective, anti-inflammatory, and anti-ulcerogenic effects. Phytochemical analysis of isolated compounds such as diosgenin, solamargine, solanine, apigenin, and lupeol has shown them to be cytotoxic in different cancer cell lines, including liver cancer (HepG2, Hep3B, SMMC-772), lung cancer (A549, H441, H520), human breast cancer (HBL-100), and prostate cancer (PC3). Since analysis of their phytochemical constituents has shown them to have a notable effect on several signaling pathways, a great deal of attention has been paid to identifying the biological targets and cellular mechanisms involved therein. Considering the promising aspects of bioactive constituents of different Solanum members, the main emphasis was on finding and reporting notable cultivars, their phytochemical contents, and their pharmacological properties. This review offers mechanistic insights into the bioactive ingredients intended to treat different ailments with the least harmful effects for potential applications in the advancement of medical research.

Keywords: Solanaceae; bioactive compounds; pharmaceuticals; phytochemistry; therapeutics

1. Introduction

Plants have long been known as an excellent source of bioactive compounds with unique pharmacological properties [1–3]. Being a good source of phytochemicals such as alkaloids, terpenoids, and phenylpropanoids, they can be used to treat different illnesses, in addition to providing a source for the development of new medicines [4–6]. *Solanum* L. (Solanaceae; also referred to as nightshades) is one of the largest genera of the Solanaceae family, distributed across tropical, subtropical, and temperate regions [7,8]. The members of the Solanaceae family, being annual, biennial, or perennial, are herbaceous and exhibit great floristic diversity, phytochemical characteristics, and ethnobotanical significance. Solanum family members have yielded a variety of pharmacologically active compounds with distinct roles such as antirheumatic, antimicrobial, antioxidant, and anti-tumor [9]. It is an important plant taxon with species known for their therapeutic value [10]. Of the 670 compounds that have been reported to be found in the Solanum genus, including steroidal saponins, alkaloids, pregnane glycosides, terpenes, flavonoids, lignans, sterols, phenols, and coumarins, steroidal alkaloids such as solasodine, solasonine, and solamargine are of particular interest due to their unique medicinal properties, which have been shown to improve human health [11,12]. The chemical composition of many of these species is still poorly understood or has yet not received much attentionm, but they have been documented to have immense pharmacological potential, which offers greater hope for the development of innovative medicine effective against a wide range of human illnesses. Against this background, the present study provides a thorough evaluation of the morphological characteristics, regional distribution, and secondary metabolites with unique pharmacological capabilities across seven different species of the Solanum genus. Special attention has been paid to elucidating the mechanisms by which they modulate signaling across diverse signaling pathways to control the course of cells toward apoptotic and/or autophagic processes.

2. Distribution and Morphology

The Solanaceae family has 97 genera and about 2700 species, distributed across all continents except Antarctica [13]. The greatest diversity has been reported in Central and South America [14,15]. It is commonly distributed in the equatorial zone of Ibero-America. It inhabits distinct habitats (wastelands, old fields, ditches, cultivated land, overgrazed grass fields, railway cuttings, areas near buildings, riverbanks, etc.), from deserts to rainforests, and establishes itself in the lower vegetation and conquers distraught areas [16,17]. Solanaceae is ranked the third most economically significant family after Poaceae and Fabaceae, encompassing a vast array of species and boasting the most abundant collection of edible varieties within its genus Solanum. Several species of Solanum are economically important at a global scale, with several major agricultural crops (*S. tuberosum* L., potato; *S. lycopersicum* L., tomato; *S. melongena* L., eggplant), and the genus also contains locally important fruit crops [18]. Furthermore, the closely related Capsicum genus features widely consumed vegetables such as peppers [19–21]. Some of the medicinally important plants of the genus Solanum are *Solanum nigrum*, *Solanum torvum*, *Solanum indicum*, *Solanum surattense*, *Solanum Villosum*, *Solanum viarum*, and *Solanum incanum* (Figure 1).

S. nigrum, an annual herb commonly found in Europe, Asia, and Africa, exhibits a widespread distribution worldwide, spanning from sea level to elevations exceeding 11,482 feet [22,23]. It particularly thrives in moist environments and demonstrates strong adaptability to soils rich in nitrogen or phosphorus [24]. *S. nigrum* typically reaches a height of 30–120 cm. The leaves are ovoid in shape with a trilateral base, measuring 4–7 cm in length and 2–5 cm in width. The upper portion of the leaves has a smooth texture. The flowers of *S. nigrum* are whitish in color, with yellow anthers [25]. The stem of the plant is usually bifurcated (branching into two) and covered with fine hairs. Its inflorescence is arranged into extra-axillary umbels and the sepals have a cup-like shape. The elliptical anthers measure approximately 2.5–3.5 mm, while the filaments are 1.5 mm in length. At maturity, the fruits of *S. nigrum* turn dull black in color and are spherical in shape, with a diameter of 8–10 mm [26].

Solanum torvum is as an erect shrub, reaching a height of 1–3 m and characterized by a prickly stem [27]. The young stem and branches exhibit a vibrant green color and are adorned with delicate trichomes, while the bark of its mature stems takes on a brown to dark gray color. The leaves of *S. torvum* maintain their green color throughout the year and possess a broadly ovate structure, measuring between 5 and 21 cm in length and 4 and 13 cm in width. The leaf margin is usually intact but may occasionally have up to seven broad triangular lobes. Both the upper and lower leaf surfaces are covered with fine stellate hairs and there are scattered prickles along the main veins. The upper surface of the leaf is darker in color compared to the lower surface. The base of the leaf lamina is truncated and oblique, while the apex is acute to acuminate. The leaves of *S. torvum* are characterized by densely stellate-pubescent petioles, measuring 1–5.5 cm in length, with some curved prickles reaching up to 10 mm in length. The inflorescences (consist of 50–100 flowers) are dense and compact and covered with trichomes that grow up to 6 cm long and branching 1–4 times. The flowers are pentameric, with slender and hairy sepals that are 2–3 mm long, forming a calyx that measures 4–6 mm in length. The petals are white to cream in color, approximately 1 cm long, and together they form a stellate corolla with a diameter of 1.5–2 cm. The androecium is composed of connivant stamens that are yellow in color, with anthers measuring 6–8 mm in length and about 1 mm in width. The gynoecium consists of a conical ovary that is glandular, topped with a 10–12 mm long glabrous style, which ends with a capitate stigma protruding over the androecium. The fruits of *S. torvum* are globular berries, measuring 1–1.5 cm in diameter. When ripe, they appear pale grayish–green and contain a few to several flat woody seeds, which are 1.5–2 mm long.

Figure 1. Representation of different members of the genus Solanum.

Solanum indicum is known to occur in regions such as India, Sri Lanka, Malaya, China, and the Philippines, spanning from sea level up to approximately 1500 m above sea level [16]. The plant is a biennial herb or small shrub with upright growth and spiky characteristics. Its sturdy stems feature large, sharp prickles that have a long compressed base, often with a slightly curved shape. The plant's leaves are oblong in shape, measuring 5–15 cm in length and 2.5–7.5 cm in width. They have a sub-entire margin with a few triangular–oval lobes and sparse prickles and hair on both sides. The leaf base is cordate or truncate, occasionally unevenly shaped, and the petioles are hairy on both sides and range from 1.3 to 2.5 cm in length. The flowers are grouped into racemose extra-axillary cymes, which are held up by short peduncles covered in stellate hairs. The corolla, with a length of about 0.8 cm, displays a pale purple hue and is adorned with darker purple stellate hairs on its outer surface. The lobes of the corolla measure roughly 5 mm. The stamens, connected to the corolla, feature brief filaments and large anthers, constituting the male reproductive components of the flower. The fruits, or berries, are globose with a diameter of ~70.8 cm when fully ripe. They are glabrous and have a dark yellow color at maturity. The seeds are about 0.4 cm in diameter and exhibit minute pits.

Solanum surattense is a perennial herbaceous plant that is widely distributed in Australia, Ceylon, India, Malaysia, Polynesia, and Southeast Asia [28]. The species typically reaches a height of about 1.2 m and has a woody base. The stem is highly branched, exhibiting a zigzag pattern, with young branches covered in dense satellite and tomentose hairs. The straight, glabrous, and shiny prickles on the stem are often 1–3 cm in length. The leaves are ovate–elliptic, deeply lobed, and tapered at the base, with spines present along the veins and margins. The flowers are bluish pink and arranged into extra-axillary racemes. The calyx has five free lobes that are ovate and prickly. The corolla is broadly ovate–triangular, with five acute lobes. The plant produces globose berries that are initially green with white stripes but turn yellow when ripe. The seeds are circular and numerous and have a smooth surface [29].

Solanum villosum is found in the Euro-Siberian, Irano-Turanian, and Mediterranean regions of the globe [30]. It shows distribution in countries such as Africa, Kenya, China, India, and Pakistan [31,32]. *S. villosum* possess a brittle stem that can reach a height of ~1 m [33]. The leaves of *S. villosum* are rhombic to ovate–lanceolate, measuring 2.0–7.0 cm in length and 1.5–4.0 cm in width. The leaf margins can be either smooth or slightly wavy with shallow teeth. The inflorescence is simple, forming umbellate or loosely arranged solitary cymes. The calyces are approximately 1.2–2.2 mm long and slightly curved and may bend downward or adhere to the base of mature berries. The berries themselves are usually longer than they are wide and orange in color and have a width of 6–10 mm. When fully ripe, the berries detach from the calyces and fall off [34].

Solanum viarum has its origin in Brazil and Argentina but exhibits widespread distribution as a weed throughout South America, Africa, India, Nepal, the West Indies, Honduras, Mexico, Cameroon, and the Democratic Republic of the Congo [35,36]. Both agricultural and natural environments have been occupied by *S. viarum* [37,38]. The plant's spiky prickles make handling them challenging. The prickles, which are dispersed on the stem and the leaves, are about 1–2 cm long and range in color from white to yellowish. The leaves are alternate, have a broad end at the base, and, especially on young leaves, have somewhat wavy borders. *S. viarum* leaves are pubescent, 10–20 cm long, 6–15 cm wide, and deeply split into broad pointed lobes. The upper side of the leaf is gray–green and the lower surface is greenish white [38]. The flowers have cream-colored filaments and are white in color [39]. The unripe fruits exhibit white mottling patterns and undergo a color transformation to yellow upon ripening. They are smooth and round, measuring approximately 2–3 cm in diameter. Each fruit contains a range of 190–385 light red–brown seeds, with a diameter of 2.2–2.8 mm [36,40].

Solanum incanum is indigenous to the Horn of Africa and has a broad distribution in that region. It is characterized by its thorny leaves, yellow fruits, and blue flowers with yellow pistils [11,41]. The plant reproduces through seed propagation and the germination process is relatively slow. Table 1 summarizes the basic features of the above-mentioned species of the genus Solanum.

Table 1. Comparison of the morphological parameters across different species of Solanaceae.

	S. viarum	S. surattense	S. nigrum	S. villosum	S. incanum	S. indicum	S. torvum
Habit	Shrub; perennial	Herb; perennial	Herb; annual	Herb; perennial	Perennial; bushy herb or shrub	Herb or small shrub	Shrub; perennial
Root system	Tap root	Tap root	Tap root	Tap root	Tap root	Tap root	Tap root
Stem	Stem densely yellow, hirsute, and with straight patent prickles. The hooked prickles are shorter than straight ones	Stem profusely branched somewhat zig zag, young branches with dense satellite and tomentose hairs, prickles compressed straight, glabrous, and shiny	Stem often angular, sparsely pubescent.	Stems decumbent, terete to ridged, green to purple; new growth densely pubescent with simple, uniseriate, translucent, and glandular trichomes	Stem is erect and covered with spines, green to purple in color	Upright, spiky and sturdy stem	Stems erect, branched and armed prickles sparse.
Leaves	Ovate–triangular, sinuate-lobed, lobes sub-obtuse or subacute, prickles straight, scattered; trichomes viscid, dense, straight or stellate	Ovate-elliptic, deeply lobed base attenuate, veins and margins with spines. Ovate or elliptic sinuate or sub-pinnatifid glabrescent, very prickly	Ovate or oblong, sinuate-toothed or lobed and glabrous	Leaves rhombic to ovate-lanceolate, margins are entire to sinuate–dentate	Leaves are simple, alternate, ovate, or elliptic	Leaves are broad, oblong in form, sub-entire with a few triangular–oval lobes, sparsely prickly and hairy on both sides; the base is cordate or truncate	Elliptic, oval, ovate or oblong, entire or irregularly lobed, acute or obtuse, highly variable; prickles straight or a few hooked; trichomes dense, stellate
Inflorescence	Extraaxillary raceme	Axillary cymose	Extraaxillary raceme	Inflorescence is simple, umbellate to slightly solitary cymes	Sometimes solitary or in clusters	Racemose extra axillary cymes, with short, stellate, hairy peduncles	Many flowered corymbose cymes
Flower	Sepals lanceolate and hirsute and corolla nearly glabrous. Corolla white, deeply lobed, petals recurved, anthers same length. Has both male and bisexual flowers and a style longer than its anthers	Bluish pink extra-axillary racemes, calyx lobes free, obovate, prickly acuminate. Corolla ovate–triangular	Calyx cup-shaped, white corolla with yellow anthers, lobes ovate–oblong, pubescent abaxially, ciliate spreading. Sepals ovate-oblong, calyx teeth small and obtuse. Corolla nearly glabrous	Calyces 1.2–2.2 mm long, slightly a crescent, deflexed or adhering to base of mature berry	Yellow or white calyx is fused, the purple corolla regular, bell- or wheel-shaped with five stamens	Corolla pale purple, covered with darker purple stellate hair on the outside; the stamens are attached to the corolla, with short filaments and anthers that are large	Calyx lanceolate, sparingly hairy; corolla lobes triangular and pubescent Corolla white, shallowly lobed, star-shaped; anthers same length
Fruit	Round, pale green with dark green veins, turning dull yellow at maturity.	Globose, yellow when ripe	Globose, dull black in color	Berries are usually longer than wide, orange, 6–10 mm broad,	The fruit is fleshy, less than 3 cm in diameter and is yellow then turn black	Globose, approximately 0.8 cm in diameter and dark yellow when ripe; glabrous	Round, green turning yellow then orange or brown at maturity. Seeds compressed, light brown
Seed	Seeds are flattened, discoid, and brown in color	Seeds discoid, smooth to faintly reticulate	Many discoid yellow seeds	Seeds are flattened and teardrop-shaped with a subapical hilum	Many pale brown seeds	Seeds are 0.4 cm in diameter and have minute pits	Seeds discoid
References	[39,42]	[43]	[44]	[33]	[11,41]	[45]	[46]

3. Secondary Metabolites from Solanaceae

Primary and secondary metabolites are the two categories of organic compounds/metabolites produced by plants [47,48]. Secondary metabolites (SMs) are small organic compounds with molecular masses less than 3000 da that develop from the primary metabolites during the embolism of plants. Different plant species have different metabolite types and compositions. In a recent study, Satish et al. [49] pointed out that almost 200,000 SMs have been identified and described. SMs are fascinating due to their structural diversity and their potency as therapeutic candidates and/or antioxidants, among other diverse reasons [50]. Almost 30% of pharmaceutical drugs arise directly or indirectly from secondary metabolites [51]. Based on the biosynthetic route, we find three main categories of plant metabolites: phenolic groups, composed mainly of simple sugars and benzene rings; terpenes and steroids, primarily consisting of carbon and hydrogen; as well as alkaloids, which incorporate nitrogen [52]. The SMs produced by different members of the Solanaceae family include diosgenin, solamargine, solanine, apigenin, etc. The SMs reported in Solanaceae exhibit a wide range of biological roles, including antioxidant, anticarcinogenic, and anti-inflammatory activities [53]. Broadly, they are known to influence several crucial cellular processes that affect cell cycle progression, alter membrane permeability, and induce apoptosis. SMs with antiproliferative properties have been found to influence several crucial cellular processes: for example, they can affect the progression of the cell cycle, leading to cell cycle arrest and preventing uncontrolled cell division. These compounds can also modulate the expression of certain genes involved in cancer cell survival and growth. The SMs present in the Solanaceae family have demonstrated the capacity to trigger apoptosis (programmed cell death) across diverse forms of cancer cells. These compounds activate different signaling pathways, which can lead to cell death in cancer cells. The differences in their effects on cancer cells may arise from variations in the chemical structure of the compounds and the specific sensitivity of different types of cancer cells. They can interfere with various signaling pathways that regulate cell survival and proliferation and can induce changes in the mitochondrial membrane, which may lead to mitochondrial dysfunction and apoptosis. They can alter the metabolic processes in cancer cells, affecting their energy production and survival. The diversity of effects observed in different cancer cell types can be attributed to the complex interactions between these SMs and the specific molecular characteristics of the cancer cells.

3.1. Diosgenin

Diosgenin, identified as 25R-spirost-5-en-3β-ol, is a derivative resulting from the hydrolysis of dioscin present in the rootstock of yam (Dioscorea). It is prevalent in natural plant sources in the glucoside form [54]. Diosgenin, a steroidal sapogenin, is present in a variety of plants, including Solanum species, namely Dioscorea nipponoca, Costus speciosus, and Trigonella foenum graecum [55,56]. This biologically active phytochemical plays a significant role in various plant actions, including its involvement in functional food properties [54]. It serves as a common starting point in the synthesis of steroids and contraceptives. Additionally, it finds application in medicinal forms for treating conditions such as leukemia, hypercholesterolemia, climacteric syndrome, and colon cancer [56]. Furthermore, diosgenin is employed as the primary material for the synthesis of oral contraceptives, sex hormones, and various other steroidal compounds [54]. Within plants, it exists in the configuration of steroidal saponins. The anti-tumor effects of diosgenin are attributed to its involvement in multiple mechanisms (inhibiting tumor cell growth, metastasis, and invasion, promoting the apoptosis of tumor cells, and blocking the cell cycle), with actions on different targets and pathways (Figure 2).

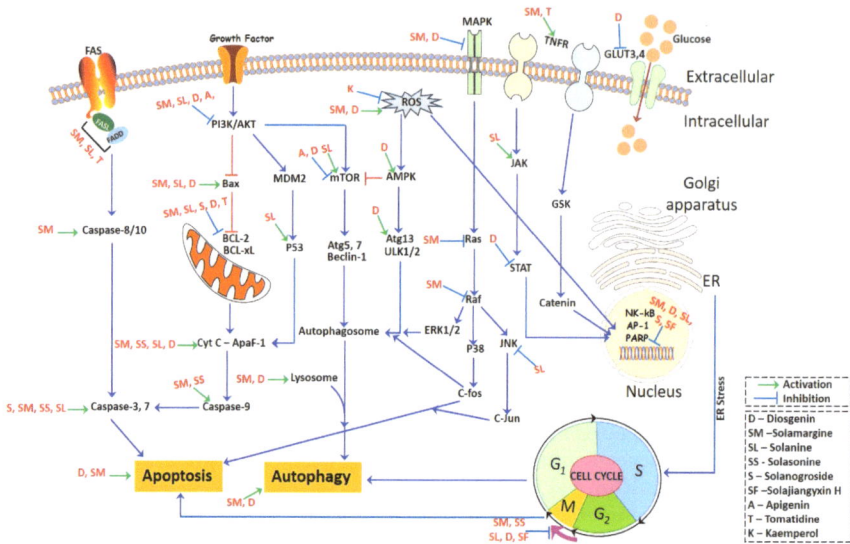

Figure 2. Molecular machinery of bioactive moieties governing different signaling pathways.

3.1.1. Inhibiting tumor cell growth, metastasis, and invasion

Diosgenin reduces the expression of matrix metalloproteinases (MMPs), including MMP-2 and -9, which are critical for tumor cell migration and invasion [57]. By inhibiting MMPs, diosgenin interferes with the metastatic process of tumor cells. Diosgenin exhibits significant effects in inhibiting tumor growth and metastasis via a reduction in the phosphorylation of IKKβ and NF-κB, leading to the inhibition of TNF-α and IL-6 production in endothelial cells. Furthermore, it inhibits the synthesis of endothelin-1 (ET-1) and plasminogen activator inhibitor 1 (PAI-1) within the endothelial cells, subsequently reinstating insulin-mediated vasodilation [58].

Angiogenesis plays a vital role in the progression of tumors, facilitating their growth, metastatic spread, and invasion into surrounding tissues [59]. Vascular endothelial growth factor (VEGF) serves as a pivotal controller of the angiogenic process and differentiation, playing a vital role in various angiogenic processes [60]. VEGF facilitates tumor growth by providing nutrition and waste excretion pathways for tumors and creating routes for tumor cells to enter the circulatory system through binding to the corresponding receptor, VEGFR. Moreover, the vascular endothelium plays a significant role in orchestrating both angiogenesis and vascular tone, constituting a specialized monolayer of cells with distinct site-specific functions and morphology [61]. Diosgenin has demonstrated the capability to impede angiogenesis by targeting several signaling pathways, encompassing HIF-1α, GRP78, VEGFR, PI3K/AKT, ERK1/2, and FAK. Additionally, it diminishes the expression of the VEGF and fibroblast growth factor 2 (FGF2) proteins, restraining the tubular formation of endothelial cells within cancer cells, consequently exerting a suppressive effect on angiogenesis [62]. These findings highlight the potential of diosgenin as a therapeutic agent to impede tumor angiogenesis, which could have significant implications in cancer treatment and management. By targeting VEGF and the vascular endothelium, diosgenin can effectively inhibit tumor angiogenesis, which is crucial for tumor growth and metastasis.

3.1.2. Regulation of the Apoptosis Pathway

Apoptosis is a well-featured process of programmed cell death, characterized by cell shrinkage and fragmentation [63]. In the apoptotic process, chromosomal DNA undergoes fragmentation (a process mediated by specific endonucleases) and the DNA is cleaved into oligonucleosomal-size fragments [64]. Apoptosis can be triggered through two principal pathways: the extrinsic (transmembrane) pathway and the intrinsic (mitochondrial) pathway. Recently, a novel lysosomal pathway of apoptosis has been identified [65,66]. Several factors, such as lysosomotropic photosensitizer agents, oxidative stress, oxidized lipids, serum starvation, DNA damage, and the activation of Fas and TNF-α, initiate the lysosomal pathway. This pathway encompasses the liberation of lysosomal cathepsins. Lysosomal membrane permeabilization, a part of this process, can result in both caspase-dependent and caspase-independent forms of apoptosis-like cell death, and in certain instances, it may lead to necrosis.

Diosgenin prompts apoptosis by elevating the presence of cytochrome C within the cytosol, along with increased levels of cleaved-caspase-3 and cleaved-PARP1. This is accompanied by an increased Bax/Bcl-2 ratio [67]. Within the HepG2 cells, diosgenin initiates apoptosis by engaging the mitochondria/caspase-3-dependent pathway, mediated by the Bcl-2 protein family. This family is implicated in tumor progression, with changes in the expression of its constituents serving as prognostic indicators across diverse lymphoid malignancies [68]. Caspases, categorized as cysteinyl aspartate-specific proteinases and belonging to the interleukin-1β-converting enzyme family, hold pivotal roles as regulators of apoptosis in eukaryotic cells. They are important for governing processes such as cell growth, differentiation, and apoptosis [69]. Their activation and function within the caspase cascade system are regulated by various molecules, including the inhibitor of apoptosis proteins, Bcl-2 family proteins, calpain, and Ca^{2+}. The p53 gene, a well-known tumor suppressor gene, is frequently subject to mutations and inactivation in the development of most human cancers [70,71]. Diosgenin triggers apoptosis in different cancer cell types, including bladder cancer, K562, rectal cancer, and glioblastoma cancer cells, through the induction of DNA damage. This outcome is linked to substantial increases in the protein levels of Bak, Bax, Bid, p53, caspase-3, and caspase-9. Simultaneously, there is a reduction in the protein levels of Bcl-2, Bcl-xL, survivin, and RNA [72,73]. One of the key targets of endoplasmic reticulum stress is calcium homeostasis, and the disruption of Ca^{2+} homeostasis and mitochondrial dysfunction play crucial roles in apoptosis [74]. Diosgenin triggers erythrocyte contraction and the disturbance of erythrocyte membrane phospholipids, leading to Ca^{2+} influx, oxidative stress, and ceramide formation [75]. By regulating calcium homeostasis, diosgenin can induce mitochondrial-dependent cell apoptosis, causing Ca^{2+} release, reducing the intracellular Ca^{2+} concentration, and inducing cell cycle arrest and apoptosis [76].

3.1.3. Blocking the Cell Cycle

Diosgenin's anti-tumor mechanism involves blocking the tumor cell cycle. It causes cell cycle arrest at various phases, including G2/M, S, and G0/G1, by regulating cyclin B1, p21cip1/Waf1, cdc2, pRb, E2F, CDK2/4/6, CyclinD1, p21, and p27 [73,77]. These regulatory effects lead to the suppression of tumor cell proliferation across various cancer types, including osteosarcoma, hepatoma, chronic myeloid leukemia, and pancreatic cancer.

3.1.4. Regulating Gene Expression and Proteins

Diosgenin has been found to upregulate the autophagy marker LC3 protein and induce autophagy in the hepatoma cells by reducing the ratio of LC3-I to LC3-II [78]. Additionally, diosgenin regulates various molecular pathways in different cancer types. In gastric, lung, and breast cancers, diosgenin modulates the expression of p21 mRNA, nuclear factor UUB activation, prostaglandin E2 synthesis, the miR-145 expression level, and the methylation status. Furthermore, it reduces the expression of cyclooxygenase (COX)-2 and microsomal prostaglandin E synthase (MPGES)-1, while inducing a substantial increase in miR-34a expression. Additionally, it downregulates the genes targeted by miR-34a, including E2F1, E2F3, and CCND1, resulting in the suppression of cancer cell proliferation [4,79]. Diosgenin inhibits the telomerase activity in lung cancer cells (A549) by downregulating the expression of the telomerase reverse transcriptase gene HTERT. Within glioblastoma cells, diosgenin prompts cell differentiation while decreasing cell dedifferentiation. This is achieved through an elevation in the expression of glial fibrillary acidic protein (GFAP), a marker of differentiation, and a concurrent reduction in the levels of dedifferentiation markers such as Id2, N-Myc, TERT, and Notch-1 [80]. Diosgenin demonstrates an anticancer impact by suppressing the expression of mesoderm posterior 1 (MESP1) within gastric cancer cells. This action stimulates processes like cell proliferation, apoptosis, and overall growth inhibition [81].

3.1.5. Regulating Cell Signaling

The signal pathways play a crucial role in mediating the anti-tumor effects of diosgenin. These pathways encompass the mitogen-activated protein kinase (MAPK) signaling pathway, which includes cascades like c-Jun N-terminal kinase (JNK) and extracellular signal-regulated kinase (ERK), as well as p38-MAPK and ERK5. The MAPK/ERK pathway is implicated in diverse cellular functions, including cell proliferation, differentiation, migration, senescence, and apoptosis. Another crucial pathway is the phosphatidylinositol-3 kinase (PI3K)/Akt pathway, which plays a role in cell survival, proliferation, and growth. The nuclear factor kappa B (NF-κB)/STAT3 pathway, which plays a role in regulating inflammation and cell survival. The Wnt/β-catenin signal pathway is responsible for cell fate determination and cell proliferation. The cAMP/PKA/CREB pathway is associated with cell signaling and gene expression regulation. By targeting these diverse signal pathways, diosgenin exhibits its anti-tumor effects by influencing various cellular processes, providing a comprehensive approach to combating tumor development and progression [82]. Diosgenin exhibits its anti-tumor effects through various mechanisms in different types of cancer cells. In colon cancer, esophageal, and osteosarcoma cells, diosgenin governs critical aspects like proliferation, apoptosis, migration, and invasion by obstructing the initiation of the epithelial–mesenchymal transition (EMT). This modulation involves the manipulation of EMT-associated proteins like transforming growth factor β1, E-cadherin, and vimentin. Diosgenin also triggers apoptosis in A549 cells through its impact on proteins within the MAPK signal pathway, including caspase 8, 9, and 3 [83]. Diosgenin downregulates the MAPK, NF-κB, and Akt signaling pathways by inhibiting the phosphorylation of NF-κB/p65, JNK, IKK-β, Akt, and ERK, resulting in its anti-tumor effect. In pancreatic tumor cells, diosgenin effectively inhibits the ERK, JNK, and PI3K/AKT signaling pathways [84]. In human hepatocellular carcinoma cells, diosgenin inhibits NF-κB activity and the activation of NF-κB/STAT3, leading to a significant decrease in the expression of various oncogene products and the inhibition of cell proliferation [85]. In MG-63 cells, diosgenin stimulates both cell proliferation and differentiation by impeding the Wnt/β-catenin signaling pathway, which leads to a decrease in the count of calcified nodules and a restraint in the expression of osteopontin and osteocalcin [86]. Diosgenin inhibits aerobic glycolysis in colorectal cancer cells by regulating glucose transporters (GLUT3 and GLUT4) and inhibiting CREB phosphorylation through the cAMP/PKA/CREB pathway, leading to apoptosis [87]. These diverse pathways and mechanisms collectively contribute to the anti-tumor effects of diosgenin, highlighting its potential as a multi-target and multi-pathway therapeutic agent in cancer treatment.

3.2. Solamargine

Solamargine is classified as a glycoalkaloid, which is a type of chemical compound that contains both sugar and an alkaloid. The compound has been isolated and identified from various plants, with structural variations depending on the specific plant source. Solamargine has been investigated for its potential pharmacological properties, including anti-inflammatory, antiviral, and anticancer activities. Some studies suggest that solamargine may exhibit cytotoxic effects on certain cancer cells, making it a subject of interest in cancer research [88,89].

3.2.1. Solamargine-Mediated Cytotoxicity

Solamargine has demonstrated remarkable cytotoxic activity against various cancer cell lines, making it one of the most potent cytotoxic compounds in this plant family [90]. Numerous studies utilizing diverse methodologies consistently demonstrate the cytotoxic effects of solamargine on an extensive array of cancer cell lines. These encompass human liver cancer lines such as HepG2, Hep3B, H22, and SMMC-7721; human lung cancer lines including A549, H441, H520, H661, H69, and H1650; human breast cancer lines like HBL-100, SK-BR-3, ZR-75-1, and MCF-7; human myelogenous leukemia line K562; squamous carcinoma line KB; human prostate cancer line PC3; osteosarcoma lines U2OS, MG-63, and Saos-2; murine melanoma line B16F10; human colon carcinoma line HT29; human cervical adenocarcinoma line HeLa; and human glioblastoma lines MO59J, U343, and U251. Notably, solamargine exhibits minimal toxicity toward normal cell lines, including Chinese hamster lung fibroblasts (V79), human lung fibroblasts (GM07492A), human liver cells (HL-7702), and the human retinal pigment epithelium (RPE1). The cytotoxic effects of solamargine have been found to be dependent on both the dosage and exposure time [91–93]. The preceding evidence supports the notion of the heightened cytotoxicity of solamargine specifically against lung cancer cell lines, while demonstrating its minimal impact on normal cells.

3.2.2. Solamargine in the Apoptotic Process

The intrinsic mitochondrial pathway of apoptosis is intricately governed by maintaining equilibrium between the proapoptotic and anti-apoptotic members within the Bcl-2 protein family [94]. The Bcl-2 family proteins play a critical role in regulating the mitochondrial membrane permeability and the intrinsic apoptotic pathway. This family is composed of two main groups: proapoptotic proteins, which promote cell death, and anti-apoptotic proteins, which inhibit apoptosis. Some of the proapoptotic Bcl-2 family proteins include Bcl-10, Bax, Bak, Bik, Bid, Bim, and Bad. These proteins are involved in promoting the mitochondrial membrane's permeability, which allows the release of proapoptotic factors, such as cytochrome c, into the cytosol. This, in turn, triggers the caspase cascade and leads to apoptosis. On the other hand, the anti-apoptotic Bcl-2 family proteins, such as Bcl-2, Bcl-XS, Bcl-XL, Bcl-x, Bcl-w, and BAG, are important in maintaining the mitochondrial membrane's integrity and preventing the release of proapoptotic factors [95]. Li et al. [96] demonstrated that solamargine induced apoptosis through mechanisms involving the upregulation of Bax and p53. The increased expression of Bax, a proapoptotic member of the Bcl-2 family, promotes apoptosis by facilitating the permeabilization of the mitochondrial membrane and releasing proapoptotic factors, such as cytochrome c, into the cytosol, where it becomes a key trigger for programmed cell death [97]. Cytochrome c oxidase (Cyt c or Complex IV, EC 1.9.3.1) is an enzyme crucially involved in ATP synthesis and is normally associated with the inner membrane of the mitochondrion, playing a vital role in supporting cellular life.

Specifically, the anti-apoptotic Bcl-xL and Bcl-2 genes were downregulated in solamargine-treated H44, H520, H661, H69, K562, A549, ZR-75-1, SK-BR-3, B16F10, HBL-100, HT29, HepG2, MCF-7, U343, HeLa, MO59J, and U251 cells. The downregulation of these anti-apoptotic genes promotes apoptosis by reducing their inhibitory effect on the process. Furthermore, an increase in the expression of cleaved caspase-3 and caspase-9 was observed in solamargine-treated K562, B16F10, HT29, MCF-7, HeLa, HepG2, MO59J, U343, and U251 cells [98,99]. The presence of cleaved caspase-3 and caspase-9 indicates the activation of the caspase cascade, which is a hallmark of the execution phase of apoptosis. Overall, these findings suggest that solamargine induces intrinsic mitochondrial apoptosis by altering the expression of Bcl-2 family members, leading to the activation of caspases and subsequent programmed cell death in the treated cancer cells.

The extrinsic (transmembrane) pathway of apoptosis is initiated when external death ligands, like the Fas ligand or TNF-related apoptosis-inducing ligand (TRAIL), attach to their respective death receptors located on the cell membrane. This interaction triggers the activation of caspase enzymes, which play a crucial role in coordinating the process of cellular self-destruction. The extrinsic pathway plays a crucial role in regulating cell death and is often implicated in the immune system's defense against abnormal or infected cells. A recent experiment demonstrated that solamargine induced apoptosis in K562 cells by causing early lysosomal destabilization, followed by subsequent mitochondrial damage. This damage was characterized by an overload of Ca^{2+}, a decrease in the membrane potential, and the release of Cyt c. Additionally, it has been observed that solamargine can induce permeabilization of the lysosomal membranes in cancer cells, leading to an increase in the influx of water. Therefore, lysosomes undergo swelling, ultimately resulting in the formation of vacuoles [98]. Enlarged lysosomes contribute to an elevation in surface tension, a reduction in lysosomal integrity, and eventual fragmentation. Furthermore, in the event of cellular damage, repair of enlarged lysosomes becomes more challenging compared to repairing smaller ones [100].

3.2.3. Cell Cycle Arrest

The cell cycle consists of four widely acknowledged stages, i.e., the G1, S, G2, and M phases [101]. Solamargine has demonstrated the capacity to elevate the ratio of the apoptotic sub-G1 peak within several cell lines, including A549 [93,102], H661, H441, H520, and H69 [103], as well as MCF-7, HBL-100, and SK-BR-3 [90], along with Hep3B cells [91,92]. This effect exhibits a dependence on both time and concentration. Conversely, the administration of solamargine led to a reduction in the G2/M phase in HBL-100, MCF-7, SK-BR-3 [90,104], Hep3B [92], SMMC-7721 [105], ZR-75-1 [104], and A549 cells [93,102]. Furthermore, investigations revealed no significant alterations in the G0/G1 phase in ZR-75-1, HBL-100, and SK-BR-3 cells following treatment with solamargine [104]. However, contrary to the previous findings, a more recent study demonstrated that solamargine did not induce cell cycle arrest in K562 cells [106]. The malignant phenotype of cancer cells is primarily attributed to uncontrolled cell cycle regulation, leading to abnormal cell growth and proliferation. Mitogens can impose inhibitory effects on the progression of the cell cycle by inducing the activation of G1-S cyclin-dependent kinase (CDK) activities. This activation subsequently initiates the phosphorylation of the retinoblastoma protein (pRB). The function of pRB is frequently impaired in cancer cells, contributing to their uncontrolled proliferation [107]. Tumorigenic mutations have been identified as the underlying factors behind the malfunctioning of the regulatory mechanisms that control the progression of the cell cycle. These mutations have been observed in diverse tumor types and can disrupt various mitogenic signaling pathways. These pathways encompass the HER2 gene and downstream signaling networks like the PI3K-Akt or Ras-Raf-MAPK pathways, as well as genes that play a role in regulating the cell cycle [108,109]. These alterations disrupt the normal regulation of the cell cycle, leading to uncontrolled cell growth and the proliferation characteristic of cancer cells. The proto-oncogene HER-2/neu plays a crucial role as a regulator of cell proliferation. Studies have shown that solamargine treatment resulted in an upregulation of HER2/neu and topoisomerase II α (Topo-IIα) expression in MCF-7, HBL-100, and SK-BR-3 cells [90].

3.2.4. Regulating Gene Expression and Proteins

Solamargine has been shown to induce apoptosis in cancer cells by modulating the expression of apoptosis-related genes and proteins. Treatment with solamargine has been found to influence the gene and protein expression of specific receptors and signal proteins associated with apoptosis in various cancer cell lines. In the HBL-100, SK-BR-3, ZR-75-1, A549, Hep3B, B16F10, HT29, MCF-7, MO59J, U343, HeLa, HepG2, and U251 cell lines, solamargine upregulated the gene expression of Hep3B [91,92,103] and TNF-R1 (tumor necrosis factor receptor 1) [93]. Additionally, in the A549 and Hep3B cell lines, it also upregulated the gene expression of TNF-R2 (tumor necrosis factor receptor 2). Furthermore, solamargine significantly increased the protein expressions of Fas (also known as CD95) and other downstream signal proteins like FADD (Fas-associated death domain protein) and TRADD (tumor necrosis factor receptor type 1-associated death domain protein) in

H441, H520, H661, B16F10, HT29, MCF-7, HBL-100, SK-BR-3, ZR-75-1, A549, HeLa, HepG2, MO59J, U343, and U251 cells. These findings suggest that solamargine activates the signal transduction of TNF receptors (TNFRs), leading to the activation of caspase-3 and -8 in A549, H69, HBL-100, ZR-75-1, SMMC-7721, H44, H520, H661, and SKBR-3 cells [104] This activation of caspases further supports the induction of apoptosis in the treated cancer cells. Solamargine can induce intrinsic apoptosis by regulating intrinsic apoptotic death mediators. In the case of solamargine treatment, it was observed that cytochrome c was released from the mitochondria into the cytosol in a dose-dependent manner in various cell lines, including A549, K562, H44, H520, H661, H69, HBL-100, ZR-75-1, SK-BR-3, and U2OS cells [93,96,98,103,104].

3.3. Solanine

A natural steroidal alkaloid—was initially identified in 1820 and subsequently recognized as prevalent in various plants, including the genus Solanum. It is categorized into α-solanine, β-solanine, and γ-solanine, with α-solanine exhibiting the highest concentration [110]. It acts as a natural defense mechanism for plants against herbivores and pests. While solanine is generally present in low concentrations in these plants, its level can increase under certain conditions, such as exposure to light and storage [111]. It has been observed to initiate the liberation of Ca^{2+} from the mitochondria, which consequently raises the cytoplasmic levels of Ca^{2+} within HepG2 cells. This outcome subsequently leads to a decrease in the potential of the mitochondrial membrane, ultimately instigating the process of apoptosis [112]. Additionally, in the HepG2 cell line, Ji et al. [113] demonstrated the induction of apoptosis as evidenced by the appearance of a sub-G0 apoptosis peak at various doses of α-solanine. Moreover, a reduction in the level of the antiapoptotic protein Bcl-2 was noted in a manner that correlated with the administered dose. In pancreatic cancer cells, α-solanine was discovered to induce an upregulation of p53 and Bax expression, concomitant with the downregulation of Bcl-2. Consequently, this event prompted the release of cytochrome c, thereby initiating the activation of the mitochondrial pathway for apoptosis. This decline in Bcl-2 and elevation in the Bax levels were likewise observed in cancerous tissue [113]. In a study by Mohsenikia et al. [114], it was demonstrated that α-solanine treatment in mice with breast cancer resulted in an increase in the proapoptotic Bax protein in breast cancer tissue. α-solanine has been found to exert another significant effect on cancer cells by inhibiting cell migration and invasion. This inhibition is achieved through the suppression of JNK, PI3K, and Akt phosphorylation, which subsequently leads to the downregulation of MMP-2 and MMP-9 expression. Moreover, α-solanine treatment has been shown to cause a reduction in the nuclear content of NF-κB in treated cells, further contributing to the inhibition of cancer metastasis [115]. In pancreatic cancer cells, Lv et al. [116] showed that α-solanine inhibits cell migration, invasion, and angiogenesis by reducing the levels of matrix metalloproteinases (MMPs) and vascular endothelial growth factor (VEGF). Additionally, α-solanine was found to modulate various signaling pathways, including JAK/STAT, Wnt/β-catenin, Akt/mTOR, and NF-κB/p65. These regulatory effects contribute to the inhibition of cell proliferation and the induction of apoptosis.

3.4. Apigenin

Formally classified as a flavone and recognized as 4′,5,7-trihydroxyflavone, stands out as one of the prevalent flavonoids found in a variety of plants. Its widespread presence is notably observed in plants within the Solanaceae and Asteraceae families, including those within the Artemisia genus [117]. In terms of biosynthesis, apigenin originates from the phenylpropanoid pathway and can be derived from both phenylalanine and tyrosine, two precursor molecules obtained through the shikimate pathway. Starting with phenylalanine, non-oxidative deamination leads to the formation of cinnamic acid, followed by oxidation at C-4. This intermediate is then transformed into p-coumaric acid. On the other hand, tyrosine undergoes direct deamination to produce p-coumaric acid. Upon CoA activation, p-coumarate combines with three malonyl-CoA units and undergoes aromatization through chalcone synthase to generate chalcone. Chalcone is subsequently isomerized by chalcone isomerase, resulting in the formation of naringenin. Finally, naringenin is oxidized into apigenin by flavanone synthase [118,119]. It manifests its impacts through the modulation of diverse kinase pathways, leading to the arrest of the cell cycle in the G2/M phase. Investigations have indicated that apigenin possesses the capability to impede cell proliferation and induce autophagy within HepG2 cells, exhibiting a reliance on both time and dose [120]. The autophagic mechanism in the HepG2 cells is facilitated through the suppression of the PI3K/Akt/mTOR pathway [121].

Similarly, other compounds from the Solanaceae follow varied mechanisms with different cellular backgrounds. Understanding these mechanisms can provide valuable insights for developing new cancer therapies that target specific pathways and exploit the vulnerabilities of cancer cells [122]. The cell cycle represents a pivotal mechanism governing cell proliferation. Notably, specific alkaloids sourced from the aerial parts of *Solanum nigrum* have exhibited antiproliferative attributes against gastric cancer cells. Some of these alkaloids, including β1-solasonine, solasonine, and solanigroside P, have shown the ability to induce apoptosis in gastric cancer cells. These alkaloids exert their antiproliferative effects by influencing the gene expression in cancer cells. They were found to increase Bax expression, which leads to the activation of apoptotic pathways, and decrease Bcl-2 expression, which is an anti-apoptotic protein that prevents cell death. By reducing Bcl-2 expression, the alkaloids promote the shift toward apoptosis, activating caspase-3, which triggers the breakdown of cellular components and leads to cell death [123].

4. Pharmacological Properties

Pharmacological properties refer to the specific effects and interactions of a drug or a compound with the biological systems of living organisms. These properties are responsible for the compound's ability to exert therapeutic or toxic effects within the body. Pharmacological properties encompass a wide range of characteristics that determine how a substance behaves in the body and how it affects physiological processes (Table 2). Some common pharmacological properties for compounds isolated from different members of the Solanaceae family are listed as follows.

4.1. Anti-Cancerous Activity

The phytochemicals found in plants of the Solanaceae are known to possess anticancer properties. *S. surattense*, a medicinal plant, exhibits an anticancer efficacy attributed to the presence of specific phytochemicals, including solamargine, diosgenin, apigenin, and lupeol. These compounds are considered valuable sources of anticancer components and contribute to the plant's potential anticancer activity [124]. In a study conducted by Kumar and Pandey, [125], it was reported that the fruit extract of *S. surattense* exhibited anticancer activity against human lung cancer cell lines (HOP-62) and leukemic cell lines (THP-1). The study demonstrated the potential anticancer efficacy of the fruit extract from *S. surattense* on these specific cell lines. Similarly, the methanolic extract of *S. nigrum* fruits was investigated for its inhibitory effect on the HeLa cell line (uterine cervix) [126]. The cell line viability was evaluated using the trypan blue dye exclusion technique. The cytotoxic impact of *S. nigrum* on the HeLa cells was gauged using both the MTT and SRB assays. The findings revealed that the methanolic extract of *S. nigrum* exhibited notable cytotoxic activity against the HeLa cell line across a concentration range spanning from 0.0196 mg/mL to 10 mg/mL, as ascertained using the SRB assay.

Table 2. Structural–activity relationship of different bioactive compounds of the family Solanaceae.

Name	Structure	Plant Parts	Medicinal Use	Other Use	References
Solasodine		Immature berries	Suppresses cancer growth, antioxidant, cytotoxic, hepatoprotective, anti-inflammatory	Used in the production of tonics, creams, and lotions	[127]
Diosgenin		Roots	Antiproliferative and anti-inflammatory activities	Used in skincare products	[128]
Solanine		Berries	Anti-inflammatory activity on LPS-activated RAW 264.7 macrophages	Used as a pesticide and fungicide	[129]
Solasonine		Berries	For production of contraceptives and steroidal anti-inflammatory drugs	Used as a pesticide	[130]
Solamargine		Berries	Antidiabetic, antifungal, antiparasitic, antibiotic, antimicrobial, and anti-cancerous properties. Cytotoxicity against skin tumors		[131]

Table 2. Cont.

Name	Structure	Plant Parts	Medicinal Use	Other Use	References
Degalactotigonin		Whole plant extract	Suppressed the growth and metastasis of osteosarcoma	Used as a wood protectant, acaricide, and pesticide	[132]
Beta-carotene		Leaves	Provides strong immune system and prevents cataracts	Used in confectionary, dairy, and packaged foods	[133]
Niacin		Leaves	Helps to keep the digestive and nervous systems healthy	Used in infant formulas, breakfast cereals, and energy drinks	[134]
Ascorbic acid		Leaves	To treat stomachache, jaundice, liver problems, and skin diseases	Used in hormone biosynthesis and as a cofactor	[135]
Citric acid		Leaves	Enhances Ca, P, and Mg absorption	Electroplating and leather tanning	[136]
Stearic acid		Seeds	Used as a tablet and capsule lubricant	Used in soaps, shaving creams, detergents, lotions, moisturizers, and candles	[137]
Palmitic acid		Seeds	Anti-inflammatory effect	Acts as an emollient, food additive, industrial mold release	[138]
Oleic acid		Seeds	Improves heart conditions by lowering cholesterol and reducing inflammation	Used as a lubricant, detergent, and surfactant	[139]

Table 2. Cont.

Name	Structure	Plant Parts	Medicinal Use	Other Use	References
Linoleic acid	linoleic acid	Seeds	To treat skin-related disorders	Used as moisturizer for skin, nails, and hair	[140]
Rutin		Berries	Antioxidant, cytoprotective, vasoprotective, anticarcinogenic, neuroprotective, and cardioprotective	Used as a colorant, antioxidant, and preservative	[141]
Protocatechuic acid		Leaves	Anti-inflammatory, antioxidant, estrogenic activity	Used to produce plastics and polymers	[142]
Gallic acid		Leaves	Neuroprotective, anti-inflammatory, decreases myocardial infarction	Used in paper manufacturing and ink dyes	[143]
Naringenin		Leaves	Anti-HCV and cardioprotective	Used as flavoring agent in carbohydrate drinks	[144]
Isoquercitrin		Berries/leaves	Antiviral, antioxidant, anti-inflammatory, stimulates mitochondrial biogenesis	Used in beverages	[145]

A series of indiosides (A–E) originating from *S. indicum* showcased a dose-dependent inhibitory influence on Bel-7402 cell proliferation, accompanied by the capability to induce cell death via the mitochondrial pathway. Furthermore, solavetivone-1, a compound present in *S. indicum*, was identified to possess cytotoxic properties against OVCAR-3 cells. These findings highlight the potential anticancer properties of these compounds derived from *S. indicum* [146]. The anticancer activity of *S. indicum* was investigated by testing the chloroform-soluble and insoluble fractions of the ethanolic extract obtained from the whole plant. In vitro techniques were employed to assess the anticancer potential against seven different cancer cell lines: HeLa, Hep-2 (laryngeal epidermoid), GBM8401/TSGH (glioma), H1477 (colon), Colo-205 (colon), KB (nasopharynx), and melanomas.

Using a dye exclusion assay (DEA) and 3-(4,5-dimethylthiazol-2-yl)-2,5-diphenyltetrazolium bromide (MTT) assays, it was determined that the purified components, dioscin and methyl protodioscin, exhibited more potent effects in terms of their anticancer activity. Moreover, in the polarographic reduction of oxygen (PRE) assay, dioscin, methyl protoprosapogenin A of dioscin, protodioscin, and methyl protodioscin exhibited cytotoxic impacts on cultured C6 glioma cells. Additionally, methyl protoprosapogenin A of dioscin, protodioscin, and methyl protodioscin demonstrated attributes of tumor suppression. Furthermore, dioscin, when administered at a concentration of 10 micrograms/mL, manifested an inhibitory influence on DNA synthesis within C6 glioma cells. These findings suggest the potential anticancer properties of these compounds, specifically in the context of C6 glioma cells [147]. Ma et al. [148] reported that several indiosides (A–E) extracted from *S. indicum* displayed a dose-dependent inhibitory effect on the proliferation of Bel7402 cells. Furthermore, these indiosides were found to induce apoptosis in the cells through the mitochondria-dependent pathway. Additionally, solavetivone-1, a specific component of *S. indicum*, demonstrated cytotoxicity against OVCAR-3 cells. These findings suggest the potential therapeutic value of these compounds in targeting cancer cells [146]. The cytotoxic influences of the chloroform-soluble and insoluble fractions from the ethanolic extract derived from the entire *S. indicum* plant were assessed via in vitro methodologies on seven distinct cancer cell lines. The tested cell lines encompassed HeLa, HA22T, Hep-2, GBM8401/TSGH, Colo-205, KB, and H1477. The refined compounds, notably dioscin and methyl protodioscin, exhibited more potent effects when subjected to evaluation using the DEA and MTT assays. These compounds demonstrated cytotoxicity in cultured C6 glioma cells.

Chiang et al. [147] reported that dioscin exhibited an inhibitory effect on DNA synthesis in C6 glioma cells at a concentration of 10 micrograms/mL. This indicates that dioscin holds promise in obstructing the replication and proliferation of C6 glioma cells. Moreover, the anticancer potential of the methanolic extract from the fruit was examined using the MTT cytotoxicity assay across diverse cancer cell lines. These encompassed prostate carcinoma (PC3 and DU145), colorectal carcinoma (HCT116), human non-small cell lung carcinoma (H1975), and malignant melanoma (A375). The evaluation of the extract's cytotoxicity using the MTT assay indicates its potential as an anticancer agent against these specific cancer cell lines.

Gopalakrishna et al. [149] observed that the fruit extract of *S. indicum* exhibited its maximum cytotoxicity in the prostate carcinoma cell line, with an IC$_{50}$ of 8.48 µg/mL in DU145 cells and 11.18 µg/mL in PC-3 cells. In H1975 cells, the extract exhibited cytotoxicity, reaching an IC$_{50}$ value of 9.03 µg/mL. Likewise, in HCT116 cells, the extract displayed cytotoxic effects, resulting in an IC$_{50}$ of 17.58 µg/mL. Furthermore, cytotoxicity was observed in A375 cells, yielding an IC$_{50}$ of 27.94 µg/mL. These findings suggest the potential of the fruit extract in inhibiting the growth of these cancer cell lines. Additionally, Rahman et al. [150] found that the fresh fruit of *S. indicum* exhibited significant brine shrimp lethality, with an LC$_{50}$ value of 4.42 ± 0.67 µg/mL. This indicates the potential bioactivity of the fruit extract against brine shrimp larvae, suggesting its possible cytotoxic properties.

The glycoalkaloids derived from the butanol extract of *S. villosum* fruit exhibited a toxic impact on the LIM-1863 human colon carcinoma cell line, contributing to cell death in these cancer cells. These naturally occurring compounds hold promise as valuable starting points for the exploration and development of prospective cancer therapeutics [151].

4.2. Antioxidant Activity

Oxidative stress is known to contribute to the development and progression of gastric ulcers, and antioxidants can help in mitigating this by reducing oxidative damage and promoting tissue healing. *S. torvum* has demonstrated significant antioxidant activity, which can help to reduce oxidative stress. Oxidative stress is associated with various health conditions, including diabetes, and can contribute to the development and progression of complications. The ability of *S. torvum* to scavenge reactive oxygen species (ROS) and mitigate oxidative stress makes it a potential natural remedy for managing oxidative-stress-related conditions like diabetes. By incorporating *S. torvum* into the diet or using its antioxidant components as a supplement, it may be possible to counteract the harmful effects of ROS and reduce the oxidative damage to cells and tissues. Indeed, in vitro studies have shown the promising antioxidant activity of this plant [152].

Antioxidants play a crucial role in preventing the oxidative damage caused by free radicals in the body, which is associated with various diseases. The restoration of antioxidant indicators such as glutathione, superoxide dismutase (SOD), glutathione reductase (GR), catalase (CAT), and lipid peroxidation (LPO) suggests that the extract possesses strong antioxidant properties. *S. surattense* has attracted attention due to its potential natural antioxidant properties. In a study by Meena et al. [153], it was discovered that methanolic and ethanolic extracts of *S. surattense* demonstrated notable antioxidant properties. This suggests that the plant contains compounds that can scavenge free radicals and protect against oxidative damage. The study suggests that the leaf extract of *S. surattense* possesses compounds that can potentially modulate the antioxidant defense system in the body [154]. By enhancing the levels of antioxidant enzymes such as SOD and GPx, the extract may help to counteract the increased oxidative stress caused by alloxan-induced diabetes [155]. By increasing the levels of these antioxidant enzymes, the leaf extract of *S. surattense* may contribute to reducing oxidative stress and protecting against cellular damage in alloxan-induced animal models.

A *S. nigrum* glycoprotein exhibited a radical scavenging activity that was dependent on the dosage. It demonstrated the ability to scavenge various radicals, including the hydroxyl radical (OH),1, 1-diphenyl-2-picrylhydrazyl (DPPH) radicals, and superoxide anions (O^{2-}), in a dose-dependent manner [156]. The antioxidant activity of the methanolic extract obtained from *S. indicum* berries was evaluated using the in vitro DPPH radical scavenging assay. Additionally, both the aqueous and ethanolic extracts of *S. indicum* leaves demonstrated DPPH scavenging potential, indicating their ability to neutralize free radicals [157]. The extract exhibited the highest inhibition rate (70.007 ± 0.841%) at a concentration of 200 µg/mL [158]. In a distinct study, IC_{50} values were established for both ethanolic and aqueous extracts of berries through the utilization of the DPPH scavenging assay and the β-carotene/linoleate model system. Notably, the ethanolic extract displayed more pronounced effectiveness (IC_{50} 37.22 ± 1.3 µg/mL) in the β-carotene assay, while the aqueous extract demonstrated superior efficacy in the DPPH assay (IC_{50} 21.83 ± 0.84 µg/mL) [159]. Furthermore, studies have indicated that the antioxidant capacity of the fruit intensifies during the ripening process, as observed using the ferric-reducing antioxidant power (FRAP) test and Folin–Ciocalteau assay. This enhancement in antioxidant potential is likely attributed to a significant increase in the concentration of β-carotene in red berries. Nevertheless, specific compounds such as caffeoylquinic acids, caffeic acid, flavonol glycosides, and naringenin demonstrated an increase in concentration as the berries reached maturity. On the other hand, the levels of p-coumaric acid and feruloylquinic acids remained constant at all stages of ripening [160].

4.3. Hepatoprotective

The investigators explored the hepatoprotective potential of aqueous and methanolic extracts of *S. nigrum* in rats subjected to a 10-day consecutive regimen of carbon tetrachloride (CCl4) injections. Carbon tetrachloride is recognized for its hepatotoxic effects. The experimental procedure involved the initial administration of CCl4 to induce liver damage in the rats. Subsequently, the rats were treated with aqueous extract of *S. nigrum* at doses spanning from 250 to 500 mg/kg. The purpose was to assess whether the extract could provide protection against the liver damage caused by CCl4. According to the findings of the study, the aqueous extracts of *S. nigrum* demonstrated a hepatoprotective effect against CCl4-induced liver damage in the rats. The evidence for this hepatoprotective effect was observed according to several parameters such as decreased serum aspartate aminotransferase (AST), alanine aminotransferase (ALT), and alkaline phosphatase (ALP) activities: elevated levels of these liver enzymes in the blood are indicative of liver damage. The administration of the aqueous extract resulted in a decrease in the activities of these enzymes, suggesting a protective effect on the liver. The study found that the administration of the aqueous extract led to a decrease in bilirubin concentration, indicating improved liver function. In terms of mild histopathological lesions, histopathological examination involves the microscopic examination of tissue samples to assess any abnormal changes. In this study, the rats injected with CCl4 alone exhibited more severe histopathological lesions in the liver tissue. However, when the rats were treated with the aqueous extract of *S. nigrum*, the severity of these lesions was reduced. This suggests that the extract helped to mitigate the damage caused by CCl4. Similarly, the administration of methanolic extracts of *S. nigrum* at doses of 250 to 500 mg/kg exhibited hepatoprotective effects. Notably, the levels of serum AST, ALT, ALP, and bilirubin demonstrated a significant reduction in animals treated with *S. nigrum* methanolic extract when compared to an untreated control group [161].

A study conducted by Bhuvaneswari and Suguna, [162], investigated the hepatoprotective role of *S. indicum* extract in a rat model of hepatotoxicity induced by carbon tetrachloride (CCl4). The researchers administered *S. indicum* extract at a dose of 200 mg/kg to the rats with CCl4-induced hepatotoxicity. They evaluated various liver markers like ALT, AST, ALP, ACP, and LDH and biochemical parameters to assess the extent of the liver damage and the effects of the plant extract. Furthermore, the investigation encompassed an analysis of diverse biochemical indicators, encompassing total bilirubin, total protein, triglycerides, total cholesterol, and urea. The findings of the study indicated that the *S. indicum* extract significantly ameliorated the damage caused by CCl4. While specific details about the extent of improvement or the exact values of the parameters were not provided, the significant amelioration suggests a protective effect of the *S. indicum* extract on the liver.

4.4. Cardioprotective

One study focused on evaluating the cardioprotective potential of the methanolic extract sourced from *S. nigrum* berries. To achieve this, an in vitro model simulating global ischemia–reperfusion injury was utilized to gauge the impact of the extract on cardiac function. The extract was administered at doses of 2.5 and 5.0 mg/kg over six days per week, spanning a total duration of 30 days. The findings of the investigation strongly suggested that the methanolic extract displayed noteworthy cardioprotective effects against global ischemia–reperfusion injury within the in vitro model. This activity was observed in a dose-dependent manner, meaning that higher doses of the extract provided a stronger protective effect [163].

The study investigated the cardiotonic activity of the methanolic extract derived from the fruits of *S. indicum* using a frog heart model. The researchers administered the extract at concentrations of 5 and 10 mg/mL to assess its effects on the force of contraction and heart rate. The results of the study indicated that the methanolic extract exhibited marked cardiotonic activity in a dose-dependent manner. When administered at a concentration of 5 mg/mL, the extract induced a modest enhancement in the force of contraction; however, no substantial alterations in heart rate were detected. Conversely, at a higher concentration of 10 mg/mL, the extract elicited a noteworthy escalation in the force of contraction, accompanied by a slight elevation in heart rate. Importantly, the study found that the methanolic extract of *S. indicum* exhibited a wide therapeutic index, meaning that it had a significant cardiotonic effect without showing any signs of cardiac toxicity at the higher tested doses, up to 5 mg/mL. These findings suggest that the methanolic extract of *S. indicum* fruits possesses cardiotonic activity, meaning it can enhance the force of contraction in frogs' hearts. The extract demonstrated a dose-dependent response, with higher concentrations resulting in more pronounced effects on cardiac function. Importantly, the extract exhibited a wide therapeutic index, indicating a relatively safe profile without causing cardiac toxicity at the tested higher doses [164].

4.5. Nephroprotective

In a study conducted by Waghulde [165], the nephroprotective effects of *S. torvum* fruits against doxorubicin (DOX)-induced nephrotoxicity were demonstrated in rats. DOX is recognized for its ability to cause harm to kidney cells in both rat and human subjects by triggering an excessive production of free radicals of the semiquinone type [166,167]. The flavonoids present in *S. torvum* have antioxidant and nephroprotective properties, as reported by Ching et al. [168]. These flavonoids are also capable of chelating free iron, which contributes to a reduction in DOX-induced toxicity in the kidneys. This suggests that the antioxidant and iron-chelating properties of *S. torvum* flavonoids play a role in mitigating the nephrotoxic effects induced by DOX [169]. Mohan et al. [170] conducted a histopathological analysis of the kidneys and observed that *S. torvum* reversed the structural damage caused by DOX, including tubular necrosis, renal lesions, and glomerular congestion. This suggests that *S. torvum* can mitigate the detrimental effects of DOX on the kidneys and restore their normal structure. Loganayaki et al. [171] also mentioned the nephroprotective action of phenolic compounds extracted from different parts of *S. torvum*. These phenolic compounds may possess properties that helps in protecting the kidneys from damage and maintain their normal function.

4.6. Hypertensive and Anti-Thrombotic Activity

Aqueous extract derived from dried fruits of *S. torvum* has demonstrated the ability to lower blood pressure. This effect may be attributed to a decrease in the sensitivity of vasorelaxant agents and an increase in hypersensitivity to contractile factors. In laboratory experiments conducted *in vitro*, the extract exhibited potent vasocontractile activity by activating both the Alpha 1-adrenergic pathway and calcium reflux [172]. Considering these observations, the potential utility of *S. torvum* in managing severe hypotension becomes apparent, especially in scenarios stemming from autonomic dysfunction that necessitate the application of vasopressor agents.

The potential anti-thrombotic impact of the aqueous extract derived from *S. torvum* was additionally examined using isolated rat platelets. Notably, the intravenous administration of both aqueous and methanolic extracts led to a noteworthy decrease in arterial blood pressure. This observed anti-thrombotic aggregation effect of *S. torvum* could hold significance for its potential cardiovascular implications in conditions such as arterial hypertension and hemostatic disorders [173]. The impact of a standardized ethanolic extract originating from *S. indicum* fruit (containing > 0.15 percent chlorogenic acids) on blood pressure was investigated in both normotensive and hypertension-induced (N(W)-nitro-L-arginine methylester (L-NAME) treated) rats. In normotensive rats, the administration of the extract (30 mg/kg) for a duration of four weeks did not exhibit any influence on blood pressure. However, after L-NAME administration, the extract effectively prevented the onset of hypertension in the animals [174]. The hypotensive effects of a standardized ethanolic extract sourced from *S. indicum* fruit (containing > 0.15% chlorogenic acids) were assessed in both normotensive and hypertensive (N(W)-nitro-L-arginine methyl ester (L-NAME)-treated) rats. The administration of the extract at a dose of 30 mg/kg for a duration of four weeks did not induce hypotensive effects in normotensive rats. However, it effectively averted the development of hypertension in the animals following L-NAME administration [175]. Similarly, an ethanolic extract of *S. villosum* exhibited mild antihypertensive activity in experimental rats. However, the study did not provide detailed information on the specific mechanisms of action or active compounds responsible for this effect [176].

4.7. Anti-Ulcerogenic

The presence of flavonoids, sterols, and triterpenes in *S. torvum* suggests that these compounds may contribute to its anti-ulcer properties. *S. torvum* is known to strengthen the mucosal barrier by promoting the production of mucus and bicarbonate. It may also reduce the volume of gastric acid secretion or neutralize gastric activity, thereby potentially preventing the development or progression of gastric ulcers [177]. Rats were subjected to different stress-inducing methods including cold restraint stress, indomethacin administration, pyloric ligation, ethanol ingestion, and acetic acid exposure, in order to induce stress ulcers in the experimental model. The extract from *S. torvum* fruits demonstrated a significant inhibition of the gastric lesions induced by these stressors, with the percentages of inhibition ranging from 70.6% to 80.1%. The potency of the extract was equal to or higher than that of omeprazole, a known anti-ulcer medication. The administered extract demonstrated a decrease in gastric secretory volume, acidity, and pepsin secretion in the rats afflicted with ulcers. Furthermore, a 7-day administration of the extract expedited the healing of the ulcers induced by acetic acid. To investigate the anti-secretory action of the extract, enzymatic studies were conducted on H^+/K^+ ATPase activity. The results showed that the *S. torvum* fruit extract significantly inhibited H^+/K^+ ATPase activity, which plays a role in gastric acid secretion. Additionally, the extract reduced the gastrin secretion in an ethanol-induced ulcer model, indicating its potential in modulating gastric secretory functions. These findings suggest that the extract from *S. torvum* fruits possesses anti-ulcer properties, inhibits gastric acid secretion, and accelerates ulcer healing [178].

The potential anti-ulcerogenic effects of the methanolic extract derived from the fruit of *S. indicum* were investigated in rats with induced ulceration caused by aspirin and ethanol. The administration of the extract at a dose of 750 mg/kg demonstrated significant effects. One of the key findings was that the extract protected the stomach mucosa from the damaging effects of aspirin and ethanol. This indicates its potential as a gastroprotective agent by preventing ulcer formation. Additionally, the extract exhibited the added benefit of promoting ulcer repair. This suggests that it can aid in the healing process of existing ulcers, which is crucial for restoring the integrity of the gastric mucosa. The observed anti-ulcerogenic effects of the extract are likely attributed to its antioxidant capacity. These effects are likely mediated through its antioxidant capacity, which helps restore the balance of oxidative stress in the stomach [179]. The objective of the rat study was to assess the choleretic activity of a suspension containing the fruit of *S. indicum*. The rats were initially anesthetized with intraperitoneal sodium pentobarbital. Then, they were exposed to bile duct cannulation, where a catheter was inserted to collect bile. Before administering the plant solution, bile was collected for one hour to establish the baseline levels. After this initial collection period, the rats were intraduodenally administered the plant solution at a dosage of 500 mg/kg. Following the administration of *S. indicum*, the researchers observed a significant increase in bile flow. Specifically, there was a 31 percent increase in bile flow compared to the baseline levels [180]. This finding indicates that the fruit of *S. indicum* exhibited choleretic activity via stimulation of the production and secretion of bile from the liver into the bile ducts.

4.8. Analgesic and Anti-Inflammatory Activity

The peripheral analgesic activity of aqueous extract obtained from *S. torvum* leaves was examined, revealing its potential in providing relief from pain. Its reported analgesic properties have been attributed to the inhibition of prostaglandin synthesis. By inhibiting their synthesis, the aqueous extract of *S. torvum* leaves may reduce pain perception and provide relief from pain. However, the dosage, mode of administration, and potential side effects should be considered before applying this plant extract for pain management [181].

The documented anti-inflammatory potential of *S. torvum* extract suggests its capacity to potentially modulate the later stages of the inflammatory response by inhibiting cyclooxygenase, an enzyme pivotal in prostaglandin synthesis, known to mediate inflammation. Correspondingly, an examination of the anti-inflammatory attributes of the methanolic extract from entire *S. nigrum* plants was conducted using animal models. The study revealed that the methanolic extract, administered at 100 mg/kg and 200 mg/kg body weight concentrations, exhibited significant and dose-dependent anti-inflammatory effects in rats afflicted with hind paw edema induced by carrageenin and egg white [182]. These findings suggest that both *S. torvum* and *S. nigrum* extracts possess anti-inflammatory properties, which can potentially be attributed to their ability to modulate inflammatory processes and inhibit the production of inflammatory mediators.

The anti-inflammatory potential of ethanolic extracts from *S. nigrum* was assessed utilizing the rat paw edema model as induced by carrageenan. Oral administration of varying doses—100 mg/kg, 250 mg/kg, and 500 mg/kg—was conducted. Notably, the study highlighted a significant anti-inflammatory effect ($p < 0.001$) of the 500 mg/kg dose of the extract in comparison to the reference drug, diclofenac sodium (50 mg/kg). Similarly, the impact of methanolic extracts sourced from *S. nigrum* berries was investigated concerning carrageenan-induced paw edema. The results indicated a notable reduction in hind paw edema due to the methanolic extract. Particularly, the anti-inflammatory potential was significant at a dose of 375 mg/kg body weight. These findings suggest that both ethanolic and methanolic extracts of *S. nigrum* possess anti-inflammatory properties [183,184].

4.9. Antibacterial and Antiviral Activity

The ethanolic leaf extract of *S. surattense* has been found to possess potential antimicrobial activity against various pathogens. This extract has shown activity against bacteria such as *Vibrio cholera, Pseudomonas aeruginosa, Staphylococcus aureus, Streptococcus species, Escherichia coli, Salmonella typhi,* and *Shigella dysenteriae*. These findings suggest that the ethanolic leaf extract of *S. surattense* may have the ability to inhibit the growth of or kill these microorganisms, indicating its potential as an antimicrobial agent [10]. Ethanol and methanol extracts of *S. surattense* have been found to possess strong antibacterial activity against *Pseudomonas aeruginosa* [185]. Additionally, the fruit extract of *S. surattense* has shown potential inhibition of the growth of bacteria such as *Escherichia coli, Salmonella typhi, Micrococcus luteus, Staphylococcus aureus, Pasteurella multifida,* and *Vibrio cholera*. These findings suggest that extracts of *S. surattense* may contain bioactive compounds with antibacterial properties, making them effective against a range of bacterial pathogens [186]. *S. surattense* extract has been studied for its antifungal effectiveness against several fungi, including *Aspergillus niger, A. flavus, A. fumigatus,* and *Trichoderma viride* [10]. Mahmood et al. [187] conducted a targeted investigation into the antifungal properties of *S. surattense*, focusing on its impact on the growth of *A. fumigatus* and *A. niger*. The study revealed noteworthy antifungal efficacy within the methanolic extract derived from *S. surattense* seeds, particularly exhibiting substantial activity against *Rhizopus oryzae* as well as *A. fumigatus*. These findings suggest that the methanolic seed extract of *S. surattense* may possess compounds that can inhibit the growth of certain fungal pathogens, making it a potential antifungal agent [188].

Research has also aimed to assess the antibacterial potential of methanol and aqueous extracts sourced from *S. nigrum* leaves. Employing the disc diffusion technique, the extracts were subjected to screening against two Gram-negative bacterial strains, *Xanthomonas campestris* (a plant pathogen) and *Aeromonas hydrophila* (an animal pathogen). The findings unveiled the notable antibacterial effectiveness of the methanol extracts from all plant samples, manifesting significant activity against both tested bacterial strains. Clear zones of inhibition were observed for the methanol extracts of *S. nigrum*, indicating their potential antibacterial properties against the tested microorganisms [189]. The ethanolic extract of

S. indicum leaves exhibited antibacterial activity against *Pseudomonas* spp., *Corynebacterium diptheriae*, and *Salmonella typhimorium*, as reported by Gavimath et al. [190]. It has also been found that the ethanolic extract of *S. indicum* leaves showed antibacterial activity against *Bacillus cereus*, *Staphylococcus aureus*, and *Escherichia coli*. Additionally, the chloroform extract, acetone extract, and ethanol extract of *S. indicum* demonstrated antibacterial activity against Pseudomonas species [191]. The fruits of *S. indicum* have been found to possess antibacterial activity. Both aqueous and ethanolic extracts of the fruits showed effectiveness against *Escherichia coli*, *Listeria innocua*, *Staphylococcus aureus*, and *Pseudomonas aeruginosa* strains. It was observed that the ethanolic extract exhibited better activity compared to the aqueous extract in terms of its antibacterial effectiveness [192]. In the study, the concentration-dependent inhibitory influence of the aqueous fraction from the ethanolic extract of *S. indicum* berries was demonstrated against diverse Pseudomonas strains, notably encompassing *Pseudomonas fluorescens*, *Pseudomonas aeruginosa*, and *Pseudomonas syringae*. The tested aqueous fraction was reported to contain flavonoids, carotenoids, and saponins, which could contribute to its antibacterial properties [192].

Ethanol, methanol, and water are commonly used as polar extraction solvents for the plant parts of *S. incanum*, and they have been found to yield phytochemicals with antibacterial and antifungal properties [193–195]. In studies, the antibacterial effects of ethanol and aqueous crude extracts derived from the leaves, fruits, and stems of *S. incanum* were evaluated against established strains of both Gram-positive and Gram-negative bacteria. The crude extracts from *S. incanum* displayed diverse degrees of growth restraint against the bacterial strains under investigation. While certain extracts demonstrated no discernible inhibition, others exhibited significant inhibitory effects. Specifically, the aqueous stem extract did not show any growth inhibition. However, the ethanol extracts of the leaves, fruits, and stems demonstrated significant growth inhibition against the tested bacterial strains. Among these, the ethanol and aqueous leaf extracts were found to be particularly effective in inhibiting bacterial growth [196,197]. Assessment of the antibacterial efficacy of *S. villosum* leaf extracts was conducted against a pair of Gram-positive bacteria, as well as two Gram-negative bacteria. It was observed that all strains displayed susceptibility to its aqueous, n-hexane, and ethanol extracts. Notably, the organic extracts exhibited superior effectiveness compared to the aqueous counterparts. The leaf extracts of *S. villosum*, encompassing aqueous, n-hexane, and ethanol compositions, demonstrated substantial effectiveness against the entire spectrum of bacterial strains examined, thus indicating their potential utility against pathogenic microorganisms affecting humans [198].

The methanolic extract of *S. torvum* fruits was examined for its antiviral activity against herpes simplex virus type 1 (HSV-1). During the extraction process, a new C4-sulfated isoflavonoid called torvanol A, a steroidal glycoside named torvoside A, and torvoside H were derived from the extract. The researchers found that these compounds exhibited antiviral activity against HSV-1. This suggests that the methanolic extract of *S. torvum* fruits, containing torvoside A, torvanolA, and torvoside H, has the potential to suppress the replication or activity of HSV-1 [199].

4.10. Anthelmintic Activity

Gunaselvi et al. [200] reported that aqueous extracts of *S. surattense* fruit exhibit anthelmintic activities. The findings suggest that aqueous extracts of *S. surattense* fruit powder possess properties that may be effective against helminth infections. A study by Priya et al. [201] indicates that various extracts of the whole plant of *S. surattense*, including aqueous, hydroethanolic, and ethanolic extracts, exhibit anthelmintic activity. Furthermore, the anthelmintic potential of the butanol and aqueous fractions obtained from the methanolic extract of *S. indicum* fruits was evaluated using the *C. elegans* bioassay. This assay involves assessing the percentage of dead nematodes after a 24 h incubation period. The fractions eluted from DEAE cellulose showed anthelmintic activity, and this activity was observed at four separate peaks based on the *C. elegans* assay. The fractions obtained from the methanolic extract of *S. indicum* fruits, specifically the butanol and aqueous fractions, exhibited eluted peaks at NaCl concentrations of 0.1, 0.28, 0.48, and 0.85 M. These eluted peaks correspond to distinct compounds possessing anthelmintic properties. Notably, the peak-associated mean death percentages were 37%, 53%, 59%, and 61%, respectively, in comparison to the negative control. These findings suggest that the *S. indicum* fruit contains at least four different anthelmintic compounds, each contributing to the observed anthelmintic activity. Further characterization and identification of these compounds would be necessary to determine their specific structures and mechanisms of action. This information could potentially contribute to the development of new anthelmintic treatments or the isolation of active compounds for further investigation [202]. A methanolic extract of *S. indicum* berries at a concentration of 100 mg/mL demonstrated paralyzing effects on the Indian earthworm (*Pheretima posthuma*) within an average time of 9.16 ± 0.12 s. Additionally, helminths died within an average time of 17.71 ± 0.21 s after exposure to the extract. These results indicate the potential anthelmintic activity of the methanolic extract of *S. indicum* berries against the Indian earthworm. The observed paralysis and subsequent death of the helminth suggest the presence of bioactive compounds in the extract that could be responsible for these effects [158]. These findings suggest that both *S. surattense* and *S. indicum* possess compounds with potential anthelmintic properties. Nevertheless, additional investigations are warranted to ascertain the precise active constituents accountable for the noted anthelmintic effects and to explore their potential application to the management of helminth infections.

4.11. Antiplasmodial Activity

Research conducted by Zirihi et al. [203] was centered on the assessment of the ethanolic fruit extract of *S. indicum* for its antiplasmodial potential against the chloroquine-resistant FcB1 strain of *Plasmodium falciparum*, the causative agent of malaria. Furthermore, the extract's impact on human MRC-5 and rat L-6 cell lines was examined to gauge its cytotoxicity. The study's outcomes revealed the noteworthy antimalarial efficacy of the ethanolic fruit extract against the *Plasmodium falciparum* strain resistant to chloroquine. The IC_{50} value, which represents the concentration required to inhibit the growth of the parasite by 50%, was determined to be 41.3 ± 7.0 g/mL for the extract. Furthermore, cytotoxicity tests were conducted on human MRC-5 and rat L-6 cell lines to assess the potential toxicity of the extract on human and rat cells. The IC_{50} value, which represents the concentration required to cause a 50% reduction in cell viability, was found to be greater than 50 g/mL for both cell lines. This suggests that the extract exhibited low cytotoxicity against these cell lines at the concentrations tested. In summary, the ethanolic fruit extract demonstrated significant antiplasmodial activity against the chloroquine-resistant FcB1 strain of *Plasmodium falciparum*. The extract exhibited an IC_{50} value of 41.3 ± 7.0 g/mL, indicating its potency in inhibiting the growth of the parasite. Additionally, the extract showed low cytotoxicity on the human MRC-5 and rat L-6 cell lines, with IC_{50} values greater than 50 g/mL. These findings highlight the potential of the ethanolic fruit extract as a source of antimalarial compounds for further exploration and development [193]. The crude extracts obtained using methanol, petroleum ether, chloroform, ethyl acetate, and aqueous solvents from this plant demonstrated considerable and suitably specific antiprotozoal efficacy against *Trypanosoma brucei*, *T. cruzi*, *Plasmodium falciparum*, and *Leishmania infantum* [204].

4.12. Antimalarial Activity

The utilization of *S. surattense*, both in in vivo and in vitro contexts, has exhibited notable antimalarial effects without any associated toxicity. Notably, dichloromethane extract from *S. surattense* has displayed robust anti-plasmodial efficacy, underscoring its potential as a promising intervention for malaria treatment. Furthermore, Zihiri et al., 2005 [203], reported that an ethyl acetate extract derived from the aerial parts of *S. surattense* exhibited efficacy against the larvae of *Plasmodium falciparum*, the parasite responsible for causing malaria. This suggests that specific compounds present in the extract possess antimalarial properties and could potentially be developed into antimalarial drugs. These findings highlight the promising potential of *S. surattense* as a natural source for discovering new antimalarial compounds.

5. Conclusions and Future Perspectives

Traditional herbal medicines, often used in curing different human ailments, have gained significant momentum for being vital in the maintenance of the health and wellbeing of humans. Being rich in phytochemical compounds, their use in medical practice owing to their pharmacological attributes is often considered to have been a living tradition since time immemorial. The existing literature on plants with medicinal properties was used to describe their effectiveness against particular or multiple human diseases. To be precise, there was previously no in-depth study performed on the clinical efficacy of Solanaceae family members in terms of the plant parts used, phytochemical production, and correlation made between phytochemicals in line with their possible clinical applications. This paper aimed to describe the common but prominent members of the Solanaceae family grown across different sub-continents with special mentions of the parts used and metabolites screened for the evaluation of their therapeutic potential in diverse clinical applications, ranging from curing different diseases to their use in the treatment of different cancers. Of the plants studied, special stress has been placed on the production of different secondary metabolites, highlighting their structural diversity and elaborating on the mechanistic insights to resolve their modi operandi against different cellular backgrounds. The health benefits are covered in terms of their broad-spectrum use as anti-cancerous, antioxidant, hepato-, cardio- and nephroprotective, antihypertensive, anti-ulcerogenic, analgesic and anti-inflammatory agents, as well as being used as antibacterial, antiviral, anthelmintic, antiplasmodial, and antimalarial treatments. Despite their remarkable pharmacological properties, a gap exists between the therapeutic potential of the bioactive moieties and the clinical outcomes, limited by fewer studies having been performed on the structural–activity relationships for the improvement of their properties, not to mention the need to perform safety and potency checks of metabolites with drug-like properties. An in-depth understanding of the structural–activity relationship will help in enhancing their role within the system for use in the prevention and treatment of different ailments. Additionally, it will help in outlining a direction for future investigations to confirm their therapeutic properties and reap the benefit in terms of products for human welfare.

Author Contributions: Conceptualization, J.K. and A.T.J.; methodology, S.J. and A.T.J.; software, O.B.; validation, S.N.S., M.A.K. and S.R.; investigation, A.T.J.; data curation, S.I., S.N.S. and S.J.; writing—original draft preparation, S.J. and S.I.; writing—review and editing, S.R., M.A.K., O.B., J.K. and A.T.J.; supervision, J.K. and A.T.J.; project administration, A.T.J.; funding acquisition, J.K. and A.T.J. All authors have read and agreed to the published version of the manuscript.

Funding: This research was funded by the Basic Science Research Program through the National Research Foundation of Korea (NRF), funded by the Ministry of Education (2020R1A6A1A03044512, 2020R1I1A3060716, and RS-2022-00167049). The research was also funded by JK ST&IC and DST under grant nos. JK ST&IC/SRE/996-998 and CRG/2019/004106.

Institutional Review Board Statement: Not applicable.

Informed Consent Statement: Not applicable.

Data Availability Statement: Not applicable.

Acknowledgments: The authors extend their appreciation to their fellow colleagues who helped in improving the contents of the manuscript.

Conflicts of Interest: The authors declare no conflicts of interest.

References

1. Dong, Y.; Hao, L.; Fang, K.; Han, X.X.; Yu, H.; Zhang, J.J.; Cai, L.J.; Fan, T.; Zhang, W.D.; Pang, K.; et al. A network pharmacology perspective for deciphering potential mechanisms of action of *Solanum nigrum* L. in bladder cancer. *BMC Complement. Med. Ther.* **2021**, *21*, 45. [CrossRef] [PubMed]
2. Przeor, M. Some common medicinal plants with antidiabetic activity, known and available in Europe (A Mini-Review). *Pharma* **2022**, *15*, 65. [CrossRef] [PubMed]
3. Bhat, M.A.; Mishra, A.K.; Kamal, M.A.; Rahman, S.; Jan, A.T. *Elaeagnus umbellata*: A miraculous shrub with potent health-promoting benefits from Northwest Himalaya. *Saudi J. Biol. Sci.* **2023**, *30*, 103662.
4. Li, Y.; Kong, D.; Fu, Y.; Sussman, M.R.; Wu, H. The effect of developmental and environmental factors on secondary metabolites in medicinal plants. *Plant Physiol. Biochem.* **2020**, *148*, 80–89. [CrossRef] [PubMed]
5. AlSheikh, H.M.A.; Sultan, I.; Kumar, V.; Rather, I.A.; Al-Sheikh, H.; Tasleem Jan, A.; Haq, Q.M.R. Plant-based phytochemicals as possible alternative to antibiotics in combating bacterial drug resistance. *Antibiotics* **2020**, *9*, 480. [CrossRef] [PubMed]
6. Hemlata; Bhat, M.A.; Kumar, V.; Ahmed, M.Z.; Alqahtani, A.S.; Alqahtani, M.S.; Jan, A.T.; Rahman, S.; Tiwari, A. Screening of natural compounds for identification of novel inhibitors against blaCTX-M-152 reported among Kluyvera georgiana isolates: An in vitro and in silico study. *Microb. Pathog.* **2020**, *150*, 104688. [CrossRef] [PubMed]
7. de Jesus Matias, L.; Rocha, J.A.; Menezes, E.V.; de Melo Júnior, A.F.; de Oliveira, D.A. Phytochemistry in medicinal species of *Solanum* L. (Solanaceae). *Pharmacogn. Res.* **2019**, *11*, 47–50.
8. Paumgartten, F.J.R.; de Souza, G.R.; da Silva, A.J.R.; De-Oliveira, A.C.A.X. Analgesic properties of plants from the genus *Solanum* L. (Solanaceae). In *Treatments, Mechanisms, and Adverse Reactions of Anesthetics and Analgesics*; Academic Press: Cambridge, MA, USA, 2022; pp. 457–471.
9. Ahmad, V.U.; Abad, M.J.; Ali, M.S.; Aly, A.H.; Green, I.R.; Hussain, J.; Nasser, R.A.; Shah, M.R.; Wray, V. *Solanaceae and Convolvulaceae: Secondary Metabolites*; Springer Sci & Business Media: Berlin/Heidelberg, Germany, 2013.
10. Sheeba, E. Antibacterial activity of *Solanum surattense* Burm. F. Kathmandu university. *J. Sci. Eng. Technol.* **2010**, *6*, 1–4.
11. Kaunda, J.S.; Zhang, Y.J. The genus solanum: An ethnopharmacological, phytochemical and biological properties review. *Nat. Prod. Bioprospect.* **2019**, *9*, 77–137. [CrossRef]
12. Chidambaram, K.; Alqahtani, T.; Alghazwani, Y.; Aldahish, A.; Annadurai, S.; Venkatesan, K.; Kandasamy, G. Medicinal plants of Solanum species: The promising sources of phyto-insecticidal compounds. *J. Trop. Med.* **2022**, *2022*, 4952221. [CrossRef]
13. Olmstead, R.G.; Bohs, L. A summary of molecular systematic research in Solanaceae: 1982–2006. *VI Int. Solanaceae Conf. Genom. Meets Biodivers.* **2006**, *745*, 255–268. [CrossRef]
14. Peralta, I.E.; Spooner, D.M.; Knapp, S. Taxonomy of wild tomatoes and their relatives (*Solanum* sect. Lycopersicoides, sect. Juglandifolia, sect. Lycopersicon; Solanaceae). *Syst. Bot.* **2008**, *84*, 186.
15. Hawkes, J.G. *The Potato: Evolution, Biodiversity and Genetic Resources*; Belhaven Press: Hoboken, NJ, USA, 1990.
16. Jayanthy, A.; Maurya, A.; Verma, S.C.; Srivastava, A.; Shankar, M.B.; Sharma, R.K. A brief review on pharmacognosy, phytochemistry and therapeutic potential of *Solanum indium* L. used in Indian Systems of Medicine. *Asian J. Res. Chem.* **2016**, *9*, 127–132. [CrossRef]
17. Wilf, P.; Carvalho, M.R.; Gandolfo, M.A.; Cúneo, N.R. Eocene lantern fruits from Gondwanan Patagonia and the early origins of Solanaceae. *Science* **2017**, *355*, 71–75. [CrossRef] [PubMed]
18. Echeverría-Londoño, S.; Särkinen, T.; Fenton, I.S.; Purvis, A.; Knapp, S. Dynamism and context-dependency in diversification of the megadiverse plant genus *Solanum* (Solanaceae). *J. Syst. Evol.* **2020**, *58*, 767–782. [CrossRef]
19. Judd, W.S.; Campbell, C.S.; Kellogg, E.A.; Stevens, P.F.; Donoghue, M.J. A Phylogenetic Approach. In *Plant Systematics*; Sinauer Associates, Inc.: Sunderland, MA, USA, 2016.
20. Spooner, D.; Spooner, D.M.; Peralta, I.E.; Knapp, S. Comparison of AFLPs with other markers for phylogenetic inference in wild tomatoes [*Solanum* L. section Lycopersicon (Mill.) Wettst.]. *Taxon* **2005**, *54*, 43–61. [CrossRef]
21. Perry, L.; Dickau, R.; Zarrillo, S.; Holst, I.; Pearsall, D.M.; Piperno, D.R.; Zeidler, J.A. Starch fossils and the domestication and dispersal of chili peppers (*Capsicum* spp. L.) in the Americas. *Science* **2007**, *315*, 986–988. [CrossRef]
22. Edmonds, J.M.; Chweya, J.A. *Black Nightshades: Solanum nigrum L. and Related Species (Vol. 15)*; Bioversity International: Roma, Italy, 1997.
23. Boulos, L. Volume three (Verbenaceae—Compositae). In *Flora of Egypt*; Al-Hadara Publishing: Cairo, Egypt, 2002; 373p.
24. Holm, L.G.; Plucknett, D.L.; Pancho, J.V.; Herberger, J.P. World's Worst Weeds. In *Dist Bio*; University Press of Hawaii: Honolulu, HI, USA, 1977.
25. Turner, N.J.; von Aderkas, P. *The North American Guide to Common Poisonous Plants and Mushrooms*; Timber Press: Portland, OR, USA, 2009.
26. Lin, H.M.; Tseng, H.C.; Wang, C.J.; Lin, J.J.; Lo, C.W.; Chou, F.P. Hepatoprotective effects of *Solanum nigrum* Linn extract against CCl4-iduced oxidative damage in rats. *Chem. Bio. Int.* **2008**, *171*, 283–293. [CrossRef]
27. Vorontsova, M.S.; Knapp, S. *Solanum torvum*. In Solanaceae Source. Knapp (ed.) 2014. Available online: http://www.solanaceaesource.org/content/solanum-torvum (accessed on 18 January 2024).
28. Parmar, K.M.; Itankar, P.R.; Joshi, A.; Prasad, S.K. Anti-psoriatic potential of *Solanum xanthocarpum* stem in imiquimod-induced psoriatic mice model. *J. Ethnopharmacol.* **2017**, *198*, 158–166. [CrossRef]
29. Singh, O.M.; Singh, T.P. Phytochemistry of *Solanum xanthocarpum*: An amazing traditional healer. *J. Sci. Ind. Res.* **2010**, *69*, 732–740.

30. Mosallam, H.A. Comparative study on the vegetation of protected and non-protected areas, Sudera, Taif, Saudi Arabia. *Int. J. Agric. Biol.* **2007**, *9*, 202–214.
31. Yousaf, Z.; Shinwari, Z.K.; Khan, M.A. Phenetic analysis of medicinally important species of the genus Solanum from Pakistan. *Pak. J. Bot.* **2010**, *42*, 1827–1833.
32. Knapp, S.; Barboza, G.E.; Bohs, L.; Särkinen, T. A revision of the Morelloid clade of *Solanum* L.(Solanaceae) in North and Central America and the Caribbean. *Phyto Keys* **2019**, *123*, 1–144. [CrossRef] [PubMed]
33. Begum, A.S.; Goyal, M. Phcog Mag.: Review Article Research and Medicinal Potential of the genus Cestrum (Solanaceae)—A Review. *Pharmacogn. Rev.* **2007**, *1*, 320–332.
34. Chowdhury, N.; Bhattacharjee, I.; Laskar, S.; Chandra, G. Efficacy of *Solanum villosum* Mill. (Solanaceae: Solanales) as a biocontrol agent against fourth instar larvae of Culex quinquefasciatus Say. *Turk. J. Zool.* **2007**, *31*, 365–370.
35. Jaeger, P.; Hepper, F.N. A review of the genus *Solanum* in Africa. In *Solanaceae: Biology and Systematics*; D'Arcy, W.G., Ed.; Columbia University Press: New York, NY, USA, 1986; pp. 41–55.
36. Nee, M. Synopsis of Solanum Section Acanthophora: A group of interest for glycoalkaloids. In *Solanaceae III, Taxonomy, Chemistry, Evolution*; Kew and Linnean Society of London: London, UK, 1991; pp. 257–266.
37. Mullahey, J.J.; Ferrell, J.; Sellers, B. Tropical soda apple: A noxious weed in Florida. In *Agronomy Department Document SS-AGR-77*; University of Florida/IFAS: Gainesville, FL, USA, 2006; p. 32611.
38. Mullahey, J.J.; Nee, M.; Wunderlin, R.P.; Delaney, K.R. Tropical soda apple (*Solanum viarum*): A new weed threat in subtropical regions. *Weed Technol.* **1993**, *7*, 783–786. [CrossRef]
39. Wunderlin, R.P.; Hansen, B.F.; Delaney, K.R.; Nee, M.; Mullahey, J.J. Solanum viarum and *S. tampicense* (Solanaceae): Two weedy species new to Florida and the United States. *SIDA Contrib. Bot.* **1993**, *15*, 605–611.
40. Nee, M.H. *A Revision of Solanum Section Acanthophora*; University of Wisconsin—Madison: Madison, WI, USA, 1980.
41. Mwonjoria, J.; Ngeranwa, J.; Kariuki, H.; Githinji, C.; Sagini, M.; Wambugu, S. Ethno medicinal, phytochemical and pharmacological aspects of *solanum incanum* (lin.). *Int. J. Pharmacol. Toxicol.* **2014**, *2*, 17–20. [CrossRef]
42. Miller, J.H. Nonnative invasive plants of southern forests: A field guide for identification and control [Revised]. In *General Technical Reports SRS-62*; US Department of Agriculture, Forest Service, Southern Research Station: Asheville, NC, USA, 2003; Volume 62, 93p.
43. Tekuri, S.K.; Pasupuleti, S.K.; Konidala, K.K.; Amuru, S.R.; Bassaiahgari, P.; Pabbaraju, N. Phytochemical and pharmacological activities of *Solanum surattense* Burm. f.—A review. *J. Appl. Pharma. Sci.* **2019**, *9*, 126–136.
44. Keerthana, S.D.; Philip, A.; Abdul, R.; Kumar, R.; Syed, S.A.; Syed, A.; Bharathi, D.R. Ethnopharmacology of *solanum nigrum*: A review. *World J. Curr. Med. Pharm. Res.* **2022**, *4*, 48–52.
45. Iqubal, M.; Sharma, S.K.S.; Hussain, M.S.; Mujahid, M. An updated ethnobotany, phytochemical and pharmacological potential of *Solanum indicum* L. *J. Drug Deliv. Ther.* **2022**, *12*, 160–172. [CrossRef]
46. Bryson, C.T.; Reddy, K.N.; Byrd, J.D. Growth, development, and morphological differences among native and nonnative prickly nightshades (*Solanum* spp.) of the southeastern United States. *Invasive Plant Sci. Manag.* **2012**, *5*, 341–352. [CrossRef]
47. Tiwari, R.; Rana, C.S. Plant secondary metabolites: A review. *Int. J. Eng. Res. Gen. Sci.* **2015**, *3*, 661–670.
48. Sharma, A.; Sharma, S.; Kumar, A.; Kumar, V.; Sharma, A.K. Plant secondary metabolites: An introduction of their chemistry and biological significance with physicochemical aspect. In *Plant Secondary Metabolites: Physico-Chemical Properties and Therapeutic Applications*; Springer Nature: Singapore, 2022; pp. 1–45.
49. Satish, L.; Shamili, S.; Yolcu, S.; Lavanya, G.; Alavilli, H.; Swamy, M.K. Biosynthesis of secondary metabolites in plants as influenced by different factors. In *Plant-Derived Bioactives: Production, Properties and Therapeutic Applications*; Springer Nature: Singapore, 2020; pp. 61–100.
50. Bourgaud, F.; Gravot, A.; Milesi, S.; Gontier, E. Production of plant secondary metabolites: A historical perspective. *Plant Sci.* **2001**, *161*, 839–851. [CrossRef]
51. Cragg, G.M.; Newman, D.J. Natural products: A continuing source of novel drug leads. *Biochim. Biophys. Acta BBA Gen. Subj.* **2013**, *1830*, 3670–3695. [CrossRef]
52. Saxena, M.; Saxena, J.; Nema, R.; Singh, D.; Gupta, A. Phytochemistry of medicinal plants. *J. Pharmacogn. Phytochem.* **2013**, *1*, 168–182.
53. Huang, W.Y.; Cai, Y.Z.; Zhang, Y. Natural phenolic compounds from medicinal herbs and dietary plants: Potential use for cancer prevention. *Nutr. Cancer* **2009**, *62*, 1–20. [CrossRef]
54. Zhang, Y.; Tang, L.; An, X.; Fu, E.; Ma, C. Modification of cellulase and its application to extraction of diosgenin from Dioscorea zingiberensis CH Wright. *Biochem. Eng. J.* **2009**, *47*, 80–86. [CrossRef]
55. Xu, L.; Liu, Y.; Wang, T.; Qi, Y.; Han, X.; Xu, Y.; Peng, J.; Tang, X. Development and validation of a sensitive and rapid non-aqueous LC–ESI-MS/MS method for measurement of diosgenin in the plasma of normal and hyperlipidemic rats: A comparative study. *J. Chromatogr. B* **2009**, *877*, 1530–1536. [CrossRef]
56. Lepage, C.; Léger, D.Y.; Bertrand, J.; Martin, F.; Beneytout, J.L.; Liagre, B. Diosgenin induces death receptor-5 through activation of p38 pathway and promotes TRAIL-induced apoptosis in colon cancer cells. *Cancer Lett.* **2011**, *301*, 193–202. [CrossRef]
57. Coleman, R.E.; Croucher, P.I.; Padhani, A.R. Bone metastases. *Nat. Rev.* **2020**, *6*, 83.
58. Liu, K.; Zhao, W.; Gao, X.; Huang, F.; Kou, J.; Liu, B. Diosgenin ameliorates palmitate-induced endothelial dysfunction and insulin resistance via blocking IKKβ and IRS-1 pathways. *Atherosclerosis* **2012**, *223*, 350–358. [CrossRef] [PubMed]

59. Chen, Y.; Xu, X.; Zhang, Y.; Liu, K.; Huang, F.; Liu, B.; Kou, J. Diosgenin regulates adipokine expression in perivascular adipose tissue and ameliorates endothelial dysfunction via regulation of AMPK. *J. Steroid Biochem. Mol. Biol.* **2016**, *155*, 155–165. [CrossRef] [PubMed]
60. Cai, H.; Gong, L.; Liu, J.; Zhou, Q.; Zheng, Z. Diosgenin inhibits tumor angiogenesis through regulating GRP78-mediated HIF-1α and VEGF/VEGFR signaling pathways. *Die Pharm. Int. J. Pharm. Sci.* **2019**, *74*, 680–684.
61. Aguirre, J.A.; Lucchinetti, E.; Clanachan, A.S.; Plane, F.; Zaugg, M. Unraveling interactions between anesthetics and the endothelium: Update and novel insights. *Anesth. Analg.* **2016**, *122*, 330–348. [CrossRef]
62. Chen, P.S.; Shih, Y.W.; Huang, H.C.; Cheng, H.W. Diosgenin, a steroidal saponin, inhibits migration and invasion of human prostate cancer PC-3 cells by reducing matrix metalloproteinases expression. *PLoS ONE* **2011**, *6*, e20164. [CrossRef] [PubMed]
63. Weerasinghe, P.; Buja, L.M. Oncosis: An important non-apoptotic mode of cell death. *Exp. Mol. Pathol.* **2012**, *93*, 302–308. [CrossRef]
64. Zhang, J.H.; Xu, M. DNA fragmentation in apoptosis. *Cell Res.* **2000**, *10*, 205–211. [CrossRef]
65. Fehrenbacher, N.; Jäättelä, M. Lysosomes as targets for cancer therapy. *Cancer Res.* **2005**, *65*, 2993–2995. [CrossRef]
66. Terman, A.; Gustafsson, B.; Brunk, U.T. The lysosomal–mitochondrial axis theory of postmitotic aging and cell death. *Chem. Biol. Interact.* **2006**, *163*, 29–37. [CrossRef]
67. Mao, X.M.; Zhou, P.; Li, S.Y.; Zhang, X.Y.; Shen, J.X.; Chen, Q.X.; Shen, D.Y. Diosgenin suppresses cholangiocarcinoma cells via inducing cell cycle arrest and mitochondria-mediated apoptosis. *Onco Targets Ther.* **2019**, *12*, 9093. [CrossRef] [PubMed]
68. Wei, M.C. Bcl-2-related genes in lymphoid neoplasia. *Int. J. Hematol.* **2004**, *80*, 205–209. [CrossRef] [PubMed]
69. Fan, T.J.; Han, L.H.; Cong, R.S.; Liang, J. Caspase family proteases and apoptosis. *Acta Biochim. Biophys. Sin.* **2005**, *37*, 719–727. [CrossRef] [PubMed]
70. Hollstein, M.; Sidransky, D.; Vogelstein, B. p53 mutations in human cancers. *Science* **1991**, *253*, 49–53. [CrossRef] [PubMed]
71. May, P.; May, E. P53 and cancers. *Pathol. Biol.* **1995**, *43*, 165–173.
72. Zhao, X.; Tao, X.; Xu, L.; Yin, L.; Qi, Y.; Xu, Y.; Peng, J. Dioscin induces apoptosis in human cervical carcinoma HeLa and SiHa cells through ROS-mediated DNA damage and the mitochondrial signaling pathway. *Molecules* **2016**, *21*, 730. [CrossRef]
73. Liu, M.J.; Wang, Z.; Ju, Y.; Wong, R.N.S.; Wu, Q.Y. Diosgenin induces cell cycle arrest and apoptosis in human leukemia K562 cells with the disruption of Ca^{2+} homeostasis. *Cancer Chemother. Pharmacol.* **2005**, *55*, 79–90. [CrossRef]
74. Bae, H.; Lee, W.; Song, J.; Hong, T.; Kim, M.H.; Ham, J.; Lim, W. Polydatin counteracts 5-fluorouracil resistance by enhancing apoptosis via calcium influx in colon cancer. *Antioxidants* **2021**, *10*, 1477. [CrossRef]
75. Mischitelli, M.; Jemaà, M.; Almasry, M.; Faggio, C.; Lang, F. Ca^{2+} entry, oxidative stress, ceramide and suicidal erythrocyte death following diosgenin treatment. *Cell Physiol. Biochem.* **2016**, *39*, 1626–1637. [CrossRef]
76. Sun, G.C.; Jan, C.R.; Liang, W.Z. Exploring the impact of a naturally occurring sapogenin diosgenin on underlying mechanisms of Ca^{2+} movement and cytotoxicity in human prostate cancer cells. *Environ. Toxicol.* **2020**, *35*, 395–403. [CrossRef]
77. Moalic, S.; Liagre, B.; Corbière, C.; Bianchi, A.; Dauça, M.; Bordji, K.; Beneytout, J.L. A plant steroid, diosgenin, induces apoptosis, cell cycle arrest and COX activity in osteosarcoma cells. *FEBS Lett.* **2001**, *506*, 225–230. [CrossRef] [PubMed]
78. Wang, X. *Mechanisms of Inhibitory Effect of Diosgenin on Human Hepatic Carcinoma Cell SMMC-7721*; Nanjing University of Chinese Medicine: Nanjing, China, 2014.
79. Tsukayama, I.; Mega, T.; Hojo, N.; Toda, K.; Kawakami, Y.; Takahashi, Y.; Suzuki-Yamamoto, T. Diosgenin suppresses COX-2 and mPGES-1 via GR and improves LPS-induced liver injury in mouse. *Prostaglandins Other Lipid Med.* **2021**, *156*, 106580. [CrossRef] [PubMed]
80. Rahmati-Yamchi, M.; Ghareghomi, S.; Haddadchi, G.; Milani, M.; Aghazadeh, M.; Daroushnejad, H. Fenugreek extract diosgenin and pure diosgenin inhibit the hTERT gene expression in A549 lung cancer cell line. *Mol. Biol. Rep.* **2014**, *41*, 6247–6252. [CrossRef] [PubMed]
81. Gu, L.; Zheng, H.; Zhao, R.; Zhang, X.; Wang, Q. Diosgenin inhibits the proliferation of gastric cancer cells via inducing mesoderm posterior 1 down-regulation-mediated alternative reading frame expression. *Hum. Exp. Toxicol.* **2021**, *40* (Suppl. 12), S632–S645. [CrossRef] [PubMed]
82. Sun, Y.; Liu, W.Z.; Liu, T.; Feng, X.; Yang, N.; Zhou, H.F. Signaling pathway of MAPK/ERK in cell proliferation, differentiation, migration, senescence and apoptosis. *J. Recept. Signal. Transduct.* **2015**, *35*, 600–604. [CrossRef] [PubMed]
83. Ren, Q.L.; Wang, Q.; Zhang, X.Q.; Wang, M.; Hu, H.; Tang, J.J.; Yang, X.T.; Ran, Y.H.; Liu, H.H.; Song, Z.X.; et al. Anticancer activity of Diosgenin and its molecular mechanism. *Chin. J. Integr. Med.* **2023**, 1–12. [CrossRef] [PubMed]
84. Yang, H.; Yue, L.; Jiang, W.W.; Liu, Q.; Kou, J.; Yu, B. Diosgenin inhibits tumor necrosis factor-induced tissue factor activity and expression in THP-1 cells via down-regulation of the NF-κB, Akt, and MAPK signaling pathways. *Chin. J. Nat. Med.* **2013**, *11*, 608–615. [CrossRef]
85. Li, F.; Fernandez, P.P.; Rajendran, P.; Hui, K.M.; Sethi, G. Diosgenin, a steroidal saponin, inhibits STAT3 signaling pathway leading to suppression of proliferation and chemosensitization of human hepatocellular carcinoma cells. *Cancer Lett.* **2010**, *292*, 197–207. [CrossRef]
86. Ge, Y.; Ding, S.; Feng, J.; Du, J.; Gu, Z. Diosgenin inhibits Wnt/β-catenin pathway to regulate the proliferation and differentiation of MG-63 cells. *Cytotechnology* **2021**, *73*, 169–178. [CrossRef]
87. Li, S.Y.; Shang, J.; Mao, X.M.; Fan, R.; Li, H.Q.; Li, R.H.; Shen, D.Y. Diosgenin exerts anti-tumor effects through inactivation of cAMP/PKA/CREB signaling pathway in colorectal cancer. *Eur. J. Pharmacol.* **2021**, *908*, 174370. [CrossRef]

88. Singh, A.; Kumar, A.; Tripathi, P.; Singh, L.H. Solasodine and its analogs: A multifunctional compound. *Med. Res. Rev.* **2019**, *39*, 103–148.
89. Murakami, R.; Hashimoto, T.; Takenaka, M. Antitumor effect of solamargine from *Solanum nigrum* Linn on human lung cancer cell lines. *Oncol. Rep.* **2014**, *31*, 231–236.
90. Shiu, L.Y.; Liang, C.H.; Chang, L.C.; Sheu, H.M.; Tsai, E.M.; Kuo, K.W. Solamargine induces apoptosis and enhances susceptibility to trastuzumab and epirubicin in breast cancer cells with low or high expression levels of HER2/neu. *Biosci. Rep.* **2009**, *29*, 35–45. [CrossRef] [PubMed]
91. Hsu, S.H.; Tsai, T.R.; Lin, C.N.; Yen, M.H.; Kuo, K.W. Solamargine Purified from *Solanum incanum* Chinese Herb Triggers Gene Expression of Human TNFR I Which May Lead to Cell Apoptosis. *Biochem. Biophys. Res. Commun.* **1996**, *229*, 1–5. [CrossRef]
92. Kuo, K.W.; Hsu, S.H.; Li, Y.P.; Lin, W.L.; Liu, L.F.; Chang, L.C.; Sheu, H.M. Anticancer activity evaluation of the Solanum glycoalkaloid solamargine: Triggering apoptosis in human hepatoma cells. *Biochem. Pharmacol.* **2000**, *60*, 1865–1873. [CrossRef] [PubMed]
93. Liang, C.H.; Liu, L.F.; Shiu, L.Y.; Huang, Y.S.; Chang, L.C.; Kuo, K.W. Action of solamargine on TNFs and cisplatin-resistant human lung cancer cells. *Biochem. Biophys. Res. Commun.* **2004**, *322*, 751–758. [CrossRef] [PubMed]
94. Dhar, S.K.; Clair, D.K.S. Nucleophosmin blocks mitochondrial localization of p53 and apoptosis. *J. Biol. Chem.* **2009**, *284*, 16409–16418. [CrossRef] [PubMed]
95. Elmore, S. Apoptosis: A review of programmed cell death. *Toxicol. Pathol.* **2007**, *35*, 495–516. [CrossRef]
96. Li, X.; Zhao, Y.; Wu, W.K.; Liu, S.; Cui, M.; Lou, H. Solamargine induces apoptosis associated with p53 transcription-dependent and transcription-independent pathways in human osteosarcoma U2OS cells. *Life Sci.* **2011**, *88*, 314–321. [CrossRef]
97. Ow, Y.L.P.; Green, D.R.; Hao, Z.; Mak, T.W. Cytochrome c: Functions beyond respiration. *Nat. Rev. Mol. Cell Bio.* **2008**, *9*, 532–542. [CrossRef]
98. Sun, L.; Zhao, Y.; Li, X.; Yuan, H.; Cheng, A.; Lou, H. A lysosomal–mitochondrial death pathway is induced by solamargine in human K562 leukemia cells. *Toxicol. In Vitro* **2010**, *24*, 1504–1511. [CrossRef] [PubMed]
99. Munari, C.C.; de Oliveira, P.F.; Campos, J.C.L.; Martins, S.D.P.L.; Da Costa, J.C.; Bastos, J.K.; Tavares, D.C. Antiproliferative activity of *Solanum lycocarpum* alkaloidic extract and their constituents, solamargine and solasonine, in tumor cell lines. *J. Nat. Med.* **2014**, *68*, 236–241. [CrossRef] [PubMed]
100. Proskuryakov, S.Y.; Konoplyannikov, A.G.; Gabai, V.L. Necrosis: A specific form of programmed cell death. *Exp. Cell Res.* **2003**, *283*, 1–16. [CrossRef] [PubMed]
101. Williams, G.H.; Stoeber, K. The cell cycle and cancer. *J. Pathol.* **2012**, *226*, 352–364. [CrossRef] [PubMed]
102. Zhou, Y.; Tang, Q.; Zhao, S.; Zhang, F.; Li, L.; Wu, W.; Hann, S. Targeting signal transducer and activator of transcription 3 contributes to the solamargine-inhibited growth and-induced apoptosis of human lung cancer cells. *Tumor Biol.* **2014**, *35*, 8169–8178. [CrossRef] [PubMed]
103. Liu, L.F.; Liang, C.H.; Shiu, L.Y.; Lin, W.L.; Lin, C.C.; Kuo, K.W. Action of solamargine on human lung cancer cells–enhancement of the susceptibility of cancer cells to TNFs. *FEBS Lett.* **2004**, *577*, 67–74. [CrossRef] [PubMed]
104. Shiu, L.Y.; Chang, L.C.; Liang, C.H.; Huang, Y.S.; Sheu, H.M.; Kuo, K.W. Solamargine induces apoptosis and sensitizes breast cancer cells to cisplatin. *Food Chem. Toxicol.* **2007**, *45*, 2155–2164. [CrossRef]
105. Xie, X.; Zhu, H.; Yang, H.; Huang, W.; Wu, Y.; Wang, Y.; Shao, G. Solamargine triggers hepatoma cell death through apoptosis. *Oncol. Lett.* **2015**, *10*, 168–174. [CrossRef]
106. Sun, L.; Zhao, Y.; Yuan, H.; Li, X.; Cheng, A.; Lou, H. Solamargine, a steroidal alkaloid glycoside, induces oncosis in human K562 leukemia and squamous cell carcinoma KB cells. *Cancer Chemother. Pharmacol.* **2011**, *67*, 813–821. [CrossRef]
107. Harbour, J.W.; Dean, D.C. The Rb/E2F pathway: Expanding roles and emerging paradigms. *Genes Dev.* **2000**, *14*, 2393–2409. [CrossRef]
108. Freier, K.; Joos, S.; Flechtenmacher, C.; Devens, F.; Benner, A.; Bosch, F.X.; Hofele, C. Tissue microarray analysis reveals site-specific prevalence of oncogene amplifications in head and neck squamous cell carcinoma. *Cancer Res.* **2003**, *63*, 1179–1182. [PubMed]
109. Zhang, J.; Yang, P.L.; Gray, N.S. Targeting cancer with small molecule kinase inhibitors. *Nat. Rev. Cancer* **2009**, *9*, 28–39. [CrossRef] [PubMed]
110. Friedman, M. Chemistry and anticarcinogenic mechanisms of glycoalkaloids produced by eggplants, potatoes, and tomatoes. *J. Agric. Food Chem.* **2015**, *63*, 3323–3337. [CrossRef] [PubMed]
111. Friedman, M. Potato Glycoalkaloids and Metabolites: Roles in the Plant and in the Diet. *J. Agric. Food Chem.* **2006**, *54*, 8655–8681. [CrossRef]
112. Gao, S.Y.; Wang, Q.J.; Ji, Y.B. Effect of solanine on the membrane potential of mitochondria in HepG2 cells and [Ca^{2+}] i in the cells. *World J. Gastroenterol. WJG* **2006**, *12*, 3359. [CrossRef] [PubMed]
113. Ji, Y.B.; Gao, S.Y.; Ji, C.F.; Zou, X. Induction of apoptosis in HepG2 cells by solanine and Bcl-2 protein. *J. Ethnopharmacol.* **2008**, *115*, 194–202. [CrossRef] [PubMed]
114. Mohsenikia, M.; Alizadeh, A.M.; Khodayari, S.; Khodayari, H.; Karimi, A.; Zamani, M.; Mohagheghi, M.A. The protective and therapeutic effects of alpha-solanine on mice breast cancer. *Eur. J. Pharmacol.* **2013**, *718*, 1–9. [CrossRef] [PubMed]
115. Lu, M.K.; Shih, Y.W.; Chien, T.T.C.; Fang, L.H.; Huang, H.C.; Chen, P.S. α-Solanine inhibits human melanoma cell migration and invasion by reducing matrix metalloproteinase-2/9 activities. *Biol. Pharm. Bull.* **2010**, *33*, 1685–1691. [CrossRef]

116. Lv, C.; Kong, H.; Dong, G.; Liu, L.; Tong, K.; Sun, H.; Zhou, M. Antitumor efficacy of α-solanine against pancreatic cancer in vitro and in vivo. *PLoS ONE* **2014**, *9*, e87868. [CrossRef]
117. Ornano, L.; Venditti, A.; Donno, Y.; Sanna, C.; Ballero, M.; Bianco, A. Phytochemical analysis of non-volatile fraction of *Artemisia caerulescens* subsp. densiflora (Viv.) (Asteraceae), an endemic species of La Maddalena Archipelago (Sardinia–Italy). *Nat. Prod. Res.* **2016**, *30*, 920–925. [CrossRef]
118. Herrmann, K.M. The shikimate pathway as an entry to aromatic secondary metabolism. *Plant. Physiol.* **1995**, *107*, 7–12. [CrossRef] [PubMed]
119. Martens, S.; Forkmann, G.; Matern, U.; Lukacin, R. Cloning of parsley flavone synthase I. *Phytochemistry* **2001**, *58*, 43–46. [CrossRef] [PubMed]
120. Sung, B.; Chung, H.Y.; Kim, N.D. Role of apigenin in cancer prevention via the induction of apoptosis and autophagy. *J. Cancer Prev.* **2016**, *21*, 216. [CrossRef]
121. Yang, J.; Pi, C.; Wang, G. Inhibition of PI3K/Akt/mTOR pathway by apigenin induces apoptosis and autophagy in hepatocellular carcinoma cells. *Biomed. Pharm. Ther.* **2018**, *103*, 699–707. [CrossRef] [PubMed]
122. Nkwe, D.O.; Lotshwao, B.; Rantong, G.; Matshwele, J.; Kwape, T.E.; Masisi, K.; Makhzoum, A. Anticancer mechanisms of bioactive compounds from Solanaceae: An update. *Cancers* **2021**, *13*, 4989. [CrossRef] [PubMed]
123. Ding, X.; Zhu, F.; Yang, Y.; Li, M. Purification, antitumor activity in vitro of steroidal glycoalkaloids from black nightshade (*Solanum nigrum* L.). *Food Chem.* **2013**, *141*, 1181–1186. [CrossRef] [PubMed]
124. Burger, T.; Mokoka, T.; Fouché, G.; Steenkamp, P.; Steenkamp, V.; Cordier, W. Solamargine, a bioactive steroidal alkaloid isolated from *Solanum aculeastrum* induces non-selective cytotoxicity and P-glycoprotein inhibition. *BMC Complement. Altern. Med.* **2018**, *18*, 137. [CrossRef]
125. Kumar, S.; Pandey, A.K. Phenolic content, reducing power and membrane protective activities of *Solanum xanthocarpum* root extracts. *Vegetos* **2013**, *26*, 301–307. [CrossRef]
126. Patel, S.; Gheewala, N.; Suthar, A.; Shah, A. In-Vitro cytotoxicity activity of *Solanum nigrum* extract against Hela cell line and Vero cell line. *Int. J. Pharm. Pharm. Sci.* **2009**, *1*, 38–46.
127. Kumar, R.; Khan, M.I.; Prasad, M. Solasodine: A Perspective on their roles in Health and Disease. *Res. J. Pharm. Technol.* **2019**, *12*, 2571–2576. [CrossRef]
128. Carrillo-Cocom, L.M.; González, B.B.V.; Santillan, R.; Soto-Castro, D.; Ocampo, P.M.S.; Zepeda, A.; Tafur, J.C. Synthesis of diosgenin prodrugs: Anti-inflammatory and antiproliferative activity evaluation. *J. Chem. Sci.* **2020**, *132*, 104. [CrossRef]
129. Zhao, L.; Wang, L.; Di, S.N.; Xu, Q.; Ren, Q.C.; Chen, S.Z.; Shen, X.F. Steroidal alkaloid solanine A from *Solanum nigrum* Linn. exhibits anti-inflammatory activity in lipopolysaccharide/interferon γ-activated murine macrophages and animal models of inflammation. *Biomed. Pharmacother.* **2018**, *105*, 606–615. [CrossRef]
130. Patel, S.S.; Savjani, J.K. Systematic review of plant steroids as potential antiinflammatory agents: Current status and future perspectives. *J. Phytopharm.* **2015**, *4*, 121–125. [CrossRef]
131. Raina, H.; Soni, G.; Jauhari, N.; Sharma, N.; Bharadvaja, N. Phytochemical importance of medicinal plants as potential sources of anticancer agents. *Turk. J. Bot.* **2014**, *38*, 1027–1035. [CrossRef]
132. Zhao, Z.; Jia, Q.; Wu, M.S.; Xie, X.; Wang, Y.; Song, G.; Shen, J. Degalactotigonin, a natural compound from *Solanum nigrum* L.; inhibits growth and metastasis of osteosarcoma through GSK3β inactivation–mediated repression of the Hedgehog/Gli1 pathway. *Clin. Cancer Res.* **2018**, *24*, 130–144. [CrossRef] [PubMed]
133. Alrawashdeh, H.M. Cataract and healthy diet, an ounce of prevention is worth a pound of cure—A mini-review. *J. Food Nutr. Health* **2020**, *1*, 1–3. [CrossRef]
134. Mikkelsen, K.; Apostolopoulos, V. Vitamin B1, B2, B3, B5, and B6 and the Immune System. In *Nutrition and Immunity*; Springer: Berlin/Heidelberg, Germany, 2019; pp. 115–125.
135. Janghel, V.; Patel, P.; Chandel, S.S. Plants used for the treatment of icterus (jaundice) in Central India: A review. *Ann. Hepatol.* **2019**, *18*, 658–672. [CrossRef]
136. Ali, N.; Rafiq, R.; Wijaya, L.; Ahmad, A.; Kaushik, P. Exogenous citric acid improves growth and yield by concerted modulation of antioxidant defense system in brinjal (*Solanum melongena* L.) under salt-stress. *J. King Saud Univ. Sci.* **2023**, *36*, 103012. [CrossRef]
137. Zhang, G.; Zhang, Y.; Huang, X.; Gao, W.; Zhang, X. Effect of pyrolysis and oxidation characteristics on lauric acid and stearic acid dust explosion hazards. *J. Loss Prev. Process. Ind.* **2020**, *63*, 104039. [CrossRef]
138. Hwangbo, H.; Ji, S.Y.; Kim, M.Y.; Kim, S.Y.; Lee, H.; Kim, G.Y.; Choi, Y.H. Anti-inflammatory effect of auranofin on palmitic acid and LPS-induced inflammatory response by modulating TLR4 and NOX4-mediated NF-κB signaling pathway in RAW264. 7 macrophages. *Int. J. Mol. Sci.* **2021**, *22*, 5920. [CrossRef]
139. Yang, Z.H.; Nill, K.; Takechi-Haraya, Y.; Playford, M.P.; Nguyen, D.; Yu, Z.X.; Remaley, A.T. Differential Effect of Dietary Supplementation with a Soybean Oil Enriched in Oleic Acid versus Linoleic Acid on Plasma Lipids and Atherosclerosis in LDLR-Deficient Mice. *Int. J. Mol. Sci.* **2022**, *23*, 8385. [CrossRef] [PubMed]
140. Cassano, R.; Serini, S.; Curcio, F.; Trombino, S.; Calviello, G. Preparation and Study of Solid Lipid Nanoparticles Based on Curcumin, Resveratrol and Capsaicin Containing Linolenic Acid. *Pharmaceutics* **2022**, *14*, 1593. [CrossRef] [PubMed]
141. Pawar, H.A. A review on extraction, analysis and biopotential of rutin. *Int. J. Chem. Pharm. Anal.* **2019**, *6*, 1.

142. Orfali, R.; Perveen, S.; Aati, H.Y.; Alam, P.; Noman, O.M.; Palacios, J.; Khan, S.I. High-performance thin-layer chromatography for rutin, chlorogenic acid, caffeic acid, ursolic acid, and stigmasterol analysis in periploca aphylla extracts. *Separations* **2021**, *8*, 44. [CrossRef]
143. Zhao, Y.; Li, D.; Zhu, Z.; Sun, Y. Improved neuroprotective effects of gallic acid-loaded chitosan nanoparticles against ischemic stroke. *Rejuvenation Res.* **2020**, *23*, 284–292. [CrossRef]
144. Salehi, B.; Fokou, P.V.T.; Sharifi-Rad, M.; Zucca, P.; Pezzani, R.; Martins, N.; Sharifi-Rad, J. The therapeutic potential of naringenin: A review of clinical trials. *Pharmaceuticals* **2019**, *12*, 11. [CrossRef]
145. Shahrajabian, M.H.; Sun, W. Survey on medicinal plants and herbs in traditional Iranian medicine with anti-oxidant, anti-viral, anti-microbial, and anti-inflammation properties. *Lett. Drug Des. Discov.* **2023**, *20*, 1707–1743. [CrossRef]
146. Syu, W.J.; Don, M.J.; Lee, G.H.; Sun, C.M. Cytotoxic and novel compounds from *Solanum indicum*. *J. Nat. Prod.* **2001**, *64*, 1232–1233. [CrossRef]
147. Chiang, H.C.; Tseng, T.H.; Wang, C.J.; Chen, C.F.; Kan, W.S. Experimental antitumor agents from *Solanum indicum* L. *Anticancer Res.* **1991**, *11*, 1911–1917.
148. Ma, P.; Cao, T.T.; Gu, G.F.; Zhao, X.; Du, Y.G.; Zhang, Y. Inducement effect of synthetic indiosides from *Solanum indicum* L. on apoptosis of human hepatocarcinoma cell line Bel-7402 and its mechanism. *Ai Zheng Aizheng Chin. J. Cancer* **2006**, *25*, 438–442.
149. Gopalakrishna, S.M.; Thimappa, G.S.; Thylur, R.P.; Shivanna, Y.; Sreenivasan, A. In-vitro anti-cancer screening of *Solanum indicum Rhus succedanea, Rheum emodi* and *Gardenia gummifera* medicinal plants in cancer cells. *Res. Rev. J. Pharm. Sci.* **2014**, *3*, 22–30.
150. Rahman, M.S.; Begum, B.; Chowdhury, R.; Rahman, K.M.; Rashid, M.A. Preliminary cytotoxicity screening of some medicinal plants of Bangladesh. *Dhaka Univ. J. Pharm. Sci.* **2008**, *7*, 47–52. [CrossRef]
151. Venkatesh, R.; Kalaivani, K.; Vidya, R. Toxicity assessment of ethanol extract of *Solanum villosum* (Mill) on wistar albino rats. *Int. J. Pharma Sci. Res.* **2014**, *5*, 406–412.
152. Gandhi, G.R.; Ignacimuthu, S.; Paulraj, M.G. *Solanum torvum* Swartz. fruit containing phenolic compounds shows antidiabetic and antioxidant effects in streptozotocin induced diabetic rats. *Food Chem. Toxicol.* **2011**, *49*, 2725–2733. [CrossRef]
153. Meena, A.K.; Rao, M.M.; Kandale, A.; Sharma, A.; Singh, U.; Yadav, A. Evaluation of physicochemical and standardization parameters of *Solanum xanthocarpum* Schrad. & Wendl. *Int. J. Chem. Anal. Sci.* **2010**, *1*, 47–49.
154. Ozsoy, N.; Can, A.; Yanardag, R.; Akev, N. Antioxidant activity of *Smilax excelsa* L. leaf extracts. *Food Chem.* **2008**, *110*, 571–583. [CrossRef]
155. Poongothai, K.; Ponmurugan, P.; Ahmed, K.S.Z.; Kumar, B.S.; Sheriff, S.A. Antihyperglycemic and antioxidant effects of *Solanum xanthocarpum* leaves (field grown & in vitro raised) extracts on alloxan induced diabetic rats. *Asian Pac. J. Trop. Med.* **2011**, *4*, 778–785. [PubMed]
156. Chen, X.; Dai, X.; Liu, Y.; Yang, Y.; Yuan, L.; He, X.; Gong, G. *Solanum nigrum* Linn.: An Insight into Current Research on Traditional Uses, Phytochemistry, and Pharmacology. *Front. Pharmacol.* **2022**, *13*, 918071. [CrossRef]
157. Narayanaswamy, N.; Balakrishnan, K.P. Evaluation of some medicinal plants for their antioxidant properties. *Int. J. Pharm. Tech. Res.* **2011**, *3*, 381–385.
158. Deb, P.K.; Das, N.; Bhakta, R.G.T. Evaluation of in-vitro antioxidant and anthelmintic activity of *Solanum indicum* Linn. berries. *Indo Am. J. Pharm. Res.* **2013**, *3*, 4123–4130.
159. Hasan, R.U.; Prabhat, P.; Shafaat, K.; Khan, R. Phytochemical investigation and evaluation of antioxidant activity of fruit of *Solanum indicum* Linn. *Int. J. Pharm. Sci.* **2013**, *5*, 237–242.
160. N'Dri, D.; Calani, L.; Mazzeo, T.; Scazzina, F.; Rinaldi, M.; Rio, D.D.; Brighenti, F. Effects of different maturity stages on antioxidant content of Ivorian Gnagnan (*Solanum indicum* L.) berries. *Molecules* **2010**, *15*, 7125–7138. [CrossRef] [PubMed]
161. Wang, H.C.; Chung, P.J.; Wu, C.H.; Lan, K.P.; Yang, M.Y.; Wang, C.J. *Solanum nigrum* L. polyphenolic extract inhibits hepatocarcinoma cell growth by inducing G2/M phase arrest and apoptosis. *J. Sci. Food Agric.* **2011**, *91*, 178–185. [CrossRef] [PubMed]
162. Bhuvaneswari, B.; Suguna, M.M.L. Hepatoprotective and antioxidant activities of *Solanum indicum* Linn. Berries. *Int. J. Clin. Toxicol.* **2014**, *2*, 71–76.
163. Bhatia, N.; Maiti, P.P.; Kumar, A.; Tuli, A.; Ara, T.; Khan, M.U. Evaluation of cardio protective Activity of Methanolic Extract of *Solanum nigrum* Linn. in Rats. *Int. J. Drug Dev. Res.* **2011**, *3*, 139–147.
164. Deb, P.K.; Das, L.; Ghosh, R.; Debnath, R.; Bhakta, T. Evaluation of laxative and cardiotonic activity of *Solanum indicum* Linn. fruits. *J. Pharm. Phytother.* **2013**, *1*, 11–14.
165. Waghulde, H.; Kamble, S.; Patankar, P.; Jaiswal, B.; Pattanayak, S.; Bhagat, C.; Mohan, M. Antioxidant activity, phenol and flavonoid contents of seeds of *Punica granatum* (Punicaceae) and *Solanum torvum* (Solanaceae). *Pharmacologyonline* **2011**, *1*, 193–202.
166. Olsen, S.; Hansen, E.S.; Jepsen, F.L. The prevalence of focal tubulo-interstitial lesions in various renal diseases. *Acta Pathol. Microbiol. Scand. Sect. A Pathol.* **1981**, *89*, 137–145. [CrossRef]
167. Sharma, M.; Kishore, K.; Gupta, S.K.; Joshi, S.; Arya, D.S. Cardioprotective potential of *Ocimum sanctum* in isoproterenol induced myocardial infarction in rats. *Mol. Cell Biochem.* **2001**, *225*, 75–83. [CrossRef]
168. Ching, T.L.; Haenen, G.R.; Bast, A. Cimetidine and other H2 receptor antagonists as powerful hydroxyl radical scavengers. *Chem. Biol. Interact.* **1993**, *86*, 119–127. [CrossRef]
169. Václavíková, R.; Kondrová, E.; Ehrlichová, M.; Boumendjel, A.; Kovář, J.; Stopka, P.; Gut, I. The effect of flavonoid derivatives on doxorubicin transport and metabolism. *Bioorg. Med. Chem.* **2008**, *16*, 2034–2042. [CrossRef] [PubMed]

170. Mohan, M.; Kamble, S.; Gadhi, P.; Kasture, S. Protective effect of *Solanum torvum* on doxorubicin-induced nephrotoxicity in rats. *Food Chem. Toxicol.* **2010**, *48*, 436–440. [CrossRef] [PubMed]
171. Loganayaki, N.; Siddhuraju, P.; Manian, S. Antioxidant activity of two traditional Indian vegetables: *Solanum nigrum* L. and *Solanum torvum* L. *Food Sci. Biotechnol.* **2010**, *19*, 121–127. [CrossRef]
172. Nguelefack, T.B.; Mekhfi, H.; Dimo, T.; Afkir, S.; Nguelefack-Mbuyo, E.P.; Legssyer, A.; Ziyyat, A. Cardiovascular and anti-platelet aggregation activities of extracts from *Solanum torvum* (Solanaceae) fruits in rat. *J. Comp. Integ. Med.* **2008**, *5*. [CrossRef]
173. Holinstat, M.; Voss, B.; Bilodeau, M.L.; McLaughlin, J.N.; Cleator, J.; Hamm, H.E. PAR4, but not PAR1, signals human platelet aggregation via Ca^{2+} mobilization and synergistic P2Y12 receptor activation. *J. Bio. Chem.* **2006**, *281*, 26665–26674. [CrossRef] [PubMed]
174. Hussain, M.S.; Fareed, S.; Ali, M. Simultaneous HPTLC-UV530 nm analysis and validation of bioactive lupeol and stigmasterol in *Hygrophila auriculata* (K. Schum) Heine. *Asian Pac. J. Trop. Biomed.* **2012**, *2*, S612–S617. [CrossRef]
175. Bahgat, A.; Abdel-Aziz, H.; Raafat, M.; Mahdy, A.; El-Khatib, A.S.; Ismail, A.; Khayyal, M.T. *Solanum indicum* ssp. distichum extract is effective against l-NAME-induced hypertension in rats. *Fundam. Clin. Pharmacol.* **2008**, *22*, 693–699. [CrossRef]
176. Kalaichelvi, K.; Sathiya, S.; Deivasigamani, K. Phytochemical analysis and antihypertensive potential of selected medicinal plants. *Int. J. Pharmacogon. Phytochem. Res.* **2016**, *8*, 1891–1896.
177. Antonio, J.M.; Gracioso, J.S.; Toma, W.; Lopez, L.C.; Oliveira, F.; Brito, A.S. Antiulcerogenic activity of ethanol extract of *Solanum variabile* (false "jurubeba"). *J. Ethnopharmacol.* **2004**, *93*, 83–88. [CrossRef]
178. Darkwah, W.K.; Koomson, D.A.; Miwornunyuie, N.; Nkoom, M.; Puplampu, J.B. Phytochemistry and medicinal properties of *Solanum torvum* fruits. *All Life* **2020**, *13*, 498–506. [CrossRef]
179. Ibrahim, A.Y.; Shaffie, N.M. Protective Effect of *Solanum indicum* Var. Distichum extract on experimentally induced gastric ulcers in rat. *Glob. J. Pharmacol.* **2013**, *7*, 325–332.
180. Al-Oqail, M.; Hassan, W.H.; Ahmad, M.S.; Al-Rehaily, A.J. Phytochemical and biological studies of *Solanum schimperianum* Hochst. *Saudi Pharm. J.* **2012**, *20*, 371–379. [CrossRef] [PubMed]
181. Atta, A.H.; Alkofahi, A. Anti-nociceptive and anti-inflammatory effects of some Jordanian medicinal plant extracts. *J. Ethnopharmacol.* **1998**, *60*, 117–124. [CrossRef] [PubMed]
182. Ndebia, E.J.; Kamgang, R.; Nkeh-ChungagAnye, B.N. Analgesic and anti-inflammatory properties of aqueous extract from leaves of *Solanum torvum* (Solanaceae). *Afr. J. Tradit. Complement. Alt. Med.* **2007**, *4*, 240–244. [CrossRef] [PubMed]
183. Arunachalam, G.; Subramanian, N.; Perumal Pazhani, G.; Karunanithi, M.; Ravichandran, V. Evaluation of anti-inflammatory activity of methanolic extract of *Solanum nigrum* (Solanaceae). *Iran. J. Pharm. Sci.* **2009**, *5*, 151–156.
184. Kaushik, D.; Jogpal, V.; Kaushik, P.; Lal, S.; Saneja, A.; Sharma, C.; Aneja, K.R. Evaluation of activities of *Solanum nigrum* fruit extract. *Arch. Appl. Sci. Res.* **2009**, *1*, 43–50.
185. Kar, D.M.; Maharana, L.; Pattnaik, S.; Dash, G.K. Studies on hypoglycaemic activity of *Solanum xanthocarpum* Schrad. & Wendl. fruit extract in rats. *J. Ethnopharmacol.* **2006**, *108*, 251–256.
186. Abbas, K.; Niaz, U.; Hussain, T.; Saeed, M.A.; Javaid, Z.; Idrees, A.; Rasool, S. Antimicrobial activity of fruits of *Solanum nigrum* and *Solanum xanthocarpum*. *Acta Pol. Pharm.* **2014**, *71*, 415–421.
187. Mahmood, A.; Mahmood, A.; Mahmood, M. In Vitro biological activities of most common medicinal plants of family Solanaceae. *World Appl. Sci. J.* **2012**, *17*, 1026–1032.
188. David, E.; Elumalai, E.K.; Sivakumar, C.; Therasa, S.V.; Thirumalai, T. Evaluation of antifungal activity and phytochemical screening of *Solanum surattense* seeds. *J. Pharm. Res.* **2010**, *3*, 684–687.
189. Upadhyay, P.; Ara, S.; Prakash, P. Antibacterial and antioxidant activity of *Solanum nigrum* stem and leaves. *Chem. Sci.* **2015**, *4*, 1013–1017.
190. Gavimath, C.C.; Kulkarni, S.M.; Raorane, C.J.; Kalsekar, D.P.; Gavade, B.G.; Ravishankar, B.E.; Hooli, R.S. Antibacterial potentials of *Solanum indicum*, *Solanum xanthocarpum* and *Physalis minima*. *Int. J. Pharma. Appl.* **2012**, *3*, 414–418.
191. Srividya, A.R.; Arunkumar, A.; Cherian, B.; Maheshwari, V.; Piramanayagam, S.; Senthoorpandi, V. Pharmacognostic, phytochemical and anti-microbial studies of *Solanum indicum* leaves. *Anc. Sci. Life* **2009**, *29*, 3. [PubMed]
192. Kouadio, A.I.; Chatigre, O.K.; Dosso, M.B. Phytochemical screening of the antimicrobial fraction of *Solanum indicum* L. berries extract and evaluation of its effect against the survival of bacteria pathogens of plants. *Int. J. Biotech. Food Sci.* **2014**, *2*, 21–30.
193. Sahle, T.; Okbatinsae, G. Phytochemical investigation and antimicrobial activity of the fruit extract of *Solanum incanum* grown in Eritrea. *Orn. Med. Plants* **2017**, *1*, 15–25.
194. Ayodele, O.A.; Yakaka, A.B.; Sulayman, T.B.; Samaila, M. Antibacterial activities of aqueous and methanol leaf extracts of *Solanum incanum* linn.(Solanaceae) against multi-drug resistant bacterial isolates. *Afr. J. Microbiol. Res.* **2019**, *13*, 70–76. [CrossRef]
195. Waithaka, P.N.; Githaiga, B.M.; Gathuru, E.M.; Dixon, M.F. Antibacterial effect of *Solanum incanum* root extracts on bacteria pathogens isolated from portable water in Egerton University, Kenya. *J. Biomed. Sci.* **2019**, *6*, 19–24. [CrossRef]
196. Beaman-Mbaya, V.; Muhammed, S.I. Antibiotic action of *Solanum incanum* Linnaeus. *Antimicrobagents Chemother.* **1976**, *9*, 920–924. [CrossRef]
197. Habtom, S.; Gebrehiwot, S. In Vitro antimicrobial activities of crude extracts of two traditionally used Ethiopian medicinal plants against some bacterial and fungal test pathogens. *Int. J. Biotech.* **2019**, *8*, 104–114. [CrossRef]
198. El-Sayed, Z.I.; Hassan, W.H. Polymethoxylated flavones from *Solanum abutiloides*, grown in Egypt (Solanaceae). *Zagazig. J. Pharm. Sci.* **2006**, *15*, 53–59. [CrossRef]

199. Kannan, M.; Dheeba, B.; Gurudevi, S.; Ranjit Singh, A.J.A. Phytochemical, antibacterial and antioxident studies on medicinal plant *Solanum torvum*. *J. Pharm. Res.* **2012**, *5*, 2418–2421.
200. Gunaselvi, G.; Kulasekaren, V.; Gopal, V. Anthelmintic activity of the extracts of *Solanum xanthocarpum* Schrad and Wendl fruits (Solanaceae). *Int. J. PharmTech Res.* **2010**, *2*, 1772–1774.
201. Priya, C.; Pal, J.A.; Aditya, G.; Gopal, R. Anti-microbial, Antioxidant and Anthelmintic Activity of Crude Extract of *Solanum xanthocarpum*, Comparative evaluation of antimicrobial potential of different extracts of *Cuscuta reflexa* growing on *Acacia arabica* and *Zizyphus jujuba*, Evaluation of Antimicrobial activity of stem bark of *Ficus bengalensis* Linn. Collected from different geographical regions. *Pharmacogn. J.* **2010**, *2*, 400–404.
202. Senaratne, U.V.; Perera, H.K.; Manamperi, A.; Athauda, S.B. *Partial Purification of Anthelmintic Compounds from Solanum Indicum*; University of Peradeniya: Galaha, Sri Lanka, 2011.
203. Zirihi, G.N.; Mambu, L.; Guédé-Guina, F.; Bodo, B.; Grellier, P. In Vitro antiplasmodial activity and cytotoxicity of 33 West African plants used for treatment of malaria. *J. Ethnopharmacol.* **2005**, *98*, 281–285. [CrossRef]
204. Sammani, A.; Shammaa, E.; Chehna, F.; Rahmo, A. The in-vitro toxic effect of the glycoalkaloids for some solanum species against the LIM-1863 cell line. *Pharmacog. J.* **2014**, *6*, 23–31. [CrossRef]

Disclaimer/Publisher's Note: The statements, opinions and data contained in all publications are solely those of the individual author(s) and contributor(s) and not of MDPI and/or the editor(s). MDPI and/or the editor(s) disclaim responsibility for any injury to people or property resulting from any ideas, methods, instructions or products referred to in the content.

Article

Phytochemical Compounds from *Laelia furfuracea* and Their Antioxidant and Anti-Inflammatory Activities

Abimael López-Pérez [1], Luicita Lagunez-Rivera [1,*], Rodolfo Solano [1], Aracely Evangelina Chávez-Piña [2,*], Gabriela Soledad Barragán-Zarate [1] and Manuel Jiménez-Estrada [3]

[1] Laboratorio de Extracción y Análisis de Productos Naturales Vegetales, Centro Interdisciplinario de Investigación para el Desarrollo Integral Regional Unidad Oaxaca, Instituto Politécnico Nacional, Hornos 1003, Santa Cruz Xoxocotlán 71230, Oaxaca, Mexico; ablopezp@ipn.mx (A.L.-P.); asolanog@ipn.mx (R.S.); gbarraganz@ipn.mx (G.S.B.-Z.)

[2] Laboratorio de Farmacología, Escuela Nacional de Medicina y Homeopatía, Instituto Politécnico Nacional, Guillermo Massieu Helguera, 239, La Escalera, Gustavo A. Madero 07320, Ciudad de México, Mexico

[3] Instituto de Química, Universidad Nacional Autónoma de México, Circuito Exterior, Insurgentes Sur, C.U., Coyoacan 04510, Ciudad de México, Mexico; manuelj@unam.mx

* Correspondence: llagunez@ipn.mx (L.L.-R.); achavezp@ipn.mx (A.E.C.-P.)

Academic Editors: Luis Ricardo Hernández, Eugenio Sánchez-Arreola and Edgar R. López-Mena

Received: 16 January 2025
Revised: 8 February 2025
Accepted: 12 February 2025
Published: 14 February 2025

Citation: López-Pérez, A.; Lagunez-Rivera, L.; Solano, R.; Chávez-Piña, A.E.; Barragán-Zarate, G.S.; Jiménez-Estrada, M. Phytochemical Compounds from *Laelia furfuracea* and Their Antioxidant and Anti-Inflammatory Activities. *Plants* **2025**, *14*, 588. https://doi.org/10.3390/plants14040588

Copyright: © 2025 by the authors. Licensee MDPI, Basel, Switzerland. This article is an open access article distributed under the terms and conditions of the Creative Commons Attribution (CC BY) license (https://creativecommons.org/licenses/by/4.0/).

Abstract: *Laelia furfuracea* is an orchid endemic to Oaxaca, Mexico, used for the treatment of cough and has anticoagulant activity. We aimed to evaluate the anti-inflammatory and antioxidant activity of the hydroethanolic extract of *L. furfuracea* leaves and identify its phytochemical compounds. The leaf material was subjected to solid–liquid extraction. Compounds were identified by UPLC-ESI-qTOF-MS/MS. The Folin–Ciocalteu and aluminum trichloride methods were used to quantify phenols and flavonoids, respectively. The DPPH method was used to determine the antioxidant activity. The anti-inflammatory activity was evaluated in a model of carrageenan-induced plantar edema induced in Wistar rats. Compounds tentatively identified in *L. furfuracea* leaves were malic, citric, succinic, hydroximethylglutaric, azelaic, eucomic, and protocatechuic acids, saponarin, luteolin-7,3′-di-O-glucoside, isoorientin, and vitexin. The contents of total phenols and flavonoids and antioxidant activity were 394.7 ± 0.1 mg EqAG/g, 129.9 ± 0.005 mg EqQ/g, and 84.6 ± 1.4%, respectively. The anti-inflammatory effect of the extract was dose-dependent, where 1000 µg/paw presented a 43.4% reduction in inflammation, similar to naproxen. The anti-inflammatory and antioxidant effect of the hydroethanolic extract of *Laelia furfuracea* leaves was demonstrated. This effect may be due to the synergy between its compounds. This orchid is a potential candidate for future pharmacological research due to its anti-inflammatory activity.

Keywords: inflammation; *in vivo* evaluation; flavonoids; traditional medicine; orchids

1. Introduction

Inflammation is a hallmark of many diseases, including cancer, neurodegenerative diseases like Alzheimer's, type II diabetes, rheumatoid arthritis, and asthma [1]. Anti-inflammatory therapy includes two groups, i.e., steroidal anti-inflammatories and nonsteroidal anti-inflammatory drugs (NSAIDs) [2]. NSAIDs have been a cornerstone in the management of various inflammatory, pain, and fever-related conditions. However, they have presented adverse effects, including those associated with cardiovascular, renal, and hepatic complications [1], as is the case with naproxen, which has presented side effects such as gastrointestinal and renal toxicity [3]. Therefore, currently a therapeutic alternative is the search for new compounds of natural origin in a safe, effective, and less harmful way to control inflammatory diseases.

Plants are an important source in the search for these new bioactive compounds, such as terpenes and phenolic compounds, mainly. They provide cellular protection against oxidative stress [4] that can otherwise lead to cell deterioration and death [5], causing acute, chronic, and degenerative inflammatory disease [6]. The discovery of specific molecules directed toward inflammatory agents, in a safe, effective, and less harmful manner, is the therapeutic strategy for the control of inflammatory diseases. Terpenes and flavonoids have beneficial effects on health due to their antioxidant [7], analgesic [8], and anti-inflammatory [9] properties.

These types of properties have been described recently in some Mexican orchids [10,11], such as the antinociceptive activity of *Cyrtopodium macrobulbon* (Lex.) G.A. Romero and Carnevali [12], vasorelaxant effect of *Trichocentrum brachyphyllum* (Lindl.) R. Jiménez [13], the anti-inflammatory and antioxidant activity of *Prosthechea michuacana* (Lex.) W.E. Higgins [14], and *P. karwinskii* (Mart.) J.M.H. Shaw [15,16]. Some species of the genus *Laelia* Lindl. are used in traditional medicine mainly by peasant communities [17]; for example, *Laelia autumnalis* (Lex.) Lindl. [18,19], *Laelia anceps* Lindl. [20,21], and *Laelia speciosa* (Kunth) Schltr. [21] are used to treat bumps and wounds. These orchids have also been studied to evaluate their biological properties. *L. anceps* presents cytotoxic activity in tumor cells of the central nervous system (U251), lung (SKLU-1) and breast (MCF-7) [22], in addition to having antihypertensive activity [20,21]. In *L. autumnalis*, its antihypertensive and vasorelaxant activity has been evaluated [18,19], while *L. speciosa* also presents vasorelaxant activity [21].

Laelia furfuracea Lindl. is an orchid endemic to Oaxaca, Mexico. In the Mixteca Alta of Oaxaca, it is known that this species is used as an herbal remedy for cough, preparing an infusion with its flowers, which is drunk as drinking water [22]. Groups of phytochemical compounds of pharmacological interest have been identified in this orchid, such as flavonoids and phenolic acids [23], and *in vitro* anticoagulant activity has been demonstrated by prolonging prothrombin (PT), activated partial thromboplastin (aPTT), and thrombin (TT) times [22,24,25]. Relating this effect to its chemical composition and its traditional use to alleviate an inflammatory process of the respiratory tract that is physiologically related to inflammation, we aimed to show evidence for the first time of the antioxidant and anti-inflammatory potential of this orchid and to identify its phytochemical composition, which alone or in synergy may be responsible for its biological effects. This is an opportunity for the search of new bioactive molecules with antioxidant and anti-inflammatory activity.

2. Results

2.1. Phytochemical Profile

Different groups of compounds were identified in the extract of *L. furfuracea*, such as tannins, flavonoids, saponins, cardiotonics, and quinones. Regarding fractions, tannins were the only group of compounds present in all fractions; flavonoids were not identified in the FC; saponins were present in the fractions of lower polarity (FH, FC, and FEA); quinones were present in the fractions of FC and FB; sesquiterpenlactones were identified only on FB; coumarins and alkaloids were not identified in the HELF and fractions (Table 1).

Table 1. Phytochemical profile of the extract and fractions of *Laelia furfuracea* leaves.

Phytochemical Tests	Sample					
	HELF	FH	FC	FEA	FB	FA
Flavonoids	+	+	−	+	+	+
Saponins	+	+	+	+	−	−
Tannins	+	+	+	+	+	+
Cardiotonics	+	−	−	+	+	+
Coumarins	−	−	−	−	−	−
Sesquiterpenlactones	−	−	−	−	+	−
Quinones	+	−	+	−	+	−
Alkaloids	−	−	−	−	−	−

Hydroethanolic extract of *Laelia furfuracea* (HELF); fractions: hexane (FH), chloroform (FC), ethyl acetate (FEA), butanol (FB), water (FA), presence (+), and absence (−).

2.2. Compounds Identified by UPLC-ESI-qTOF-MS/MS

Chromatographic analysis tentatively identifies the presence of metabolites such as carboxylic acids, particularly citric, malic, succinic, hydroxymethylglutaric, and azelaic acids; a terpene identified was the iridoid glycoside; as to phenolic compounds of the phenolic acid groups, eucomic and protocatechuic acid were present. Some flavonoids identified were saponarin, luteolin-7,3′-di-O-glycoside, isoorientin, and vitexin, and flavonoid glucoside derived from isoorientin and vitexin (Figure 1, Table 2).

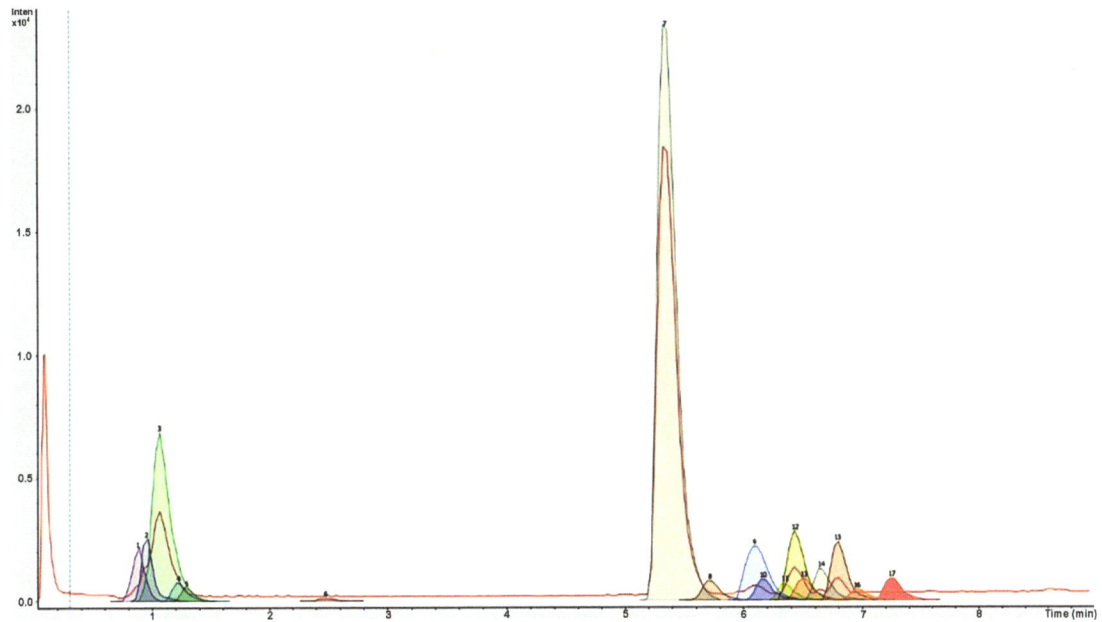

Figure 1. UPLC-ESI-qTOF-MS/MS chromatogram of the *Laelia furfuracea* leaf extract. The red line represents the baseline chromatogram. The identified compounds are listed as follows: 1 (unidentified), 2 (malic acid), 3 (citric acid), 4 (succinic acid), 5 (hydroxymethylglutaric acid), 6 (procatechuic acid), 7 (eucomic acid), 8 (luteolin-7,3′-di-O-glucoside), 9 (saponarin), 10 (isoorientin), 11 (vitexin), 12 (unidentified), 13 (flavonoid glucoside derived from isoorientin), 14 (unidentified), 15 (flavonoid glucoside derived from vitexin), 16 (iridoid glucoside), and 17 (azelaic acid).

Table 2. Tentative identification of compounds in the hydroethanolic extract of *Laelia furfuracea* leaves by UPLC-ESI-qTOF-MS/MS.

PN	RT (min)	Error (ppm)	M/Z [M-H]$^-$	Collision Energy (ev)	Fragments MS/MS *	Chemical Formula	Compound	Chemical Class	IR
1	0.9	5.6	165.0405	16.6	75.0082, 129.0180, 147.0265	$C_9H_{10}O_3$	Unknown	Organic acid	[26]
2	0.9	5.0	133.0142	15.8	115.0040, 71.0137	$C_4H_6O_5$	Malic acid	Hydroxydicarboxylic acid	[26–29]
3	1.1	0.1	191.0197	17.3	111.0090, 87.0089, 85.0296, 129.0194, 173.0085	$C_6H_8O_7$	Citric Acid	Tricarboxylic acid	[26,29]
4	1.2	0.8	117.0192	15.4	73.0292, 99.0097	$C_4H_6O_4$	Succinic acid	Dicarboxylic acid	[26]
5	1.3	0.7	161.0457	16.5	99.0448	$C_6H_{10}O_5$	Hydroxymethylglutaric acid	Dicarboxylic acid	[26,29]
6	2.5	1.4	153.0191	16.3	109.0297	$C_7H_6O_4$	Procatechuic acid	Phenol	[26,28,30]
7	5.3	2.4	239.0555	18.5	179.0346, 177.0554, 149.0604, 133.0660, 107.0496, 195.0662	$C_{11}H_{12}O_6$	Eucomic acid	Phenolic monocarboxylic acid	[27,31]
8	5.7	1.2	609.1452	28.3	447.0918, 448.0962	$C_{27}H_{30}O_{16}$	Luteolin-7,3'-di-O-glycoside	Glycoside flavone	[26]
9	6.0	1.9	593.1501	28.1	311.0551, 431.0980, 297.0406	$C_{27}H_{30}O_{15}$	Saponarin	Flavonoid	[26,32]
10	6.1	3.1	447.0925	23.7	357.0618, 327.0504, 429.0813, 285.0405, 297.0416	$C_{21}H_{20}O_{11}$	Isoorientin	Hydroxyflavone	[26]
11	6.3	1.5	431.0977	23.3	311.0557, 283.0612, 341.0634	$C_{21}H_{20}O_{10}$	Vitexin	Hydroxyflavone	[26,29]
12	6.4	2.1	449.2019	23.7	269.1394, 270.1417, 225.1492, 209.1185, 251.1284, 89.0232, 287.1479	$C_{20}H_{34}O_{11}$	Unknown	Flavonoid glycoside	[29,30,32]
13	6.5	1.8	591.1336	28.1	327.0500, 357.0610, 447.0921, 489.1032	$C_{27}H_{28}O_{15}$	Flavonoid glycoside and isoorientin derivative	Flavonoid	[26,28,33]
14	6.5	2.6	413.1442	22.8	269.1018, 99.0455, 101.0244, 125.0250	$C_{19}H_{26}O_{10}$	Unidentified	-	-
15	6.6	2.0	575.1395	27.3	311.0555, 341.0660, 431.0982, 473.1075, 513.1362, 263.0798, 161.0449	$C_{27}H_{28}O_{14}$	Flavonoid glycoside and vitexin derivative	Flavonoid	[26,28]
16	6.8	n.c.	435.2225	23.4	389.2201	Unknown	Iridoid glycoside	Terpene	[34,35]
17	7.0	6.8	187.0989	17.2	125.0962, 141.8669, 169.0867	$C_9H_{16}O_4$	Azelaic acid	Dicarboxylic acid	[26]

PN: peak number, RT: retention time, IR: identification reference, n.c.: not calculated. * The fragments were ordered according to their intensity, starting with those of greater height.

2.3. Total Flavonoids

The HELF presented a flavonoid content of 129.9 ± 0.01 mg EqQ/g with respect to the fractions. Those with the highest concentrations were the FC and FEA, with 198.1 ± 0.1 and 503.5 ± 0.2 mg EqQ/g, respectively (Table 3).

Table 3. Antioxidant activity, flavonoids, and total phenolics of *Laelia furfuracea* leaf extract and fraction.

Tests	Total Flavonoids (mg EqQ/g)	Total Phenolics (mg EqAG/g)	Antioxidant Activity	
			Inhibition (%)	AAI (mg EqAA/g)
HELF	129.900 ± 0.010	394.700 ± 0.200	84.600 ± 1.400	1.300 ± 0.020
FH	81.900 ± 0.020	201.000 ± 0.200	50.100 ± 1.400	0.800 ± 0.010
FC	198.100 ± 0.100	282.800 ± 0.100	82.900 ± 1.700	1.400 ± 0.006
FEA	503.500 ± 0.200	612.500 ± 0.300	91.600 ± 0.200	2.100 ± 0.001
FB	118.100 ± 0.040	275.200 ± 0.100	52.100 ± 0.400	0.900 ± 0.027
FA	40.700 ± 0.020	207.800 ± 0.100	23.400 ± 1.100	0.300 ± 0.010

Hydroethanolic extract of *Laelia furfuracea* (HELF); fractions: hexane (FH), chloroform (FC), ethyl acetate (FEA), butanol (FB), water (FA), and AAI (antioxidant activity index); mg EqAA (milligram equivalents of ascorbic acid); mg eqQ (milligram equivalents of quercetin); and mg eqAG (milligram equivalents of gallic acid).

2.4. Total Phenols

In the same way, the FC and FEA fractions showed the highest values of total phenols, whose concentrations were 282.8 ± 0.1 and 612.5 ± 0.3 mg EqAG/g, while the HELF presented 394.7 ± 0.2 mg EqAG/g (Table 3).

2.5. Antioxidant Activity

Regarding the *in vitro* antioxidant activity by the DPPH technique, HELF presented 84.6 ± 1.4% inhibition of free radicals, with an AAI of 1.3 ± 0.02 mg EqAA/g. The FC and FEA fractions showed the best antioxidant activity. These values show *L. furfuracea* as an orchid with high antioxidant capacity (Table 3).

2.6. Anti-Inflammatory Activity

2.6.1. Oral Administration

In the anti-inflammatory evaluation, HELF, administered orally to rats at different doses, did not present a significant dose–response effect (Figure 2a).

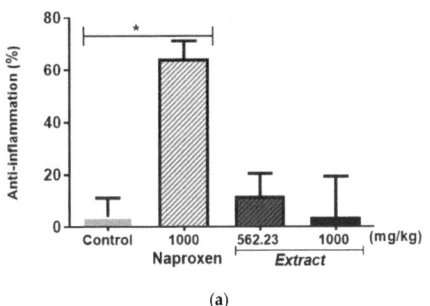
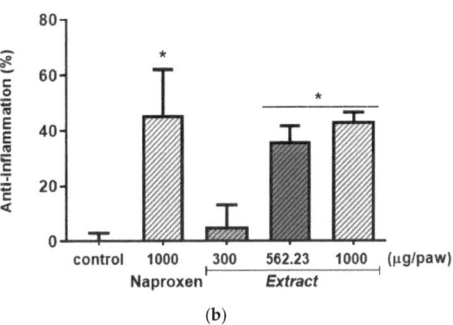

Figure 2. Anti-inflammatory effect at different doses of the hydroethanolic extract of *Laelia furfuracea* leaves in the model of plantar edema with carrageenan. (**a**) oral administration, the values were reported in mg/kg; (**b**) local administration, the values were reported in μg/paw. Each column represents the mean ± SD (n = 6); * indicates a significant difference between the treatments $p < 0.05$ of the control group (saline solution), compared to the positive control group (naproxen) and the extract at different doses.

2.6.2. Local Administration

On the other hand, the local administration of HELF, after six hours of evaluation, presented a significant dose–response anti-inflammatory effect compared to the control group. This anti-inflammatory effect of 43.7% was similar to the inhibition of inflammation caused by the administration of naproxen used as the reference drug, which had an effect of 45.7% (Figure 2b).

2.7. In Vivo Antioxidant Activity by Measurement of NO and GSH

The groups of rats treated with HELF with doses of 562.23 and 1000 µg/paw presented $NaNO_2$ concentrations of 34.1 ± 1.8 µM and 28.2 ± 4.8 µM, respectively. Meanwhile, the concentration of $NaNO_2$ in the animals in the control group was 28.0 ± 3.9 µM, with no significant differences between the values recorded for these groups.

In the measurement of GSH, the groups of animals that received the HELF doses did not present significant differences compared to the value presented by the control group.

3. Discussion

Chromatographic analysis identifies the presence of citric, malic, succinic, hydroxymethylglutaric, azelaic, eucomic, and protocatechuic acids, iridoid glycoside, saponarin, luteolin-7,3′-di-O-glycoside, isoorientin, and vitexin, and flavonoids glucoside derived from isoorientin and vitexin.

Out of the compounds identified, malic acid, eucomic acid, saponarin, and vitexin have been studied for their biological activities related to inflammation. Malic acid dependently decreases the content of tumor necrosis factor alpha (TNF-α), interferon γ (IFN-γ), and interleukins 6 (IL6) and IL10 [36]; besides, it reduces symptoms in rheumatic diseases [37]. Eucomic acid stimulates cytochrome *c* oxidase activity in the immortalized human keratinocyte cell line (HaCaT) [38]. Saponarin reduces the inflammatory response by inhibiting mitogen-activated protein kinase (MAPK) signaling, nuclear factor kappa β (NF-κB) activity, cytokine production, and expressions of marker genes specific for M1 polarization [39]; also, it presents hepatoprotective effects, increases the cellular antioxidant defense system, and levels of reduced glutathione (GSH) [40]. In the same way, it reduces the expression of inflammatory mediators such as IL-4 cytokines, IL-5, IL-13, phosphorylation of extracellular signal-regulated kinase (ERK), and p38 involved in the mitogen-activated protein kinase signaling pathway in RAW264.7 cells [41]. Vitexin inhibits cytokines IL-8, IL-17, IL-33, nitric oxide (NO), and monocyte chemoattractant protein-1 (MCP-1) and increases IL-10 [42]. We consider that the presence of these compounds in *L. furfuracea* is intervening to give the anti-inflammatory biological effect and antioxidant activity.

Regarding the antioxidant capacity of *L. furfuracea*, similar values were reported in the orchids *Dendrobium thyrsiflorum* Rchb. f. ex André [43], *D. tosaense* Makino, *D. moniliforme* (L.) Sw. [44], and *Eulophia ochreata* Lindl. [45]. This antioxidant percentage was higher in *L. furfuracea* than those reported in the extract by soxhlet with 56.6% [46] and by maceration 16.6% [47] of *Prosthechea karwinskii* and *Anacamptis pyramidalis* (L.) Rchb. f. with 54.1% [48]. With respect to the fractions, the percentage of inhibition of free radicals is higher; the FEA and FC of *L. furfuracea* presented 91.6 ± 0.2 and 82.9 ± 1.7%, and AAI of 2.1 ± 0.001 and 1.4 ± 0.006 mgEqAA/g, respectively. These fractions even turned out to be better antioxidants than the compound 4-hydroxy-3-methoxybenzyl alcohol isolated from *Gastrodia elata* Blume, which was 70% [49].

In this sense, it can be observed that there is a relationship in terms of antioxidant activity, total flavonoids, and phenols, both in the extract and in the fractions. This could be due to their composition, which has an impact on the fact that the antioxidant activity is a function of the number and position of the -OH groups, as is the case of phenols such as flavonoids, which have the highest number of hydroxyl groups and the highest rates of free radical inhibition [50].

Acute inflammation is a response of the body to repair a damaged area, which is characterized by redness, edema, heat, pain, and functional impotence [1]. This process triggers intracellular signals produced by immune cells, macrophages, and neutrophils responsible for the main inflammatory events. These are activated and bind to specific receptors to regulate cellular functions such as the expression of adhesion molecules, phagocytosis, cell death, and secretion of proinflammatory enzymes and chemotoxins, which are signs of an acute inflammatory response [1–3]. Therefore, considering that the anti-inflammatory effect in this study may be given by the phytochemical compounds of the leaf extract of *L. furfuracea*, their bioactivity depends not only on their molecular structure and concentration but also on other physiological factors in experimental animals [51]. Furthermore, structurally, phenolic compounds behave like acids, affecting solubility, glycosylation, and methylation [52], as is the case with some flavonoids linked to sugars that cause variations in their susceptibility to being digested, fermented, and absorbed in the gastrointestinal tract. However, only a part of these compounds can cross the intestinal wall to be located in the areas of the injury to act as anti-inflammatory [53].

Regarding the safety of consuming *L. furfuracea*, its use is only known in traditional medicine, where peasant communities prepare an infusion using three flowers of the plant to treat cough, but an *in vivo* study has not been carried out to demonstrate its biosafety.

On the other hand, in local administration, *L. furfuracea* showed an anti-inflammatory effect. Comparing this anti-inflammatory effect with other orchids such as *Eulophia ochreata* [54] and *Prosthechea karwinskii* [15], they also present a similar dose–response anti-inflammatory effect. Still, the leaf extract of *L. furfuracea* inhibits inflammation as a protective response, perhaps enhanced by the antioxidant action of its compounds. These compounds are likely to act on the inflammation system, since they also act on the coagulation model [23], and this model converges with inflammation [55], thus supporting its anti-inflammatory effect.

With respect to the fractions evaluated, none of these (FH, FC, FAE, FB, and FA) presented inhibition of edema in the animals, so it is inferred that the synergy of one or more compounds present plays an important role in achieving or potentiating their bioactivity [56] when the crude extract was evaluated.

In the same way, in the group of animals treated with HELF, it was observed that the concentrations of NO and GSH did not increase, so these endogenous antioxidant systems were not directly responsible for the inhibition of edema. This may be due to other enzymatic mechanisms that could work together to counteract oxidative stress, such as superoxide dismutase (SOD), catalase (CAT), and glutathione peroxidase (GPx) [57]. However, it could be that the phenolic compounds in *L. furfuracea* extract play an important role as exogenous antioxidants that complement crucial endogenous systems, maintaining a balance that contributes to antioxidant and anti-inflammatory protection, as well as the prevention of related conditions with oxidative stress [58].

4. Materials and Methods

4.1. Vegetal Material

The leaf material of *L. furfuracea* was collected in the municipality Santo Domingo Yanhuitlán, Oaxaca, México, in an oak forest with *Juniperus* and *Arbutus*. For this purpose, permission was obtained from the authorities of the Commissariat of Communal Property, as well as a scientific collection permit granted by the Mexican Ministry of Environment and Natural Resources to RS (02228/17). A voucher specimen (R. Solano 4244) was herborized and deposited in the OAX Herbarium of the Centro Interdisciplinario de Investigación para el Desarrollo Integral Regional (CIIDIR) of Oaxaca.

4.2. General Experimental Procedures

4.2.1. Reagents

Ascorbic acid (CAS 50-81-7), 1,1-Diphenyl-2-picryl-hydrazyl (CAS 1898664), $AlCl_3$ (CAS 7446-70-0), Folin–Denis Reagent (CAS 47742), gallic acid (CAS 149-91-7), quercetin (CAS 7446-70-0), naproxen (CAS 26159-34-2), and carrageenan (CAS 9000-07-1) were purchased from Sigma Aldrich (Toluca, Mexico).

4.2.2. Extraction Process

The leaf material was dehydrated at 40 °C until constant weight, then pulverized in a mill (IKA® M20, Northchase Parkway, Wilmington, DE, USA) and passed through a physical test sieve with an opening of 250 microns, mesh number 60 (MOTINOX, Toluca de Lerdo, Ciudad de México, Mexico), to have a homogeneous particle size and facilitate its extraction. The material was subjected to a solid–liquid extraction in a Soxhlet equipment (Puebla, Puebla, Mexico), using 5 g of leaf material with water–ethanol (1:1) as a solvent, at 78.2 °C for two hours. The extract was filtered and concentrated in a rotary evaporator (BÜCHI, R-210, Darmstadt, Germany). Finally, the hydroethanolic extract of *L. furfuracea* (HELF) was subjected to a liquid–liquid extraction by increasing polarities to obtain fractions of hexane (FH), chloroform (FC), ethyl acetate (FEA), butanol (FB), and water (FA) [23] (Appendix A).

4.2.3. Characterization of the Phytochemical Profile

The phytochemical profile of the extract and fractions was obtained following the procedure to determine its main groups of metabolites [59] (Appendix A).

4.2.4. Analysis of Compounds by UPLC-ESI-qTOF-MS/MS

One mg of *L. furfuracea* leaves extract was dissolved in 1 mL of water with 10% methanol. Subsequently, it was filtered through a 0.45 µm nylon filter. Ten µL of the filtered solution was taken and mixed with 90 µL of acetonitrile. The injection volume for the analysis was 10 µL.

The analysis of compounds was performed using an ultra-high-performance liquid chromatography system (UPLC, Thermo Scientific, Ultimate 3000) coupled to an Impact II mass spectrometer (Bruker Daltonics, Billerica, MA, USA), equipped with electrospray ionization and quadrupole with time-of-flight (UPLC-ESI-qTOF-MS/MS). This analysis was based on a previous study [10]. The column used was a Thermo Scientific Acclaim 120 C18 (2.2 µm, 120 Å, 50 × 2.1 mm). The mobile phase was A: 0.1% formic acid in water and B: acetonitrile. The gradient system was set up as follows: 0% B (0–2 min), 1% B (2–3 min), 3% B (3–4 min), 32% B (4–5 min), 36% B (5–6 min), 40% B (6–8 min), 45% B (8–9 min), 80% B (9–11 min), and 0% B (12–14 min) at a flow rate of 0.35 mL/min. The mass spectrometer was operated in negative electrospray mode at 0.4 Bar (5.8 psi), with a mass range of 50–1000 m/z, in MS/MS mode, and an ionization capillary voltage of 2700 V. Data were processed using the Data Analysis software 3.1 for mass spectrometry (Bruker Daltonics). The compounds were tentatively identified by comparing their mass spectra with Bruker MetaboBase, Plant metabolites, MassBank libraries, and those reported in previous articles.

4.2.5. Quantification of Total Flavonoids

The determination of total flavonoids was carried out with the aluminum trichloride ($AlCl_3$) colorimetric method [60], using quercetin as a standard for the calibration curve. The results were expressed as mg quercetin equivalents/g sample (mg EqQ/g).

4.2.6. Quantification of Total Phenols

The concentration of total phenols was determined by the method of Folin and Ciocalteau [61], using gallic acid as a standard for the calibration curve. The results were expressed as mg equivalent of gallic acid/g of sample (mg EqAG/g).

4.2.7. *In Vitro* Antioxidant Activity

The antioxidant activity was determined with the free radical scavenging method by 1,1-diphenyl-2-picryl-hydrazyl (DPPH), using ascorbic acid as a standard for the calibration curve, calculating the antioxidant activity index (IAA) with the concentration of DPPH between IC50, as well as the percentage of inhibition of the DPPH radical, as recommended by Ibarra [62].

4.2.8. Evaluation of Anti-Inflammatory Activity

Experimental Animals

Female Wistar rats, 7–9 weeks old and weighing 180 ± 20 g, were used from the vivarium of the Center for Research and Advanced Studies (CINVESTAV, Mexico City). The animals were housed under regular conditions at a temperature of from 22 to 24 °C in a 12-h light–dark cycle. Their diet consisted of standard laboratory food with free access to food, and purified water *ad libitum*, kept in 43 × 53 × 20 glass acrylic boxes with removable iron mesh lid. The designated area was 187 cm^2 per animal, and between 4 and 6 individuals were placed in each cage. Efforts were made to minimize animal suffering and to reduce the number of animals used. Each rat was used in only one experiment and euthanized in a CO_2 chamber at the end of the assay [63]. The conditions of the rats were according to the Official Mexican Standard for the Care and Management of Experimental Animals (NOM-062-ZOO-1999) and the regulations of the International Council for Laboratory Animal Science (ICLAS) [64].

Design of Experiments

The animals were randomly divided into groups for oral and local administration of HELF. In oral administration, groups of rats with 11 individuals were used: the positive control received 300 mg/kg p.o. naproxen (reference drug); the negative control received distilled water (vehicle); and the treatments received 562.23 mg/kg p.o. and 1000 mg/kg p.o. of the HELF. Treatment was administered with a gastric cannula one hour before inducing inflammation. Before oral administration, the animals were fasted for 8 h and had free access to water.

In local administration, ten groups of rats with 6 subjects were used: 1000 µg/paw of naproxen (positive control group), distilled water (vehicle control group), and three groups that received HELF in doses of 300, 562.23, and 1000 µg/paw; in five other groups, each of the extract fractions (FH, FC, FEA, FB, and FA) was administered in doses of 1000 µg/paw. The doses evaluated were selected based on an increased logarithmic scale. Each fraction was dissolved in saline with 10% Tween 80 as a vehicle. The administration of the volume was parenteral and plantar, one hour before inducing inflammation in the rats. The animals did not require prior treatment; two hours after administering the study agents, the animals received water and, three hours later, food. Inflammation was induced using the plantar edema model, applying an injection of 100 µL of 1% carrageenan in saline to the right hind paw of each rat. The volume of the paw where the inflammatory process was induced was measured with a digital plethysmometer (LE 7500 Mca. Panlab Harvard/Apparatus, Hill Road Holliston, USA) at 0, 30, 60, 120, 180, 240, 300, and 360 min after administering carrageenan. The percentage of inhibition of inflammation was calculated from the differences in the basal volume between the different times, using the following equation:

$$\% \text{ inhibition} = (ABC_{control} - ABC_{treated} / ABC_{control})100 \quad (1)$$

The area under the curve (AUC) was calculated through the following equation:

$$ABC = \sum ((\Delta vol \, B_{greater} + \Delta vol \, b_{smaller}) \, h/2) \quad (2)$$

$$\Delta vol \text{ Inflamed} - vol \text{ basal} \quad (3)$$

In Vivo Antioxidant Activity (NO and GSH)

Nitric oxide (NO) in the paw

The concentration of NO was determined in the homogenates of the paws of the rats from the negative control and basal groups and those that presented anti-inflammatory activity when the extract was administered at doses of 562.23 and 1000 µg/paw locally. The Cayman colorimetric test for nitrates/nitrites and the Griess technique were used to determine NO [65]. The concentration of sodium nitrite ($NaNO_2$) in the paw was expressed as µmol/g of tissue.

Reduced glutathione (GSH) in the paw

GSH was quantified in the same way in the rats in the negative control and basal groups and in those where there was inhibition of inflammation with HELF at doses of 562.23 and 1000 µg/paw locally, using the Ellman reagent method (5,5′-dithio-bis-(2-nitrobenzoic acid) known as DTNB (Cayman, 69-78-3, Ellsworth Rd. Ann Arbor, MI, USA), which allows detecting thiols. The readings were carried out at 412 nm. The GSH concentration in the paw was expressed as nM/g of tissue [65]. All determinations were performed in triplicate.

4.3. Statistical Analysis

The results of the anti-inflammatory evaluation were expressed as the mean and standard error for each group of animals. An analysis of variance (ANOVA) implemented in GraphPad Prism 5 was applied, as well as a Newman–Keuls multiple comparison test of means to determine the differences between them. Values were considered statistically significant when $p < 0.05$.

5. Conclusions

Laelia furfuracea is an orchid with a dose-dependent antioxidant and anti-inflammatory effect. It is possible that its activity depends on the synergy of its compounds or any of these, such as carboxylic acids, terpenes, and phenolic compounds, mainly flavonoids and phenolic acids present in its leaf structures.

A relationship was observed between the antioxidant activity *in vitro* with the concentration of phenols and flavonoids, both in the crude extract and in the ethyl acetate fraction of the leaves.

The antioxidant and anti-inflammatory effects, chemical composition, and traditional use for alleviating inflammatory process of the respiratory tract–physiologically related to inflammation– support our findings on the antioxidant and anti-inflammatory potential of *L. furfuracea* for the first time. This represents an alternative for future research studies in the search for new bioactive compounds with pharmacological activity

Author Contributions: A.L.-P.: conceptualization, methodology, validation, formal analysis, investigation, and writing—original draft, review, and editing. L.L.-R. and A.E.C.-P.: validation, resources, and writing—review, editing, supervision, and project administration. R.S.: resources and writing—original draft, review, and editing. G.S.B.-Z.: formal analysis. M.J.-E.: resources. All authors have read and agreed to the published version of the manuscript.

Funding: The financial support provided by Consejo Nacional de Ciencia y Tecnologia (CONACYT, project 270428) and Instituto Politécnico Nacional (IPN; projects SIP 2016-RE/50, SIP-20161270, SIP-20170263, SIP-20180382, SIP-20195897, and SIP-20240944).

Data Availability Statement: Data are contained within the article.

Acknowledgments: The authors acknowledge the grants provided by CONACYT and IPN for Ms.D. studies. They also acknowledge the community of Santo Domingo Yanhuitlán, Oaxaca, for their permission to collect plant materials. The comments of three anonymous reviewers were valuable in improving the manuscript.

Conflicts of Interest: The authors declare that they have no conflicts of interest.

Appendix A. Graphical Representation of the Experimental Protocol

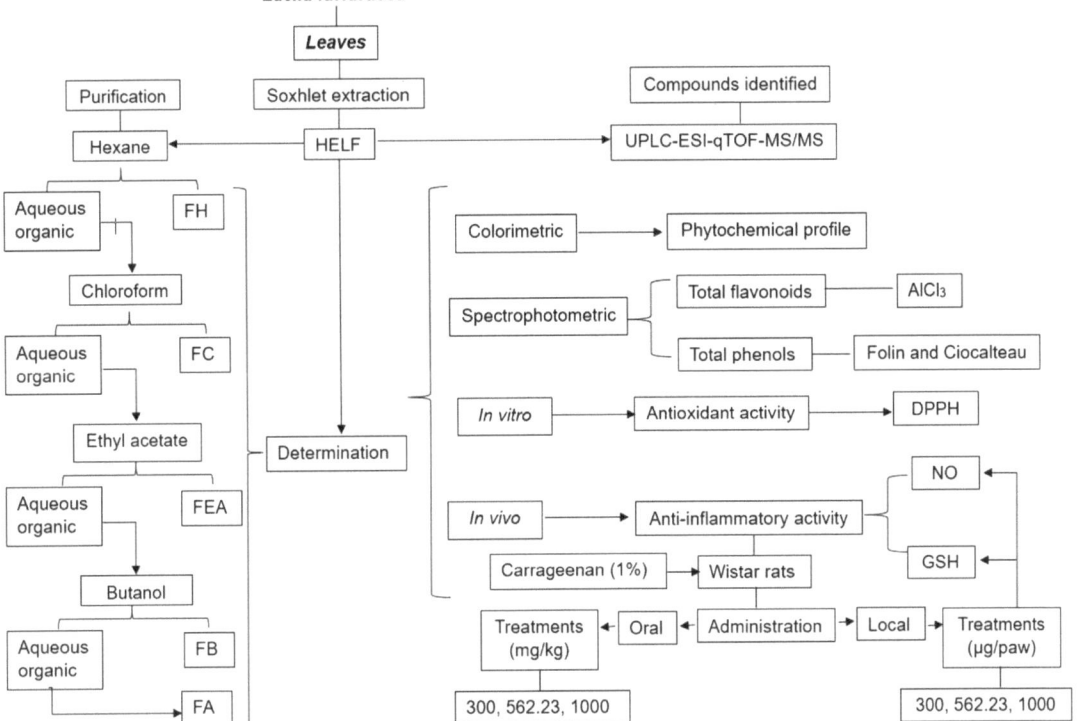

Hydroethanolic extract of *Laelia furfuracea* (HELF), fractions: hexane (FH), chloroform (FC), ethyl acetate (FEA), butanol (FB), water (FA), aluminum trichloride (AlCl$_3$), 1,1-diphenyl-2-picryl-hydrazyl (DPPH), nitric oxide (NO), reduced glutathione (GSH).

References

1. Arfeen, M.; Srivastava, A.; Srivastava, N.; Khan, R.A.; Almahmoud, S.A.; Mohammed, H.A. Design, classification, and adverse effects of NSAIDs: A review on recent advancements. *Bioorganic Med. Chem.* **2024**, *112*, 117899. [CrossRef] [PubMed]
2. Huo, L.; Liu, G.; Deng, B.; Xu, L.; Mo, Y.; Jiang, S.; Tao, J.; Bai, H.; Wang, L.; Yang, X.; et al. Effect of use of NSAIDs or steroids during the acute phase of pain on the incidence of chronic pain: A systematic review and meta-analysis of randomised trials. *Inflammopharmacology* **2024**, *32*, 1039–1058. [CrossRef] [PubMed]

3. Xie, R.; Li, J.; Jing, Y.; Tian, J.; Li, H.; Cai, Y.; Wang, Y.; Chen, W.; Xu, F. Efficacy and safety of simple analgesics for acute treatment of episodic tension-type headache in adults: A network meta-analysis. *Ann. Med.* **2024**, *56*, 2357235. [CrossRef] [PubMed]
4. Carvajal, C.C. Reactive oxygen species: Training, function and oxidative stress. *Med. Leg. Costa Rica* **2019**, *36*, 92–99.
5. Andersen, J.K. Oxidative stress in neurodegeneration: Cause or consequence? *Nat. Med.* **2004**, *10*, S18–S25. [CrossRef]
6. Vino, D. Anti-inflammatory activity. In *In Vitro Methods for Anti-Inflammatory Activity H-Bradykinin Receptor Binding*; Springer: Berlin/Heidelberg, Germany, 2014; pp. 2–137.
7. Aslani, B.A.; Ghobadi, S. Studies on oxidants and antioxidants with a brief glance at their relevance to the immune system. *Life Sci.* **2016**, *146*, 163–173. [CrossRef]
8. Alamgeer, U.M.; Muhammad, N.M.; Hafeez, U.K.; Zafirah, M.; Muhammad, N.H.M.; Taseer, A.; Fouzia, L.; Nazia, T.; Abdul, Q.K.; Haseeb, A.; et al. Evaluation of anti-inflammatory, analgesic, and antipyretic activities of *Thimus serphyllum* Linn. in mice. *Acta Pol. Pharm.* **2015**, *72*, 113–118.
9. Cuong, T.D.; Hung, T.M.; Lee, J.S.; Weon, K.Y.; Woo, M.H.; Min, B.S. Anti-inflammatory activity of phenolic compounds from the whole plant of *Scutellaria indica*. *BMCL* **2015**, *25*, 1129–1134. [CrossRef]
10. Pomini, A.M.; Sahyun, S.A.; De Oliveira, S.M.; de Faria, T.R. Bioactive natural products from orchids native to the America—A review. *An. Acad. Bras. Ciências* **2023**, *95*, e20211488. [CrossRef]
11. Castillo-Pérez, L.J.; Ponce-Hernández, A.; Alonso-Castro, A.J.; Solano, R.; Fortanelli-Martínez, J.; Lagunez-Rivera, L.; Carranza-Álvarez, C. Medicinal Orchids of Mexico: A Review. *Pharmaceuticals* **2024**, *17*, 907. [CrossRef]
12. Yañez-Barrientos, E.; González-Ibarra, A.A.; Wrobel, K.; Wrobel, K.; Corrales-Escobosa, A.R.; Alonso-Castro, A.J.; Carranza-Álvarez, C.; Ponce-Hernández, A.; Isiordia-Espinoza, M.A.; Ortiz-Andrade, R.; et al. Antinociceptive effects of *Laelia anceps* Lindl. and *Cyrtopodium macrobulbon* (Lex.) G.A. Romero & Carnevali, and comparative evaluation of their metabolomic profiles. *J. Ethnopharmacol.* **2022**, *291*, 115172. [CrossRef] [PubMed]
13. Pérez-Barrón, G.; Estrada-Soto, S.; Arias-Duran, L.; Cruz-Torres, K.C.; Ornelas-Mendoza, K.; Bernal-Fernandez, G.; Perea-Arango, I.; Villalobos-Molina, R. Calcium channel blockade mediates the vasorelaxant activity of dichloromethane extract from roots of *Oncidium cebolleta* on isolated rat aorta. *Biointerface Res. Appl. Chem.* **2023**, *13*, 80. [CrossRef]
14. Gutierrez, R.M.P.; Solis, R.V. Anti-inflammatory and wound healing potential of *Prosthechea michuacana* in rats. *Pharmacogn. Mag.* **2009**, *5*, 219–225. Available online: http://www.phcog.com/text.asp?2009/5/19/219/58163 (accessed on 21 December 2024).
15. Barragán-Zarate, G.S.; Lagunez-Rivera, L.; Solano, R.; Pineda-Peña, E.A.; Landa-Juárez, A.J.; Chávez-Piña, A.E.; Carranza-Álvarez, C.; Hernández-Benavidez, D.M. *Prosthechea karwinskii*, an orchid used as traditional medicine, exerts anti-inflammatory activity and inhibits ROS. *J. Ethnopharmacol.* **2020**, *253*, 112632. [CrossRef]
16. Barragán-Zarate, G.S.; Alexander-Aguilera, A.; Lagunez-Rivera, L.; Solano, R.; Soto-Rodríguez, I. Bioactive compounds from *Prosthechea karwinskii* decrease obesity, insulin resistance, pro-inflammatory status, and cardiovascular risk in Wistar rats with metabolic syndrome. *J. Ethnopharmacol.* **2021**, *279*, 114376. [CrossRef]
17. Hágsater, E.; Soto-Arenas, M.A.; Salazar-Chávez, G.A.; Jiménez-Machorro, R.; López-Rosas, M.A.; Dressler, R.L. *Las Orquídeas de México*; REdacta: Mexico City, Mexico, 2015.
18. Vergara-Galicia, J.; Ortiz-Andrade, R.; Castillo-España, P.; Ibarra-Barajas, M.; Gallardo-Ortiz, I.; Villalobos-Molina, R.; Estrada-Soto, S. Antihypertensive and vasorelaxant activities of *Laelia autumnalis* are mainly through calcium channel blockade. *Vasc. Pharmacol.* **2008**, *49*, 26–31. [CrossRef]
19. Aguirre, C.F.; Castillo, E.P.; Villalobos, M.R.; López, G.J.G.; Estrada, S. Vasorelaxant effect of Mexican medicinal plants on isolated rat aorta. *Pharm. Biol.* **2005**, *43*, 540–546. [CrossRef]
20. Vergara-Galicia, J.; Ortiz-Andrade, R.; Rivera-Leyva, J.; Castillo-España, P.; Ibarra, M.B.; Villalobos-Molina, R.; Ibarra-Barajas, M.; Gallardo-Ortiz, I.; Estrada-Soto, S. Vasorelaxant and antihypertensive effects of methanolic extract from roots of *Laelia anceps* are mediated by calcicum-channel antagonism. *Fitoterapia* **2010**, *81*, 350–357. [CrossRef]
21. Vergara-Galicial, J.; Castillo-España, P.; Villalobos-Molina, R.; Estrada-Soto, S. Vasorelaxant effect of *Laelia speciosa* and *Laelia anceps*: Two orchids as potential sources for the isolation of bioactive molecules. *J. Appl. Pharm. Sci.* **2013**, *3*, 034–037.
22. Jímarez-Montiel, M. Fitoquímica y Determinación de la Actividad Antiinflamatoria y Citotóxica de Compuestos y Extractos Orgánicos de Las Hojas de *Laelia anceps* Lindl. (Orchideceae). Bachelor's Thesis, Universidad Nacional Autónoma de México, Mexico City, Mexico, 2009; pp. 30–43.
23. López-Pérez, A.; Hernández-Juárez, J.; Solano, R.; Majluf-Cruz, A.; Hernández-Cruz, P.A.; Lagunez-Rivera, L. *Laelia furfuracea* Lindl.: An Endemic Mexican Orchid. with Anticoagulant Activity. *J. Mex. Chem.* **2022**, *66*, 1–16. [CrossRef]
24. López-Pérez, A.; Barragán-Zárate, G.S.; Lagunez-Rivera, L.; Solano, R. Capacidad antioxidante y perfil fitoquímico de dos especies de orquídeas mexicanas. *J. CIM* **2016**, *4*, 605–610.
25. Fernández-Rojas, B.; López-Pérez, A.; Lagunez-Rivera, L.; Solano, R.; Bernal-Martínez, A.K.; Majluf-Cruz, A.; Hernández-Juárez, J. Antiplatelet, Anticoagulant, and Fibrinolytic Activity of Orchids: A Review. *Molecules* **2024**, *29*, 5706. [CrossRef] [PubMed]
26. Available online: https://massbank.eu/MassBank/ (accessed on 28 November 2024).

27. El-Hawary, S.S.; Sobeh, M.; Badr, W.K.; Abdelfattah, M.A.O.; Ali, Z.Y.; El-Tantawy, M.E.; Rabeh, M.A.; Wink, M. HPLC-PDA-MS/MS profiling of secondary metabolites from *Opuntia ficus-indica* cladode, peel and fruit pulp extracts and their antioxidant, neuroprotective effect in rats with aluminum chloride induced neurotoxicity. *Saudi J. Biol. Sci.* **2020**, *27*, 2829–2838. [CrossRef] [PubMed]
28. Ghareeb, M.A.; Sobeh, M.; Rezq, S.; El-Shazly, A.M.; Mahmoud, M.F.; Wink, M. HPLC-ESI-MS/MS Profiling of Polyphenolics of a Leaf Extract from *Alpinia zerumbet* (Zingiberaceae) and Its Anti-Inflammatory, Anti-Nociceptive, and Antipyretic Activities *In Vivo*. *Molecules* **2018**, *23*, 3238. Available online: https://www.mdpi.com/1420-3049/23/12/3238 (accessed on 28 November 2024). [CrossRef]
29. Ragheb, A.Y.; Masoud, M.A.; El Shabrawy, M.O.; Farid, M.M.; Hegazi, N.M.; Mohammed, R.S.; Marzouk, M.M.; Aboutabl, M.E. MS/MS-based molecular networking for mapping the chemical diversity of the pulp and peel extracts from *Citrus japonica* Thunb.; *in vivo* evaluation of their anti-inflammatory and anti-ulcer potential. *Sci. Afr.* **2023**, *20*, e01672. [CrossRef]
30. Ma, T.; Lin, J.; Gan, A.; Sun, Y.; Sun, Y.; Wang, M.; Wan, M.; Yan, T.; Jia, Y. Qualitative and quantitative analysis of the components in flowers of *Hemerocallis citrina* Baroni by UHPLC–Q-TOF-MS/MS and UHPLC–QQQ-MS/MS and evaluation of their antioxidant activities. *J. Food Compos. Anal.* **2023**, *120*, 105329. [CrossRef]
31. Llorach, R.; Favari, C.; Alonso, D.; Garcia-Aloy, M.; Andres-Lacueva, C.; Urpi-Sarda, M. Comparative metabolite fingerprinting of legumes using LC-MS-based untargeted metabolomics. *Int. Food Res.* **2019**, *126*, 108666. [CrossRef]
32. Zhao, M.; Linghu, K.G.; Xiao, L.; Hua, T.; Zhao, G.; Chen, Q.; Xiong, S.; Shen, L.; Yu, J.; Hou, X.; et al. Anti-inflammatory/antioxidant properties and the UPLC-QTOF/MS-based metabolomics discrimination of three yellow *Camellia* species. *Int. Food Res.* **2022**, *160*, 111628. [CrossRef]
33. Zhang, Q.Q.; Dong, X.; Liu, X.G.; Gao, W.; Li, P.; Yang, H. Rapid separation and identification of multiple constituents in Danhong Injection by ultra-high performance liquid chromatography coupled to electrospray ionization quadrupole time-of-flight tandem mass spectrometry. *CJNM* **2016**, *14*, 147–160. [CrossRef]
34. Heffels, P.; Müller, L.; Schieber, A.; Weber, F. Profiling of iridoid glycosides in *Vaccinium* species by UHPLC-MS. *Int. Food Res.* **2017**, *100*, 462–468. [CrossRef]
35. Sun, X.; Xue, S.; Cui, Y.; Li, M.; Chen, S.; Yue, J.; Gao, Z. Characterization and identification of chemical constituents in Corni Fructus and effect of storage using UHPLC-LTQ-Orbitrap-MS.514. *Int. Food Res.* **2023**, *164*, 112330. [CrossRef] [PubMed]
36. Zhang, P.; Jiang, G.; Wang, Y.; Yan, E.; He, L.; Guo, J.; Yin, J.; Zhang, X. Maternal consumption of l-malic acid enriched diets improves antioxidant capacity and glucose metabolism in offspring by regulating the gut microbiota. *Redox Biol.* **2023**, *67*, 102889. [CrossRef] [PubMed]
37. de Carvalho, J.F.; Lerner, A. Malic Acid for the Treatment of Rheumatic Diseases. *Mediterr. J. Rheumatol.* **2023**, *30*, 592–593. [CrossRef] [PubMed]
38. Simmler, C.; Anteaume, C.; André, P.; Bonté, F.; Lobstein, A. Glucosyloxybenzyl eucomate derivatives from *Vanda teres* stimulate HaCaT cytochrome c oxidase. *J. Nat. Prod.* **2011**, *74*, 949955. [CrossRef]
39. Yu, G.R.; Lim, D.W.; Karunarathne, W.A.H.M.; Kim, G.Y.; Kim, H.; Kim, J.E.; Park, W.H. A non-polar fraction of *Saponaria officinalis* L. acted as a TLR4/MD2 complex antagonist and inhibited TLR4/MyD88 signaling *in vitro* and *in vivo*. *FASEB J.* **2022**, *36*, e22387. [CrossRef]
40. Simeonova, R.; Vitcheva, V.; Kondeva-Burdina, M.; Krasteva, I.; Manov, V.; Mitcheva, M. Hepatoprotective and antioxidant effects of saponarin, isolated from *Gypsophila trichotoma* Wend. on paracetamol-induced liver damage in rats. *Biomed. Res. Int.* **2013**, *2013*, 757126. [CrossRef]
41. Min, S.Y.; Park, C.H.; Yu, H.W.; Park, Y.J. Anti-Inflammatory and Anti-Allergic Effects of Saponarin and Its Impact on Signaling Pathways of RAW 264.7, RBL-2H3, and HaCaT Cells. *Int. J. Mol. Sci.* **2021**, *5*, 8431. [CrossRef]
42. He, M.; Min, J.W.; Kong, W.L.; He, X.H.; Li, J.X.; Peng, B.W. A review on the pharmacological effects of vitexin and isovitexin. *Fitoterapia* **2016**, *115*, 74–85. [CrossRef]
43. Bhattacharyya, P.; Kumaria, S.; Job, N.; Tandon, P. Phyto-molecular profiling and assessment of antioxidant activity within micropropagated plants of *Dendrobium thyrsiflorum*: A threatened, medicinal orchid. *Plant Cell Tiss. Organ. Cult.* **2015**, *122*, 535–550. [CrossRef]
44. Lo, S.F.; Mulabagal, V.; Chen, C.L.; Kuo, C.L.; Tsay, H.S. Bioguided fractionation and isolation of free radical scavenging components from *in vitro* propagated Chinese medicinal plants *Dendrobium tosaense* Makino and *Dendrobium moniliforme* SW. *J. Agric. Food Chem.* **2004**, *52*, 6916–6919. [CrossRef]
45. Kshirsagar, R.; Kanekar, Y.; Jagtap, S.; Upadhyay, S.; Rao, R.; Bhujbal, S.; Kedia, J. Phenanthrenes of *Eulophia ochreata* Lindl. *IJGP* **2010**, *4*, 147. [CrossRef]
46. Barragán-Zarate, G.S.; Lagunez-Rivera, L.; Solano, R.; Carranza-Álvarez, C.; Hernández-Benavides, D.M.; Vilarem, G. Validation of the traditional medicinal use of a Mexican endemic orchid (*Prosthechea karwinskii*) through UPLC-ESI-qTOF-MS/MS characterization of its bioactive compounds. *Heliyon* **2022**, *8*, e09667. [CrossRef]

47. Rojas-Olivos, A.; Solano, R.; Alexander-Aguilera, A.; Jiménez-Estrada, M.; Zilli-Hernández, S.; Lagunez-Rivera, L. Effect of *Prosthechea karwinskii* (*Orchidaceae*) on obesity and dyslipidemia in Wistar rats. *Alex. J. Med.* **2017**, *53*, 311–315. [CrossRef]
48. Stajner, D.; Popović, B.M.; Kapor, A.; Boza, P.; Stajner, M. Antioxidant and scavenging capacity of *Anacamptis pyramidalis* L.—pyramidal orchid from Vojvodina. *Phytother. Res.* **2010**, *24*, 759–763. [CrossRef]
49. Lee, J.Y.; Jang, Y.W.; Kang, H.S.; Moon, H.; Sim, S.S.; Kim, C.J. Anti-inflammatory action of phenolic compounds from *Gastrodia elata* root. *Arch. Pharm. Res.* **2006**, *29*, 849–858. [CrossRef]
50. Bijak, M.; Ziewiecki, R.; Saluk, J.; Ponczek, M.; Pawlaczyk, I.; Krotkiewski, H.; Wachowicz, B.; Nowak, P. Thrombin inhibitory activity of some polyphenolic compounds. *Med. Chem. Res.* **2014**, *23*, 2324–2337. [CrossRef]
51. Palafox-Carlos, H.; Ayala-Zavala, J.F.; González-Aguilar, G.A. The role of dietary fiber in the bioaccessibility and bioavailability of fruit and vegetable antioxidants. *J. Food Sci.* **2011**, *76*, R6–R15. [CrossRef]
52. Bowsher, C.; Steer, M.; Tobin, A. *Plant Biochemistry*; Garland Science: New York, NY, USA, Taylor & Francis Group, LLC: Oxford, UK; 2008.
53. Gutiérrez-Grijalva, E.P.; Ambriz-Pere, D.L.; Leyva-López, N.; Castillo-López, R.I.; Heiedia, J.B. Review: Dietary phenolic compounds, health benefits and bioaccessibility. *Arch. Latinoam. Nutr.* **2016**, *66*, 87–100. [PubMed]
54. Jagtap, S.; Gilda, P.; Bhondave, A.; Pradkar, P.; Pawar, A.; Harsulkar, A. Validation of the potential of *Eulophia ochreata* L. tubers for its anti-inflammatory and antioxidant activity. *Pharmacol.* **2009**, *2*, 307–316.
55. Yong, J.; Toh, C.H. The convergent model of coagulation. *J. Thromb. Haemost.* **2024**, *22*, 2140–2146. [CrossRef]
56. Saura-Calixto, F. Dietary Fiber as a Carrier of Dietary Antioxidants: An Essential Physiological Function. *J. Agric. Food Chem.* **2010**, *59*, 43–49. [CrossRef] [PubMed]
57. Jomova, K.; Raptova, R.; Alomar, S.Y.; Alwasel, S.H.; Nepovimova, E.; Kuca, K.; Valko, M. Reactive oxygen species, toxicity, oxidative stress, and antioxidants: Chronic diseases and aging. *Arch. Toxicol.* **2023**, *97*, 2499–2574. [CrossRef] [PubMed]
58. Sahoo, D.K.; Heilmann, R.M.; Paital, B.; Patel, A.; Yadav, V.K.; Wong, D.; Jergens, A.E. Oxidative stress, hormones, and effects of natural antioxidants on intestinal inflammation in inflammatory bowel disease. *Front. Endocrinol.* **2023**, *14*, 1217165. [CrossRef]
59. Valencia, G.T.; Garín, M.E.A. *Manual de Prácticas de Productos Naturales*, 1st ed.; Instituto Politécnico Nacional: Mexico City, Mexico, 2010; pp. 28–31.
60. Woisky, R.; Salatino, A. Analysis of propolis: Some parameters and procedures for chemical quality control. *J. Apic. Res.* **1998**, *37*, 99–105. [CrossRef]
61. Agbor, G.A.; Vinson, J.A.; Donnelly, P.E. Folin-Ciocalteau Reagent for Polyphenolic Assay. *Int. J. Food Sci. Nutr. Diet. (IJFS)* **2014**, *3*, 801. [CrossRef]
62. Scherer, R.; Godoy, H.T. Antioxidant activity index (AAI) by the 2,2-diphenyl-1-picrylhydrazyl method. *Food Chem.* **2009**, *112*, 654–658. [CrossRef]
63. Miranda-Lara, C.A.; Ortiz, M.I.; Rodríguez-Ramos, F.; Chávez-Piña, A.E. Synergistic interaction between docosahexaenoic acid and diclofenac on inflammation, nociception, and gastric security models in rats. *Drug Dev. Res.* **2018**, *79*, 239–246. [CrossRef]
64. Alcalde, B.S.M.; Flores, M.L.; Cordoba, O.L.; Taira, C.A.; Gorzalczany, S. Antinociceptive and anti-inflammatory activities of an aqueous extract of *Chiliotrichum diffusum*. *Rev. Bras. Farmacog* **2013**, *23*, 699–705. [CrossRef]
65. Tarpey, M.M.; Fridovich, I. Methods off detection of vascular reactive species. Nitric oxide, superoxide, hydrogen peroxide and peroxynitrite. *Circ. Res.* **2001**, *89*, 224–236. [CrossRef]

Disclaimer/Publisher's Note: The statements, opinions and data contained in all publications are solely those of the individual author(s) and contributor(s) and not of MDPI and/or the editor(s). MDPI and/or the editor(s) disclaim responsibility for any injury to people or property resulting from any ideas, methods, instructions or products referred to in the content.

Article

Application of Microwave-Assisted Water Extraction (MAWE) to Fully Realize Various Physiological Activities of *Melaleuca quinquenervia* Leaf Extract

Ting-Kang Lin [1], Jyh-Yih Leu [1,2], Yi-Lin Lai [3], Yu-Chi Chang [3], Ying-Chien Chung [3,*] and Hsia-Wei Liu [1,2,*]

1. Graduate Institute of Applied Science and Engineering, Fu Jen Catholic University, New Taipei City 242062, Taiwan; dingganglin@gmail.com (T.-K.L.); yihleu@gmail.com (J.-Y.L.)
2. Department of Life Science, Fu Jen Catholic University, New Taipei City 242062, Taiwan
3. Department of Biological Science and Technology, China University of Science and Technology, Taipei City 115311, Taiwan; yilinlai519@gmail.com (Y.-L.L.); yuchi0618@cc.cust.edu.tw (Y.-C.C.)
* Correspondence: ycchung@cc.cust.edu.tw (Y.-C.C.); 079336@mail.fju.edu.tw (H.-W.L.); Tel.: +886-2-27821862 (Y.-C.C.); +886-2-29053740 (H.-W.L.)

Abstract: *Melaleuca quinquenervia* is widely grown in tropical areas worldwide. Studies have demonstrated that extracts of its buds, leaves, and branches obtained through hydrodistillation, steam distillation, or solvent extraction exhibit physiological activities, including anti-melanogenic, antibacterial, and antioxidant properties; nevertheless, such extracts are mostly not effectively collected or adequately utilized. Accordingly, this study applied a rapid, effective, and easy-to-operate microwave-assisted water extraction (MAWE) technique for the first time to prepare *M. quinquenervia* leaf extract (MLE) with improved physiological activities. The results indicated that the optimal irradiation time and liquid/solid ratio for the production of the MLE were 180 s and 20 mL/g, respectively. Under optimal conditions, the freeze-dried MLE achieved a high yield (6.28% ± 0.08%) and highly effective broad-spectrum physiological activities. The MLE exhibited strong antioxidant, antiaging, and anti-inflammatory activities and excellent antityrosinase and antimicrobial activities. Additionally, the MLE was noncytotoxic at concentrations of $\leq$300 mg/L, at which it exhibited pharmacological activity. The results also indicated that the MLE comprised a total of 24 chemical compounds and 17 phenolic compounds. Among these compounds, luteolin contributed to antityrosinase activity. The extract's antiaging activity was attributed to ellagic acid and quercetin, its anti-inflammatory activity resulted from ellagic acid and kaempferol, and its antimicrobial activity resulted from quercetin and 3-*O*-methylellagic acid. In conclusion, the MAWE-derived MLE may be useful as a functional ingredient in cosmetic products, health foods, and botanical drugs.

Keywords: antioxidant; cytotoxicity; microwave-assisted extraction; *Melaleuca quinquenervia*; molecular docking

Citation: Lin, T.-K.; Leu, J.-Y.; Lai, Y.-L.; Chang, Y.-C.; Chung, Y.-C.; Liu, H.-W. Application of Microwave-Assisted Water Extraction (MAWE) to Fully Realize Various Physiological Activities of *Melaleuca quinquenervia* Leaf Extract. *Plants* **2024**, *13*, 3362. https://doi.org/10.3390/plants13233362

Academic Editors: Luis Ricardo Hernández, Eugenio Sánchez-Arreola and Edgar R. López-Mena

Received: 3 November 2024
Revised: 27 November 2024
Accepted: 28 November 2024
Published: 29 November 2024

Copyright: © 2024 by the authors. Licensee MDPI, Basel, Switzerland. This article is an open access article distributed under the terms and conditions of the Creative Commons Attribution (CC BY) license (https://creativecommons.org/licenses/by/4.0/).

1. Introduction

Melaleuca quinquenervia is extensively grown in Taiwan and is often used as a street tree, landscape tree, and windbreak [1]. *M. quinquenervia* belongs to a class of large evergreen trees characterized by protruding nodules on the trunk and brown or off-white bark. The leaves and buds of *M. quinquenervia* contain essential oils, which are often extracted for use as antibacterial agents, preservatives, analgesics, pesticides, tranquilizers, and treatments for atopic dermatitis or eczema [2–4]. Therefore, *M. quinquenervia* leaf and bud extracts are widely used as the main raw material in daily chemicals, beauty products, and health-care products. Although *M. quinquenervia* leaf extract (MLE) exhibits physiological activities, including skin-whitening, antibacterial, antiaging, and anti-inflammatory effects, this extract is mostly not effectively collected or adequately utilized [1]. This is primarily attributed to the lack of effective and appropriate extraction techniques. Hence, developing appropriate extraction techniques can improve the commercial value and pharmacological activity of MLEs.

Common techniques for extracting essential oils from plants include hydrodistillation, steam distillation, cold pressing, enfleurage, and solvent extraction [5]. Hydrodistillation and steam distillation are easy to operate, but these techniques provide low yields of essential oils [6]. Cold pressing is suitable only for extracting essential oils from citrus peels [7]. Enfleurage is a traditional extraction technique with complicated and labor-intensive procedures [8]; hence, it is mainly used to extract essential oils from flowers for producing perfumes and balms. Solvent extraction entails placing plants in organic solvents (e.g., ether, hexane, toluene, formaldehyde, acetaldehyde, or alcohol) in a closed container for extraction; this technique achieves a higher yield of essential oils than does distillation, but it requires a longer extraction time and can extract only chemical compounds of specific polarity [5]. In recent years, scholars have developed supercritical fluid extraction, ultrasound-assisted extraction, and microwave-assisted extraction (MAE) techniques to improve the extraction efficiency and shorten the extraction time [9]. Among these techniques, MAE is particularly noteworthy for its effectiveness in minimizing extraction time while simultaneously improving the yield of phenolic compounds [10]. Additionally, MAE is recognized as an energy-saving, environmentally friendly, and easy-to-operate technique [10]. Compared with hydrodistillation, MAE yields extracts with lower degrees of thermal decomposition and oxidation; thus, in MAE-derived extracts, the active ingredients in natural products are preserved. Furthermore, relative to hydrodistillation, MAE can shorten the extraction time by 5–20 times and can produce a greater yield of active compounds through the rapid breakdown of cell walls and extraction of intracellular substances [10].

Steam distillation and solvent (80% ethanol, water, and methanol) extraction techniques have been used to derive extracts from different parts of *M. quinquenervia* trees (e.g., buds, leaves, and branches) [1,2,11]. Such extracts have been reported to demonstrate various physiological activities, including antioxidant, anti-melanogenic, antigenotoxic, hypoglycemic, antifungal, and anti-inflammatory effects, in addition to exhibiting insecticidal activity against mosquitoes [1,2,11]. However, the types and concentrations of active ingredients in the extracts can vary according to the region of harvest, type of extraction technique used, or part of the plant used for extraction. Accordingly, the aim of the present study was to determine the physiological activities of an MLE derived through microwave-assisted water extraction (MAWE) in order to maximize the commercial value of the extract.

2. Results and Discussion

2.1. Optimization of MAE Conditions

Previous studies have reported a positive correlation between total phenolic content (TPC) and antioxidant activity [12]. Therefore, in this study, the TPC of the *M. quinquenervia* leaf extract (MLE) was used as an indicator to optimize MAWE conditions. Figure 1A illustrates the effect of various liquid/solid ratios (LSRs) on the TPC of the MLE. As the LSR changed, the TPC of the MLE first increased and then decreased, and the highest TPC (265.4 ± 2.8 mg gallic acid equivalent (GAE)/g dry weight (DW)) was observed at an LSR of 20 mL/g. Figure 1B presents the effects of irradiation time on the TPC and yield of the MLE. The TPC and yield of the MLE varied with irradiation time. The optimal TPC and yield of the MLE (312.7 ± 5.2 mg GAE/g DW and 6.28 ± 0.08%, respectively) were observed at an irradiation time of 180 s. The TPC of the MLE in this study was determined to be superior to those of methanol leaf extract of *M. cajuputi* (37 ± 0.02 mg GAE/g DW) [13], ethanol extract of *M. bracteata* (88.6 ± 1.3 mg GAE/g DW) [14], ethanol flower extract of *M. leucadendron* (153.8 ± 1.9 mg GAE/g DW) [15], and butanol extract of *M. leucadendron* (289.23 ± 5.21 mg GAE/g DW) [16]. In addition, the yield of the MLE in this study was determined to be considerably superior to that of butanol extract of *M. leucadendron* (2.22%) [16]; it was also superior to those of hydrodistillation extract of *M. leucadendra* (0.75%) [17], hydrodistillation extract of *M. leucadendra* (0.7%) [18], and hydrodistillation extract of young (1.22%) and old (1.43%) *M. leucadendra* leaves [19]. Although experimental data for the same species were not available for comparison with our results, we clearly determined the application potential of the MAWE technique for obtaining *M. quinquenervia* extracts with high antioxidant activity and yield.

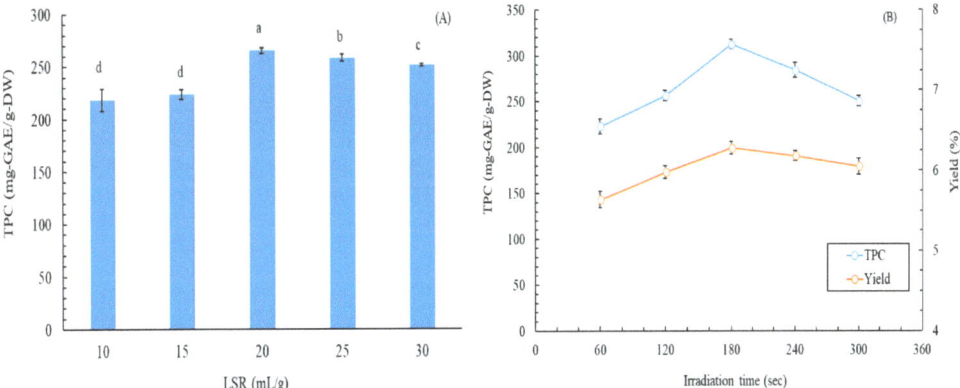

Figure 1. (**A**) Effects of LSR on the TPC of the MLE when the irradiation time was 120 s and extraction temperature was 80 °C. (**B**) Effects of irradiation time on the TPC and yield of the MLE when the LSR was 20 and the extraction temperature was 80 °C. Data are expressed as means and standard deviations of three independent experiments. The lowercase letters in the subfigure A indicate significant differences at the $p < 0.05$ level.

2.2. Extracellular and Intracellular Antityrosinase Activity

The antityrosinase activity of the MLE was evaluated under optimal extraction conditions (LSR: 20 mL/g; irradiation time: 180 s). The extracellular and intracellular antityrosinase activities of the MLE are displayed in Figure 2A,B, respectively. The antityrosinase activity of the MLE gradually increased with the concentration of the extract. The half-maximal inhibitory concentration (IC_{50}) value derived for the intracellular antityrosinase activity of the extract was 124.5 ± 3.6 mg/L, which was noted to be superior to those of α-arbutin (161.8 ± 5.4 mg/L), aqueous methanolic leaf extract of *Melaleuca subulata* (287.04 ± 4.19 mg/L) [20], and polyphenol-rich fraction of *Melaleuca rugulosa* leaves (200.56 ± 2.08 mg/L) [21]. However, the IC_{50} value derived for the antityrosinase activity of the MLE was noted to be inferior to that of kojic acid (positive control; 14.5 ± 0.8 mg/L). Figure 2B reveals a nonproportional relationship between a decrease in melanin content in the HEMn cells and an increase in the intracellular antityrosinase activity of the MLE. The results indicated that the mechanism underlying the whitening effects of the MLE (in terms of the melanin content in the HEMn cells) was complex and did not involve only the inhibition of tyrosinase activity. The mechanism might also involve the inhibition and regulation of other melanogenic genes or enzymes [22]. In addition, the IC_{50} value derived for the intracellular antityrosinase activity of the extract was 102.1 ± 1.8 mg/L, which was superior to the value found for intracellular antityrosinase activity. In this study, the MAWE-derived MLE exhibited significantly effective antityrosinase activity; this is because the melanin content decreased to 9.8 ± 1.2% in the HEMn cells treated with the MLE at a low concentration (120 mg/L).

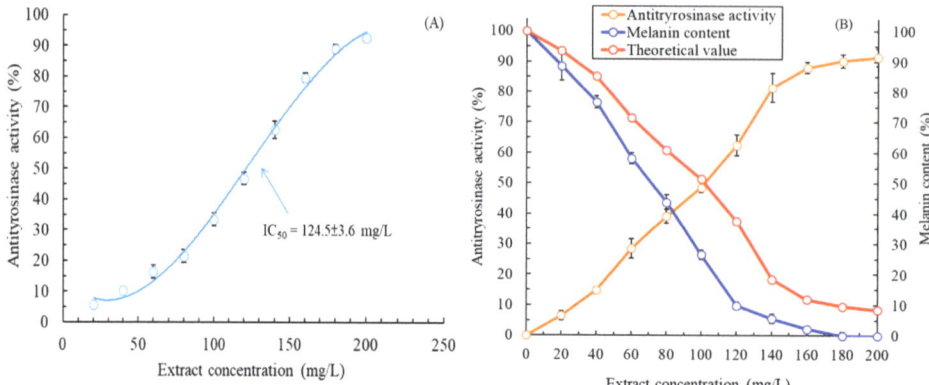

Figure 2. (**A**) Extracellular antityrosinase activity and (**B**) intracellular antityrosinase activity and melanin content in HEMn cells after treatment with the MLE obtained through MAWE (operating conditions: LSR, 20 mL/g; extraction temperature, 80 °C; microwave irradiation time, 180 s; and microwave irradiation power, 700 W). Data are expressed as means and standard deviations of three independent experiments.

2.3. Cytotoxicity Assay

In product development, product functionality is crucial, but product safety also warrants consideration. Accordingly, this study assessed the viability of the HEMn, CCD966SK, and RAW264.7 cells treated with the MLE at different concentrations, and the results are illustrated in Figure 3. Cell viability decreased as the MLE concentration increased. Among the tested cells, the HEMn cells were the most sensitive to the cytotoxic effects of the MLE, followed by the CCD966SK and RAW264.7 cells. Higher cell viability indicates that the extract is safer. Generally, cell viability of >80% implies that the extract is nontoxic [23]. In this study, the safe doses or concentrations of the MLE for the HEMn, CCD966SK, and RAW264.7 cells were found to be 300, 300, and 350 mg/L, respectively. The IC_{50} values derived for the cytotoxic effects of the MLE against the HEMn, CCD966SK, and RAW264.7 cells were 392.4 ± 5.1, 431.8 ± 7.6, and 482.3 ± 2.5 mg/L, respectively. According to these findings, the safety of the MAWE-derived MLE can be considered to be much superior to that of *M. quinquenervia* extract obtained through steam distillation [1], *M. leucadendron* extract obtained through hydrodistillation [17], ethanol leaf extract of *M. leucadendron* [24], and butanol extract of *M. Leucadendron* [16]. The low cytotoxicity of the MLE may be because the MAWE technique uses water as the extraction solvent, indicating the promising safety of MAWE for plant extraction.

2.4. Evaluation of Antioxidant Activity

A single indicator cannot be used for examining antioxidant activity/capacity, particularly for multifunctional or complex multiphase systems, because several variables affect antioxidant activity [25]. In addition, the MLE may contain hydrophilic and hydrophobic compounds. Hence, in this study, the TPC, total flavonoid content (TFC), 2,2-diphenyl-1-picrylhydrazyl (DPPH) radical scavenging activity, 2,2′-azino-bis(3-ethylbenzothiazoline-6-sulfonic acid) (ABTS) radical scavenging activity, and β-carotene bleaching (BCB) activity of the MLE were evaluated to provide comprehensive insights into the mechanisms underlying the antioxidant activity of the extract. The TPC and TFC of the MLE were 312.7 ± 5.2 mg GAE/g DW and 71.3 ± 0.2 mg rutin equivalents (RE)/g DW, respectively. As mentioned, the TPC of the MLE derived through the MAWE technique was superior to that derived through solvent extraction. Similarly, the TFC of the MLE was much superior to that of ethanol extract of *M. bracteate* (19.4 mg RE/g DW) [14]. Figure 4 presents the DPPH radical scavenging activity, ABTS radical scavenging activity, and BCB activity of the MLE.

These antioxidant activities of the MLE increased with the concentration of the extract. The IC_{50} values derived for DPPH scavenging, ABTS scavenging, and BCB activities were 132.6 ± 3.1, 78.6 ± 1.2, and 174.1 ± 5.2 mg/L, respectively, which were inferior to the values derived for the positive control (ascorbic acid: 18.6 mg/L for DPPH scavenging activity) and butylated hydroxytoluene (BHT) (82.6 mg/L for ABTS scavenging activity and 12.4 mg/L for BCB activity).

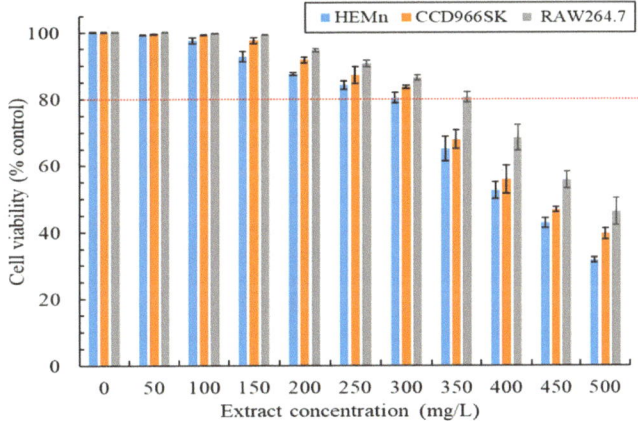

Figure 3. Cytotoxic effects of the MLE obtained through MAWE on HEMn, CCD966SK, and RAW264.7 cells after exposure for 24 h (operating conditions: LSR, 20 mL/g; extraction temperature, 80 °C; microwave irradiation time, 180 s; and microwave irradiation power, 700 W). Data are expressed as means and standard deviations of three independent experiments. The red dashed line indicates 80% cell viability.

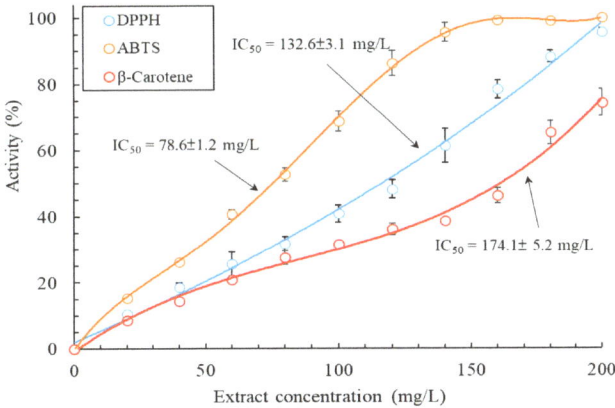

Figure 4. Antioxidant activity of the MLE obtained through MAWE (operating conditions: LSR, 20 mL/g; extraction temperature, 80 °C; microwave irradiation time, 180 s; and microwave irradiation power, 700 W). Data are expressed as means and standard deviations of three independent experiments.

The IC_{50} values derived for DPPH scavenging activity for the MLE were noted to be inferior to those observed for a polyphenol-rich fraction of *M. rugulosa* leaves (22.05 mg/L) [21], aqueous methanolic leaf extract of *M. subulata* (12.0 mg/L) [20], ethanol leaf extract of *M. 1eucadendron* (7.32 mg/L) [24], and butanol extract of *M. 1eucadendron*

(5.1 mg/L) [16]. However, the IC_{50} values derived for the MLE were significantly superior to those derived for hydrodistillation-based leaf extract of *M. leucadendron* (2400 mg/L) [18]. These results indicate that polar solvent extraction is beneficial for extracting substances with antioxidant activity from *Melaleuca* leaves. In addition, the IC_{50} values derived for BCB activity for the MLE were determined to be inferior to those derived for a polyphenol-rich fraction of *M. rugulosa* leaves (11.31 mg/L) [21] and aqueous methanolic leaf extract of *M. subulata* (4.31 mg/L) [20]. These findings demonstrate that the MLE derived through MAWE exhibited hydrophilic and hydrophobic antioxidant activities, indicating that the MAWE technique can be used in various applications. The chemical composition and relative content of the MLE are provided in the subsequent sections.

2.5. Evaluation of Antiaging Activity

Antioxidant activity/capacity is a comprehensive indicator of the physiological activities of plant extracts [25]. To assess the antiaging activity of extracts, the activity of aging-related enzymes must be evaluated. Matrix metalloproteinase-1 (MMP-1), collagenase, elastase, and hyaluronidase catalyze the degradation of the main components of the extracellular matrix. Thus, their activities in the skin indicate the current condition of the skin (wrinkles, elasticity, and even luster). Table 1 presents the IC_{50} values derived for the antiaging activities of the MLE. The IC_{50} values derived for the inhibitory effects of the MLE on MMP-1, collagenase, elastase, and hyaluronidase activities were 114.8 ± 8.1, 187.2 ± 5.4, 73.4 ± 2.4, and 60.4 ± 3.1 mg/L, respectively. The IC_{50} value derived for the inhibitory effect of the MLE on MMP-1 activity was inferior to that of EGCG (positive control; 42.3 ± 3.1 mg/L). Moreover, the IC_{50} value derived for the inhibitory effect of the MLE on collagenase activity was slightly inferior to those of EGCG (positive control; 113.7 ± 9.1 mg/L) and gallic acid (positive control; 126.8 ± 3.7 mg/L). The IC_{50} value derived for the inhibitory effect of the MLE on elastase activity was relatively close to those of EGCG (93.5 ± 7.2 mg/L) and oleanolic acid (positive control; 78.2 ± 1.8 mg/L). However, the IC_{50} value derived for the inhibitory effect of the MLE on hyaluronidase activity was significantly superior to those of the commercial antiaging agent epigallocatechin gallate (EGCG) (382 ± 12.6 mg/L) and oleanolic acid (98.6 ± 5.2 mg/L). The results may be attributed to the various active phytoconstituents of the MLE.

Table 1. Antiaging activity (IC_{50} values) of the MLE obtained through MAWE under optimal conditions *.

Tested Sample	MMP-1 Activity	Collagenase Activity	Elastase Activity	Hyaluronidase Activity
M. quinquenervia leaf extract	114.8 ± 8.1	187.2 ± 5.4	73.4 ± 2.4	60.4 ± 3.1
EGCG	42.3 ± 3.1	113.7 ± 9.1	93.5 ± 7.2	382 ± 12.6
Gallic acid	–	126.8 ± 3.7	–	–
Oleanolic acid	–	–	78.2 ± 1.8	98.6 ± 5.2

* LSR, 20 mL/g; extraction temperature, 80 °C; microwave irradiation time, 180 s; and microwave irradiation power, 700 W.

We compared our results with those of previous studies and observed that the IC_{50} values derived for the inhibitory effect of the MLE on collagenase activity were significantly superior to those of a polyphenol-rich fraction of *M. rugulosa* leaves (410 mg/L) [21] and aqueous methanolic leaf extract of *M. subulata* (382.16 mg/L) [20]. However, the IC_{50} values derived for the inhibitory effect of the MLE on elastase activity were slightly inferior to those of a polyphenol-rich fraction of *M. rugulosa* leaves (68.18 mg/L) [21] and aqueous methanolic leaf extract of *M. subulata* (59.18 mg/L) [20]. Although limited data on the antiaging activity of *M. quinquenervia* extracts were available for comparison, the data for the extracts from the same genus of *Melaleuca* indicate that the MLE obtained through MAWE has considerable potential for commercial application owing to its antiaging activity.

2.6. Evaluation of Anti-Inflammatory Activity

Extracts with anti-inflammatory activity could be a potential ingredient in skin care products, health products, and even pharmaceutical products. Accordingly, this study examined the anti-inflammatory activity of the MLE, and the results are displayed in Figure 5. Specifically, the inhibitory effects of the MLE on cyclooxygenase-1 (COX-1) activity, cyclooxygenase-2 (COX-2) activity, and tumor necrosis factor-α (TNF-α) production increased exponentially as the extract's concentration was increased. The inhibitory effects of the MLE at a concentration of 200 mg/L on COX-1 activity, COX-2 activity, and TNF-α production were 98.1 ± 0.8%, 100 ± 0.2%, and 92.3 ± 2%; this concentration was not cytotoxic to skin cells. In addition, the IC_{50} values derived for the inhibitory effects of the MLE on COX-1 activity, COX-2 activity, and TNF-α production were 27.4 ± 4.1, 8.5 ± 1.3, and 43.8 ± 2.6 mg/L, respectively. The inhibitory effects of the MLE were strongest for COX-2 activity. The IC_{50} value derived for the inhibitory effects of the anti-inflammatory drug indomethacin (positive control) on COX-1 activity, COX-2 activity, and TNF-α production were 5.2 ± 0.4, 43.6 ± 3.7, and 6.52 ± 0.8 mg/L, respectively. The anti-inflammatory activity of the MLE was slightly inferior to that of the highly purified polyphenol-rich fraction of *M. rugulosa* leaves [21]. Nevertheless, the anti-inflammatory activity of the MLE meets the requirements for commercial medicinal use.

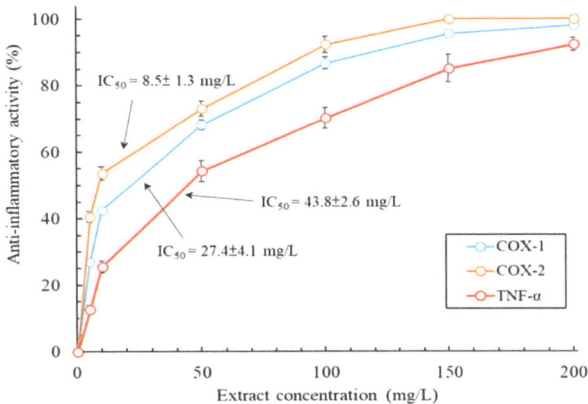

Figure 5. Anti-inflammatory activity of the MLE obtained through MAWE (operating conditions: LSR, 20 mL/g; extraction temperature, 80 °C; microwave irradiation time, 180 s; and microwave irradiation power, 700 W). Data are expressed as means and standard deviations of three independent experiments.

2.7. Evaluation of Antimicrobial Activity

In accordance with regulatory guidelines for USP 51 antimicrobial effectiveness testing and considering the microbes associated with skin diseases, this study evaluated the antimicrobial activity of the MLE. Table 2 lists the minimum inhibitory concentration (MIC) values derived for the activity of the MLE against *S. aureus*, *E. coli*, *P. aeruginosa*, and *C. acnes* strains and the minimum fungicidal concentration (MFC) values derived for the activity of the MLE against *C. albicans* and *A. brasiliensis* strains. The MIC and MFC values derived for the activity of the MLE against the tested bacterial and fungal strains were 64–128 and 128–256 mg/L, respectively. The concentrations at which the MLE exhibited antimicrobial activity were not cytotoxic to skin cells (Figure 3). The MLE exhibited stronger antibacterial activity against *S. aureus* than it did against *C. acnes*. This finding is consistent with that of Shakeel et al. (2021), who used essential oils extracted from *M. quinquenervia* [3]. The MLE exhibited strong antifungal activity against the *C. albicans* and *A. brasiliensis* strains. Valková et al. (2022) reported that essential oils extracted from *M. quinquenervia* leaves through steam distillation were effective against various *Penicillium* strains [4]. Notably, the antimicrobial activity of the MLE was significantly stronger than that of leaf extracts of similar species (*M. leucadendra*) [26–28]. These results indicate that the MLE obtained through MAWE has excellent and broad-spectrum antimicrobial activity. A possible reason for the excellent antimicrobial activity of the MLE is the use of microwaves in MAWE to improve the effectiveness and efficiency of cell breakdown [9], which could result in the release of various antimicrobial ingredients. These findings demonstrate that the MLE has favorable cosmeceutical properties. In addition to its high antimicrobial activity, the MLE has the potential to become a valuable ingredient in skin care products; its use in such products can ameliorate the necessity for the addition of artificial preservatives. In addition, it can be applied as an antimicrobial agent in cleaning products, the food industry, and even in medicine.

Table 2. Antimicrobial activity of the MLE obtained through MAWE under optimal conditions *.

Tested Sample	MIC				MFC	
	S. aureus	*E. coli*	*P. aeruginosa*	*C. acnes*	*C. albicans*	*A. brasiliensis*
M. quinquenervia leaf extract	64	64	128	128	256	128
Streptomycin	64	32	32	–	–	–
Erythromycin	–	–	–	8	–	–
Nystatin	–	–	–	–	32	16

* LSR, 20 mL/g; extraction temperature, 80 °C; microwave irradiation time, 180 s; and microwave irradiation power, 700 W.

2.8. Chemical Composition of MLE

This study subjected the MLE to gas chromatography–mass spectrometry (GC-MS) and high-performance liquid chromatography (HPLC) to determine its primary constituents and major phenolic compounds, respectively. Table 3 lists the primary constituents and their relative content associated with the volatile part of the MLE. This table lists only constituents whose relative content in the volatile part of the MLE was >0.5%; these constituents constituted 98.44% of the volatile part of the MLE. The volatile part of the MLE had a total of 24 components, and compounds with a relative content of >7% were as follows: 1,8-cineole (16.71%), viridiflorol (13.27%), α-pinene (10.61%), α-terpineol (10.12%), ledene (8.52%), limonene (7.94%), and β-pinene (7.52%). The main components in the MLE, especially 1,8-cineole, viridiflorol, and α-terpineol, were noted to be similar to those identified by Chao et al. (2017) [1], Valková et al. (2022) [4], and Vázquez et al. (2023) [29], but the relative content of these components was different from those identified by them. In this study, monoterpene was the dominant component (63.18%), followed by sesquiterpene (32.07%), aromatic compounds (2.45%), and hydrocarbons (0.74%). Studies have also reported that monoterpene compounds were the most abundant (25.42–62.4%) components of MLEs, regardless of the applied extraction technique [1,4,29].

The main phenolic compounds in plants are phenolic acids and flavonoids, which include flavonols, anthocyanins, and isoflavones [30]. Table 4 lists the main phenolic compounds and their relative content in the MLE. A total of 17 phenolic compounds were identified in the MLE: 7 phenolic acids and 10 flavonoids. The predominant compounds in the MLE were gallic acid, ellagic acid, 3-O-methylellagic acid, luteolin, kaempferol, and quercetin. The total concentrations of phenolic acids (304.1 mg GAE/g DW) and flavonoids (70.1 mg RE/g DW) were close to those of the TPC (312.7 mg GAE/g DW) and TFC (71.3 mg RE/g DW) of the MLE. These findings demonstrate that most phenolic compounds in the MLE were detected in this study. Gallic acid, ellagic acid, and 3-O-methylellagic acid were also detected in ethanol leaf extract of *M. quinquenervia* [2]. The possible roles of these components in the pharmacological activities of the MLE are discussed in the subsequent section.

Table 3. Composition of the main chemicals and their relative content in the volatile part of the MLE obtained through MAWE under optimal conditions *.

No.	RI	Chemical Compounds	Categories	Relative Content (%)
1	934	α-pinene	monoterpene	10.61
2	952	camphene	monoterpene	0.62
3	961	benzaldehyde	hydrocarbons	0.74
4	978	β-pinene	monoterpene	7.52
5	990	myrcene	monoterpene	0.52
6	1025	p-cymene	aromatic compounds	2.45
7	1029	o-cymene	monoterpene	0.75
8	1034	limonene	monoterpene	7.94
9	1042	1,8-cineole	monoterpene	16.71
10	1062	γ-terpinene	monoterpene	1.52
11	1092	terpinolene	monoterpene	2.14
12	1102	linalool	monoterpene	0.67
13	1178	4-terpineol	monoterpene	1.55
14	1205	α-terpineol	monoterpene	10.12
15	1352	α-terpinyl acetate	monoterpene	2.51
16	1428	caryophyllene	sesquiterpene	2.17
17	1465	α-humulene	sesquiterpene	1.05
18	1469	β-humulene	sesquiterpene	0.61
19	1501	ledene	sesquiterpene	8.52
20	1507	α-selinene	sesquiterpene	0.74
21	1567	nerolidol	sesquiterpene	1.81
22	1596	viridiflorol	sesquiterpene	13.27
23	1610	ledol	sesquiterpene	1.23
24	1615	globulol	sesquiterpene	2.67

* LSR, 20 mL/g; extraction temperature, 80 °C; microwave irradiation time, 180 s; and microwave irradiation power, 700 W.

Table 4. Composition of the main phenolic compounds and their content in the MLE obtained through MAWE under optimal conditions *.

Chemical Compounds	Contents
Phenolic acids (mg GAE/g DW)	
gallic acid	104.2
vanillic acid	10.2
caffeic acid	28.6
ferulic acid	8.1
rosmarinic acid	6.3
ellagic acid	88.1
3-O-methyl ellagic acid	58.6
Flavonoids (mg RE/g DW)	
rutin	1.7
luteolin	23.1
catechin	3.8
quercetin-3-O-glucuronopyranoside	2.8
kaempferol-3-O-glucoside	3.2
quercetin	13.8
apigenin	2.4
naringin	1.2
kaempferol	16.3
hesperidin	1.8

* LSR, 20 mL/g; extraction temperature, 80 °C; microwave irradiation time, 180 s; and microwave irradiation power, 700 W.

2.9. Molecular Docking Analysis

In this investigation, a total of 41 distinct compositions within MLE were identified. Previous studies indicated that extracts with a higher concentration of active components exhibited a more pronounced inhibitory effect on enzyme activities. Consequently, this study conducted a molecular docking analysis of the 10 main phytocompounds in the MLE against representative enzymes or proteins. The evaluation of whitening activity was conducted through the assessment of binding affinity between the compounds and tyrosinase. Similarly, antiaging activity was gauged by the binding affinities of the compounds with elastase, collagenase, hyaluronidase, and MMP-1. Anti-inflammatory activity was assessed based on the binding affinities with COX-1, COX-2, and TNF-α, while antimicrobial activity was evaluated through the binding affinities with tyrosyl-tRNA synthetase and sterol 14α-demethylase. Table 5 lists the analysis results. A low binding energy between compounds and proteins typically indicates a high binding affinity between compounds and proteins or high inhibitory effects of the compounds on enzymes [31]. In this study, luteolin exhibited the highest binding energy (-106.9 kcal/mol) against tyrosinase, followed by gallic acid and quercetin. Manzoor et al. (2019) revealed that luteolin exhibited strong anti-tyrosinase activity; it inhibited the expression of cyclic adenosine monophosphate and the activity of adenyl cyclase through the α-MSH pathway [32]. In the present study, ellagic acid exhibited the highest binding energies against elastase, hyaluronidase, and COX-1 (-96.5, -102.5, and -104.8 kcal/mol, respectively). Moon et al. (2018) reported that ellagic acid exhibited strong antiaging activity; it activated both TGF-β1 and Wnt signaling pathways [33]. Baradaran Rahimi et al. (2020) also revealed that ellagic acid was the major active compound in pomegranate with antiaging and anti-inflammatory activities [34]. In the present study, quercetin exhibited the highest binding energies against collagenase, MMP-1, and tyrosyl-tRNA synthetase (-98.8, -109.2, and -118.2 kcal/mol, respectively). Quercetin is regarded as an excellent antiaging active ingredient in plant extracts [35], and it possesses broad-spectrum antibacterial properties (an antibacterial agent inhibiting tyrosyl-tRNA synthetase) [36]. In the present study, kaempferol exhibited the highest binding energies against COX-2 and TNF-α (-109.4 and -113.8 kcal/mol, respectively). Its anti-inflammatory properties are well documented [37]. Moreover, 3-O-methylellagic acid exhibited the highest binding energy (-105.3 kcal/mol) against sterol 14α-demethylase, thus demonstrating high antifungal activity [38].

Figure 6 illustrates the docking interactions and docking complexes formed by selected phytochemical compounds in the MLE against tested enzymes. This study investigated the docking interactions of luteolin, ellagic acid, kaempferol, quercetin, and 3-O-methylellagic acid against tyrosinase, elastase, COX-2, tyrosyl-tRNA synthetase, and sterol 14α-demethylase, respectively. Among the compounds in the MLE, luteolin exhibited antityrosinase properties owing to the formation of H bonds with Ser106, Cys101, Phe105, Gly103, His100, Thr69, and Val68; π-cation bonds with Arg114; unfavorable acceptor–acceptor interactions with Glu66; π-anion bonds with Glu451; alkyl bonds with Pro446; and π-alkyl bonds with Pro445 and Cys101 (Figure 6A). Luteolin formed various molecular bonds with tyrosinase, facilitating the formation of protein–ligand complexes. Ellagic acid exhibited anti-elastase properties owing to the formation of H bonds with I1e8, Leu20, Ile22, Ala99, Val99, Asp98, and Asp97; π-σ bonds with Ile19; and π-alkyl bonds with Ile21, Ala99, and Ile19 (Figure 6B). Kaempferol exhibited anti-COX-2 properties owing to the formation of H bonds with His207, π-π bonds with His386, and π-alkyl bonds with Ala202 (Figure 6C). Quercetin exhibited anti-tyrosyl-tRNA synthetase activity mainly owing to the formation of H bonds with Arg86, Lys82, Tyr169, Thr73, Asn123, Tyr34, Asp176, and Gln195 and the formation of π-anion bonds with Asp78 (Figure 6D). In addition, 3-O-methylellagic acid exhibited anti-sterol 14α-demethylase activity owing to the formation of H bonds with Lys226, Thr229, Met508, Val509, Pro230, and Ser507; π-anion bonds with Asp225; π-σ bonds with Leu511; and π-π bonds with His310 (Figure 6E). On the basis of these results, we confirmed the relationship between the various physiological activities of the MLE and its active constituents. Nonetheless, it is important to note that these findings will require validation through future experiments utilizing pure chemical substances to inhibit the enzymes.

Table 5. Results of the molecular docking analysis of the 10 main phytocompounds in the MLE obtained through MAWE under optimal conditions *.

	Gallic Acid	Ellagic Acid	3-O-Methylellagic Acid	Luteolin	Quercetin	Kaempferol	α-Pinene	1,8-Cineole	α-Terpineol	Viridiflorol
	Total binding energy (kcal/mol)									
tyrosinase	−101.6	−92.3	−94.2	−106.9	−98.2	−96.2	−49.7	−52.3	−62.9	−83.2
elastase	−80.7	−96.5	−83.3	−88.9	−87.4	−91.8	−55.5	−62.3	−61.3	−64.7
collagenase	−78.4	−91.8	−92.6	−94.2	−98.8	−95.1	−56.6	−60.7	−69.8	−73.6
hyaluronidase	−78.0	−102.5	−96.2	−90.1	−99.9	−95.6	−55.2	−57.9	−59.2	−64.6
MMP-1	−93.3	−95.6	−87.9	−98.6	−109.2	−103.6	−56.3	−55.7	−70.1	−72.6
COX-1	−85.8	−104.8	−100.2	−92.3	−96.8	−93.8	−57.6	−60.5	−66.5	−77.5
COX-2	−81.7	−103.4	−104.1	−102.1	−101.7	−109.4	−55.6	−57.4	−70.4	−72.8
TNF-α	−80.3	−100.7	−102.6	−90.6	−90.4	−113.8	−51.2	−60.9	−65.6	−77.3
tyrosyl-tRNA synthetase	−82.0	−110.3	−108.0	−95.5	−118.2	−88.9	−49.2	−54.8	−57.5	−66.7
sterol 14α-demethylase	−88.1	−95.8	−105.3	−98.4	−98.1	−90.2	−50.3	−54.1	−62.3	−67.8

* LSR, 20 mL/g; extraction temperature, 80 °C; microwave irradiation time, 180 s; and microwave irradiation power, 700 W.

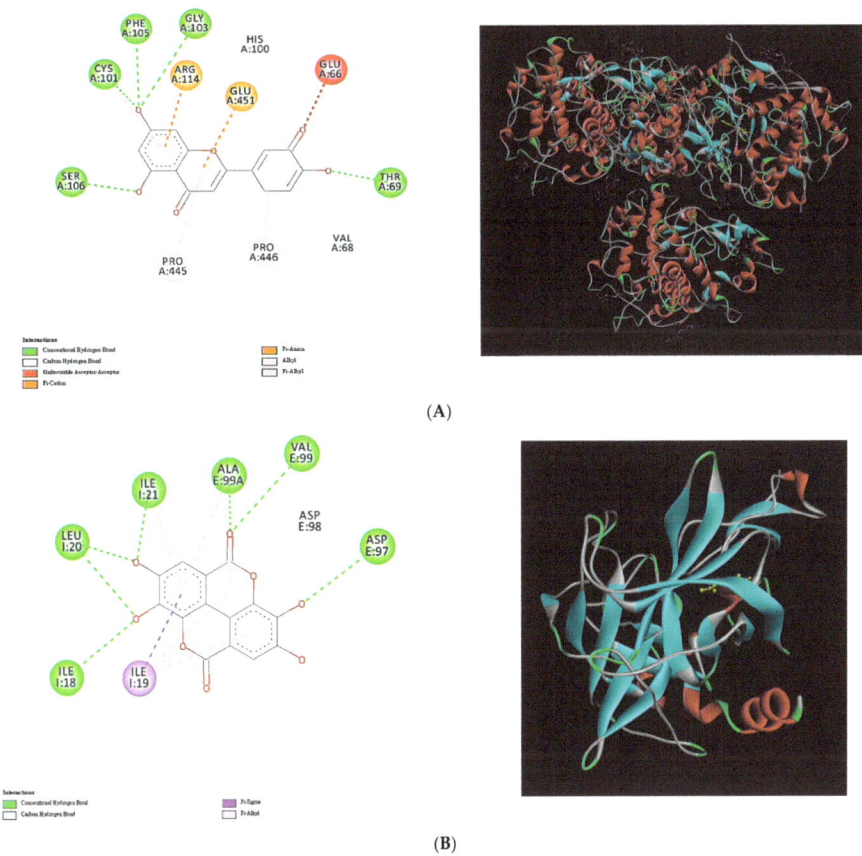

Figure 6. *Cont.*

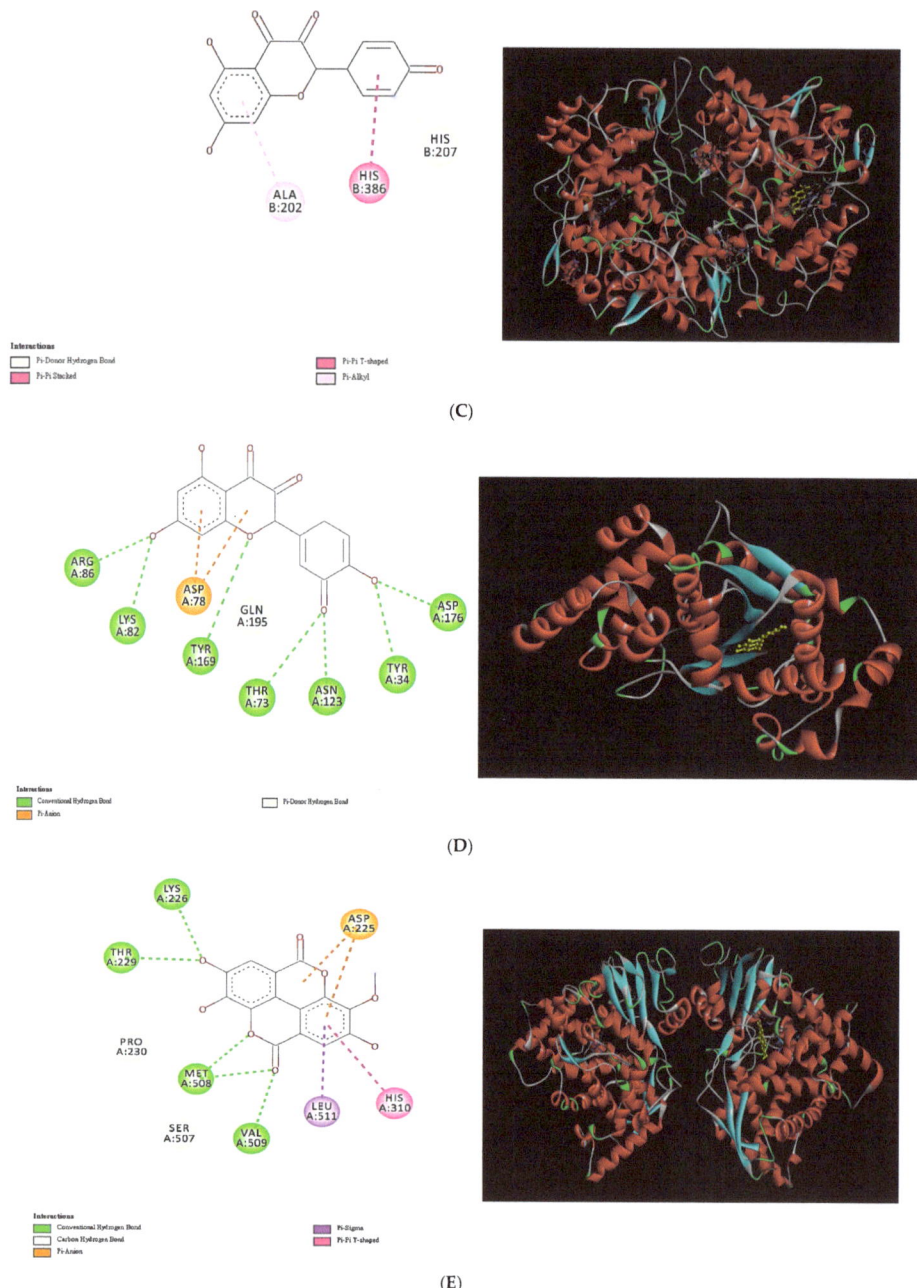

Figure 6. Docking interactions and docking complex of the optimal phytochemical compounds in the MLE (obtained through MAWE) against tested enzymes or proteins: (**A**) molecular docking of the interactions between luteolin and tyrosinase; (**B**) molecular docking of the interactions between ellagic acid and elastase; (**C**) molecular docking of the interactions between kaempferol and COX-2; (**D**) molecular docking of the interactions between quercetin and tyrosyl-tRNA synthetase; (**E**) molecular docking of the interactions between 3-*O*-methylellagic acid and sterol 14α-demethylase.

3. Materials and Methods

3.1. Plant Material and Extraction Procedure

M. quinquenervia leaves were collected from Nangang District, Taipei City, Taiwan (25°03′22″ N, 121°60′96″ E); the collected leaves were identified by Professor Bau-Yuan Hu. Voucher specimens (accession no. 20230215) were deposited in the herbarium of China University of Science and Technology, Taiwan. The collected leaves were washed with distilled water and dried in an oven (Eyela, Tokyo, Japan) at 50 °C for 2 h. The dried leaves were pulverized to a powder, which was passed through a 0.5 mm mesh. A sample of the powder (10 g) was extracted with various volumes of distilled water at LSRs of 10–30 mL/g. This extraction process was conducted in a microwave digestion instrument (SINEO, Shanghai Sineo Microwave Chemistry Technology Co., Ltd., Shanghai, China) operated at a power/frequency of 700 W/2.45 GHz; the extraction temperature was set to 80 °C and the irradiation time was 120 s. The optimal irradiation time was determined by extracting the sample with distilled water at an LSR of 20 mL/g in the microwave digestion instrument operated at an extraction temperature of 80 °C for various irradiation periods (60–300 s). The crude extracts were filtered through a Whatman filter (0.45 μm) and lyophilized using a shelf freeze-dryer (Uniss Corp., Taipei City, Taiwan) for the subsequent analysis of physiological activities.

3.2. Microbial Strains, Cells, and Reagents

Four bacterial strains, namely, ATCC 6538 (*Staphylococcus aureus*), ATCC 8739 (*Escherichia coli*), ATCC 9027 (*Pseudomonas aeruginosa*), and ATCC 6919 (*Cutibacterium acnes*), and two fungal strains, namely, ATCC 10231 (*Candida albicans*) and ATCC 16404 (*Aspergillus brasiliensis*), were employed for the current study; they were purchased from the Bioresource Collection and Research Center (BCRC; Hsinchu, Taiwan). The *S. aureus*, *E. coli*, and *P. aeruginosa* strains were cultured in tryptic soy broth (TSB) (DIFCO, Tucker, GA, USA) under aerobic conditions. The *C. acnes* strain was cultured in brain heart infusion broth (BHI) (DIFCO, Tucker, GA, USA) under anaerobic conditions. The *C. albicans* and *A. brasiliensis* strains were cultured in potato dextrose broth (PDB) (DIFCO, Tucker, GA, USA). The human skin fibroblast cell line CCD966SK (BCRC 60153) and the murine macrophage cell line RAW264.7 (BCRC 60001) were obtained from the BCRC. Normal human primary epidermal melanocytes neonatal (HEMn; C-102-5C) were obtained from Cascade Biologics (Portland, OR, USA). CCD966SK and HEMn cells were cultured in minimum essential medium containing 10% fetal bovine serum (FBS) and in Medium 254 with human melanocyte growth supplement (HMGS) (Thermo, Waltham, MA, USA), respectively. RAW264.7 cells were cultured in RPMI 1640 supplemented with 10% FBS and 1% penicillin–streptomycin (Thermo, Waltham, MA, USA). The chemicals used in this study were of analytical grade (purity $\geq$99.2%) and were purchased from Sigma-Aldrich (St. Louis, MO, USA).

3.3. Evaluation of Antioxidant Activity

The TPC of the derived MLE was determined using the Folin–Ciocalteu method, and TPC is expressed as gallic acid equivalents, as estimated using the method of Kujala et al. (2000) [39]. The TFC of the extracts was determined using the aluminum chloride colorimetric method [40] and is expressed as milligrams of rutin equivalents per gram of dry weight.

The antioxidant activity of the MLE was assessed in accordance with the protocols of Wu et al. (2018), Merchán-Arenas et al. (2011), and Lee et al. (2012) by using the DPPH free radical, ABTS free radical, and BCB assays, respectively [41–43]. For the DPPH, ABTS, and BCB assays, absorbance was recorded at 517, 734, and 470 nm on an ultraviolet–visible spectrophotometer (UV-2600i, Shimadzu, Kyoto, Japan), respectively. Ascorbic acid was used as a positive control for the DPPH assay. BHT was used as a positive control for the ABTS and BCB assays.

3.4. Extracellular and Intracellular Antityrosinase Activity

The extracellular antityrosinase activity of the MLE was assessed using the method of Zheng et al. (2012) [44]. In brief, the freeze-dried leaf extract was dissolved in dimethyl sulfoxide (DMSO) to obtain an MLE with concentrations of 0–200 mg/L. Moreover, 30 µL of the sample was mixed with 970 µL of 0.05 mM phosphate-buffered saline (PBS). Subsequently, 1 mL of 100 mg/L L-tyrosine and 1 mL of 350 U/mL mushroom tyrosinase solution were added to the sample and reacted in the dark for 20 min. After completion of the reaction, the absorbance of the solution was measured at 490 nm. Positive controls were α-arbutin and kojic acid, a commercial whitening agent. The antityrosinase activity of the MLE was calculated using the following formula:

$$\text{Antityrosinase activity (\%)} = \frac{[(A - B) - (C - D)]}{(A - B)} \times 100 \quad (1)$$

where A is the OD_{490} of the control (without the MLE), B is the OD_{490} of the blank of A (without the MLE or tyrosinase), C is the OD_{490} of the experimental group (with the MLE and tyrosinase), and D is the OD_{490} of the blank of C (without tyrosinase).

The intracellular antityrosinase activity of the MLE and the melanin content of the HEMn cells were analyzed using the method of Wu et al. (2018) [41]. In brief, the HEMn cells were seeded in 24-well plates at a density of 3×10^5 cells/well and cultured with 2.5 mL Medium 254 supplemented with 1% HMGS at 37 °C under 5% CO_2. After 24 h of cultivation, the cells were treated with the MLE (0–200 mg/L) for another 24 h. Subsequently, the cells were washed with PBS, lysed with lysis buffer, and sonicated using an ultrasonic sonicator (Qsonica, Newtown, CT, USA). To determine the intracellular antityrosinase activity of the MLE, the lysate was collected and reacted with 1.25 mM l-dopa for 3 h, and the absorbance of the solution was measured at 475 nm on an Epoch ELISA reader (BioTek Instruments, Santa Clara, CA, USA). To measure the melanin content of the HEMn cells, the sonicated product was further centrifuged at $8000 \times g$ for 10 min by using a micro-ultracentrifuge (Thermo Fisher Scientific, Waltham, MA, USA). The pellets were then dissolved in 1 N NaOH solution containing 10% DMSO and reacted for 30 min at 80 °C. Absorbance was measured at 405 nm on the ELISA reader. We quantified the melanin content of the HEMn cells by using a calibration curve plotted with OD_{405} values of synthetic melanin versus varying concentrations of synthetic melanin.

3.5. Cytotoxicity Assay

The viability of the HEMn, CCD966SK, and RAW264.7 cells was determined using the 3-(4,5-dimethylthiazol-2-yl)-2,5-diphenyltetrazolium bromide (MTT) colorimetric assay [45]. These cells were seeded in 96-well microplates at a density of 5×10^5 cells/well and cultured at 37 °C under 5% CO_2. The next day, the culture medium was removed, and 0–500 mg/L MLE in fresh medium was added to the wells. After 24 h of incubation, MTT solution (5 mg/mL in PBS) was added to the cells and reacted for 2 h. Subsequently, 0.1 mL of DMSO was added to each well to stop the reaction and solubilize the resulting formazan crystals. The absorbance of the final solution was measured at 570 nm on the ELISA reader. Cell viability (%) was estimated as the percentage of the absorbance of the sample with the MLE relative to the absorbance of the blank (without the MLE).

3.6. Evaluation of Antiaging Activity

The antiaging activity of the MLE was evaluated by assessing its inhibitory effects on MMP-1, collagenase, elastase, and hyaluronidase activities. The activity of MMP-1 in the CCD966SK cells was assessed using the human MMP-1 ELISA kit (RayBiotech, Norcross, GA, USA), as described by Chen et al. (2022) [46]. In brief, CCD-966SK cells were cultured in 96-well plates for 24 h in a 5% CO_2 atmosphere at 37 °C. Subsequently, the culture media were discarded, and varying concentrations of MLE were introduced to each well for an additional 24 h. The ELISA kit facilitated the mixing of components, and the reaction was conducted at 25 °C. After a 2 h incubation period, the resultant mixture was analyzed spectrophotometrically at 420 nm. Additionally, extracellular collagenase activity was assessed using the modified fluorogenic DQ gelatin degradation assay, in accordance with the method of Li et al. (2020) [47]. Briefly, varying concentrations of MLE were added to a 96-well plate, along with 1 U/mL collagenase (100 μL per well) and 15 μg/mL DQ gelatin. The mixture was allowed to react for 20 min, after which the absorbance was recorded at 485 nm and 530 nm (excitation and emission wavelengths, respectively) to assess the rate of gelatin proteolysis. Extracellular elastase activity was evaluated using porcine pancreatic elastase as a model enzyme, with N-succinyl-Ala-Ala-Pro-Val-p-nitroanilide (Suc-Ala) serving as the substrate [41]. In this procedure, 50 μL of MLE at varying concentrations was combined with 125 μL of 7 mM Suc-Ala (pH 8.0, prepared in 0.1 M Tris–Cl buffer) and incubated for 15 min at 25 °C in a 96-well plate. Following this, 25 μL of 0.3 U/mL neutrophil elastase was added, and the reaction continued for an additional 15 min. The absorbance was subsequently measured at 405 nm using the Epoch ELISA reader. Finally, extracellular hyaluronidase activity was evaluated using the Hyaluronidase Enzymatic Assay Kit (Sigma-Aldrich, St. Louis, MO, USA), employing a spectrophotometric method with hyaluronidase as the enzyme and hyaluronic acid as the substrate [48]. The hyaluronidase activity was quantified by measuring the absorbance at 600 nm with the Epoch ELISA reader. EGCG, gallic acid, or oleanolic acid was used as a positive control in the antiaging assay if required.

3.7. Evaluation of Anti-Inflammatory Activity

The anti-inflammatory activity of the MLE was evaluated through both enzyme-based and cell-based assays. The inhibitory effects of the MLE on the activities of the proinflammatory enzymes COX-1 and COX-2 were assessed using a colorimetric COX inhibitor screening assay kit (Cayman, Ann Arbor, MI, USA) according to the manufacturer's instructions. In the cell-based assay, the anti-inflammatory activity of the MLE in RAW264.7 cells was examined by inducing inflammatory responses in the cells by using a lipopolysaccharide. Moreover, TNF-α was used as an indicator of anti-inflammatory activity, and its levels were determined using commercial ELISA kits (R&D systems Inc., Minneapolis, MN, USA) according to the manufacturer's instructions. Indomethacin was used as a positive control in the assays.

3.8. Evaluation of Antimicrobial Activity

The MIC and MFC of the MLE were employed as indicators of its antibacterial and antifungal activities, respectively. The antibacterial activity of the MLE in the *E. coli*, *S. aureus*, *P. aeruginosa*, and *C. acnes* strains was evaluated using the tube dilution method [49]. In brief, a test tube was prepared by combining 1 mL of MLE, with its concentration being adjusted as necessary; 2 mL of TSB for *E. coli*, *S. aureus*, and *P. aeruginosa* or BHI for *C. acnes*; and 1 mL of inoculum at a concentration of 2×10^7 cfu/mL. The incubation conditions varied according to the bacterial species; the test tubes containing *E. coli*, *S. aureus*, or *P. aeruginosa* were incubated for 18 h at 37 °C under aerobic conditions, while the test tube containing *C. acnes* was incubated for 48 h at 37 °C under anaerobic conditions. MIC refers to the lowest concentration of a chemical/compound that prevents cell growth. The antifungal activity of the MLE in the *C. albicans* and *A. brasiliensis* strains was determined using the conventional plate count method [46]. In brief, a mixture was prepared by combining 1 mL of MLE, with its concentration being adjusted, 1 mL of inoculum at a concentration of 8×10^6 cfu/mL, and 100 mL of PDB in a conical flask. This mixture was subsequently incubated for 5 days at 25 °C. MFC refers to the lowest extract concentration of a chemical/compound at which no visible growth of the subculture occurs. Streptomycin, a broad-spectrum bactericidal agent, was used as a positive control against *E. coli*, *S. aureus*, and *P. aeruginosa*. Erythromycin, which is frequently used for acne vulgaris treatment, was used as a positive control against *C. acnes*. Nystatin, an antifungal agent, was used as a positive control against *C. albicans* and *A. brasiliensis*.

3.9. Quantification of Chemical Compounds in MLE

The primary chemical compounds in the MLE were analyzed through GC-MS (Shimadzu, Kyoto, Japan), in accordance with the method of Chao et al. (2017) [1]. In brief, GC-MS was applied in the EI mode at 70 eV with a mass range of m/z 35–500. A DB-5 fused capillary column (30 m × 0.25 mm i.d.) with a thickness of 0.25 µm for the coated material was used. The injector and detector temperatures were set at 280 and 300 °C, respectively. The temperature program was as follows: the temperature was initially maintained at 35 °C for 3 min, increased at a rate of 5 °C/min to 300 °C, and then maintained at 300 °C for 10 min. The flow rate of the carrier gas (helium) was maintained at 1.5 mL/min. The chromatographic retention index (RI) was calculated by referring to a homologous series of n-alkanes (C_6–C_{22}), and GC-MS was conducted in accordance with the previously specified conditions. The chemical compounds were quantified using the percentage relative peak area and identified by referring to the chromatographic peaks in a standard library from the National Institute of Standards and Technology (NIST 20) MS spectral database. The chemical compounds in the MLE were confirmed by comparing their RI values with those of authentic compounds.

The main phenolic acids and flavonoids in the MLE were analyzed through HPLC (Hitachi, Tokyo, Japan), in accordance with the method of Trabelsi et al. (2013) [50], with slight modifications. In brief, a prontosil C18 column (250 mm × 4.0 mm × 5 µm) was used for HPLC. The mobile phase was composed of two solvents: 0.025% trifluoroacetic acid in H_2O (A) and acetonitrile (B). The elution program at a flow rate of 1 mL/min was as follows: 15% B, followed by 20% B at 5 min, 26% B at 10 min, 38% B at 15 min, 50% B at 20 min, 100% B at 25 min, and 15% B at 30 min. The injection volume was 20 µL and peaks in the chromatogram were monitored at 280 nm. The peaks were identified through comparison with those of standard samples under the same conditions. To compare the total concentrations of phenolic acids and flavonoids identified by HPLC with TPC and TFC in the MLE, the concentrations of individual phenolic acids and flavonoids were initially quantified using standard curves that correlate the concentrations of standard samples with their corresponding peak area values. Following this, the concentrations of each phenolic acid and flavonoid were assessed utilizing the Folin–Ciocalteu method [39] and the aluminum chloride colorimetric method [40].

3.10. Molecular Docking Study

To examine the possible mechanisms underlying the various physiological activities of the MLE, the 10 main phytocompounds found in the MLE were subjected to a molecular docking study. The molecular docking study was conducted using iGEMDOCK V2.1 software with the following parameters: population size = 200, generations = 70, and number of solutions = 2. The 3D chemical structures of selected compounds were docked against the active sites of tyrosinase, elastase, collagenase, hyaluronidase, MMP-1, COX-1, COX-2, TNF-α, tyrosyl-tRNA synthetase, and sterol 14α-demethylase (CYP51). The 3D chemical structures of selected compounds and the crystal structures of target proteins were obtained from the PubChem database and the Protein Data Bank database, respectively. The best match was chosen on the basis of the total binding energy, which is the sum of the energy of the hydrogen bond, van der Waals forces, and electrostatic interactions.

3.11. Statistical Analysis

The study data were assessed using a one-way analysis of variance, followed by Duncan's multiple range test. The data are expressed as means ± standard deviations ($n = 3$). A p value of <0.05 was considered indicative of statistical significance. The IC_{50} was calculated using GraphPad Prism software (version 9) (GraphPad Software, San Diego, CA, USA). All statistical analyses were performed using IBM SPSS (version 26) (IBM Corp., Armonk, NY, USA).

4. Conclusions

This study applied MAWE to produce an MLE and comprehensively demonstrated its physiological or pharmacological activities. The MLE obtained under optimal extraction conditions exhibited a higher yield and superior physiological activities compared with extracts obtained through conventional extraction techniques. Moreover, our study is the first to reveal the antiaging and anti-inflammatory activities of the MAWE-derived MLE. The MAWE technique is environmentally friendly because it uses water as the extraction solvent. Furthermore, the MLE was noted to exhibit low cytotoxicity and broad-spectrum physiological activities, indicating that it has potential for use in cosmetics, food, medicine, and other products. A molecular docking analysis of the primary constituents of the MLE revealed that luteolin exhibited optimal skin-whitening effects, ellagic acid exhibited excellent antiaging and anti-inflammatory activities, quercetin exhibited favorable antiaging and antibacterial activities, kaempferol exhibited excellent anti-inflammatory activity, and 3-O-methylellagic acid exhibited the highest antifungal activity. Thus, these potential active ingredients can be further extracted and purified for use in various industries.

Author Contributions: Conceptualization, Y.-C.C. (Ying-Chien Chung) and H.-W.L.; methodology, J.-Y.L.; validation, Y.-C.C. (Yu-Chi Chang), T.-K.L. and Y.-L.L.; formal analysis, T.-K.L.; investigation, T.-K.L., J.-Y.L., Y.-C.C. (Yu-Chi Chang) and H.-W.L.; resources, T.-K.L. and J.-Y.L.; data curation, Y.-L.L. and Y.-C.C. (Yu-Chi Chang); writing—original draft preparation, T.-K.L., Y.-C.C. (Ying-Chien Chung) and H.-W.L.; writing—review and editing, Y.-C.C. (Ying-Chien Chung) and H.-W.L.; supervision, Y.-C.C. (Ying-Chien Chung) and H.-W.L.; funding acquisition, Y.-C.C. (Ying-Chien Chung). All authors have read and agreed to the published version of the manuscript.

Funding: This research was funded by the National Science and Technology Council, grant number NSTC 112-2622-E-157-001 and NSTC 112-2313-B-157-001-MY3.

Data Availability Statement: All data generated or analyzed during this study are included in this published article.

Acknowledgments: The authors would like to thank Chi-Hsiang Tang for their help with analytical measurements.

Conflicts of Interest: The authors declare no conflicts of interest.

References

1. Chao, W.W.; Su, C.C.; Peng, H.Y.; Chou, S.T. *Melaleuca quinquenervia* essential oil inhibits α-melanocyte-stimulating hormone-induced melanin production and oxidative stress in B16 melanoma cells. *Phytomedicine* **2017**, *34*, 191–201. [CrossRef] [PubMed]
2. Moharram, F.A.; Marzouk, M.S.; El-Toumy, S.A.A.; Ahmed, A.A.E.; Aboutabl, E.A. Polyphenols of *Melaleuca quinquenervia* leaves—Pharmacological studies of grandinin. *Phytother. Res.* **2003**, *17*, 767–773. [CrossRef] [PubMed]
3. Shakeel, F.; Salem-Bekhit, M.M.; Haq, N.; Alshehri, S. Nanoemulsification improves the pharmaceutical properties and bioactivities of niaouli essential oil (*Melaleuca quinquenervia* L.). *Molecules* **2021**, *26*, 4750. [CrossRef] [PubMed]
4. Valková, V.; Ďuranová, H.; Vukovic, N.L.; Vukic, M.; Kluz, M.; Kačániová, M. Assessment of chemical composition and anti-*Penicillium* activity of vapours of essential oils from *Abies alba* and two *Melaleuca* species in food model systems. *Molecules* **2022**, *27*, 3101. [CrossRef]
5. Aziz, Z.A.A.; Ahmad, A.; Setapar, S.H.M.; Karakucuk, A.; Azim, M.M.; Lokhat, D.; Rafatullah, M.; Ganash, M.; Kamal, M.A.; Ashraf, G.M. Essential oils: Extraction techniques, pharmaceutical and therapeutic potential-a review. *Curr. Drug. Metab.* **2018**, *19*, 1100–1110. [CrossRef]
6. Wang, Y.; Yan, M.; Qin, R.; Gong, Y. Enzymolysis-microwave-assisted hydrodistillation for extraction of volatile oil from *Atractylodes chinensis* and its hypoglycemic activity in vitro. *J. AOAC Int.* **2021**, *104*, 1196–1205. [CrossRef]
7. Yilmaz, E.; Güneşer, B.A. Cold pressed versus solvent extracted lemon (*Citrus limon* L.) seed oils: Yield and properties. *J. Food Sci. Technol.* **2017**, *54*, 1891–1900. [CrossRef]
8. Paibon, W.; Yimnoi, C.A.; Tembab, N.; Boonlue, W.; Jampachaisri, K.; Nuengchamnong, N.; Waranuch, N.; Ingkaninan, K. Comparison and evaluation of volatile oils from three different extraction methods for some Thai fragrant flowers. *Int. J. Cosmet. Sci.* **2011**, *33*, 150–156. [CrossRef]
9. Haro-González, J.N.; Castillo-Herrera, G.A.; Martínez-Velázquez, M.; Espinosa-Andrews, H. Clove essential oil (*Syzygium aromaticum* L. Myrtaceae): Extraction, chemical composition, food applications, and essential bioactivity for human health. *Molecules* **2021**, *26*, 6387. [CrossRef]

10. Bagade, S.B.; Patil, M. Recent advances in microwave assisted extraction of bioactive compounds from complex herbal samples: A review. *Crit. Rev. Anal. Chem.* **2021**, *51*, 138–149. [CrossRef]
11. Cock, I.E.; Winnett, V.; Sirdaarta, J.; Matthews, B. The potential of selected Australian medicinal plants with anti-*Proteus* activity for the treatment and prevention of rheumatoid arthritis. *Pharmacogn Mag.* **2015**, *11* (Suppl. S1), S190–S208. [CrossRef] [PubMed]
12. Li, X.; Wu, X.; Huang, L. Correlation between antioxidant activities and phenolic contents of radix *Angelicae sinensis* (Danggui). *Molecules* **2009**, *14*, 5349–5361. [CrossRef] [PubMed]
13. Al-Abd, N.M.; Mohamed Nor, Z.; Mansor, M.; Azhar, F.; Hasan, M.S.; Kassim, M. Antioxidant, antibacterial activity, and phytochemical characterization of *Melaleuca cajuputi* extract. *BMC Complement. Altern. Med.* **2015**, *15*, 385. [CrossRef] [PubMed]
14. Hou, W.; Zhang, W.; Chen, G.; Luo, Y. Optimization of extraction conditions for maximal phenolic, flavonoid and antioxidant activity from *Melaleuca bracteata* leaves using the response surface methodology. *PLoS ONE* **2016**, *11*, e0162139. [CrossRef] [PubMed]
15. Bianchini Silva, L.S.; Perasoli, F.B.; Carvalho, K.V.; Vieira, K.M.; Paz Lopes, M.T.; Bianco de Souza, G.H.; Henrique Dos Santos, O.D.; Freitas, K.M. *Melaleuca leucadendron* (L.) flower extract exhibits antioxidant and photoprotective activities in human keratinocytes exposed to ultraviolet B radiation. *Free Radic. Biol. Med.* **2020**, *159*, 54–65. [CrossRef] [PubMed]
16. Surh, J.; Yun, J.M. Antioxidant and anti-inflammatory activities of butanol extract of *Melaleuca leucadendron* L. *Prev. Nutr. Food Sci.* **2012**, *17*, 22–28. [CrossRef]
17. Zhang, J.; Wu, H.; Jiang, D.; Yang, Y.; Tang, W.; Xu, K. The antifungal activity of essential oil from *Melaleuca leucadendra* (L.) L. grown in China and its synergistic effects with conventional antibiotics against *Candida*. *Nat. Prod. Res.* **2019**, *33*, 2545–2548. [CrossRef]
18. Pino, J.A.; Regalado, E.L.; Rodríguez, J.L.; Fernández, M.D. Phytochemical analysis and in vitro free-radical-scavenging activities of the essential oils from leaf and fruit of *Melaleuca leucadendra* L. *Chem. Biodivers.* **2010**, *7*, 2281–2288. [CrossRef]
19. An, N.T.G.; Huong, L.T.; Satyal, P.; Tai, T.A.; Dai, D.N.; Hung, N.H.; Ngoc, N.T.B.; Setzer, W.N. Mosquito larvicidal activity, antimicrobial activity, and chemical compositions of essential oils from four species of *Myrtaceae* from central Vietnam. *Plants* **2020**, *9*, 544. [CrossRef]
20. Mady, M.S.; Elsayed, H.E.; El-Sayed, E.K.; Hussein, A.A.; Ebrahim, H.Y.; Moharram, F.A. Polyphenolic profile and ethno pharmacological activities of *Callistemon subulatus* (Cheel) Craven leaves cultivated in Egypt. *J. Ethnopharmacol.* **2022**, *284*, 114698. [CrossRef]
21. Ebrahim, H.Y.; Mady, M.S.; Atya, H.B.; Ali, S.A.; Elsayed, H.E.; Moharram, F.A. *Melaleuca rugulosa* (Link) Craven Tannins: Appraisal of anti-inflammatory, radical scavenging activities, and molecular modeling studies. *J. Ethnopharmacol.* **2022**, *298*, 115596. [CrossRef] [PubMed]
22. Wang, G.H.; Lin, Y.M.; Kuo, J.T.; Lin, C.P.; Chang, C.F.; Hsieh, M.C.; Cheng, C.Y.; Chung, Y.C. Comparison of biofunctional activity of *Asparagus cochinchinensis* (Lour.) Merr. extract before and after fermentation with *Aspergillus oryzae*. *J. Biosci. Bioeng.* **2019**, *127*, 59–65. [CrossRef] [PubMed]
23. López-García, J.; Lehocký, M.; Humpolíček, P.; Sáha, P. HaCaT keratinocytes response on antimicrobial atelocollagen substrates: Extent of cytotoxicity, cell viability and proliferation. *J. Funct. Biomater.* **2014**, *5*, 43–57. [CrossRef] [PubMed]
24. Hashim, A.N.; Swilam, N.F.; Moustafa, E.S.; Bakry, S.M.; Labib, R.M.; Barakat, H.H.; Singab, A.B.; Linscheid, M.W.; Nawwar, M.A. A cytotoxic flavonol glycoside from *Melaleuca leucadendra* leaves extract with immunostimulant activity. *Pharmazie*. **2018**, *73*, 61–64.
25. Capanoglu, E.; Cekic, S.D.; Baskan, K.S.; Avan, A.N.; Uzunboy, S.; Apak, R. Antioxidant activity and capacity measurement. In *Plant Antioxidants and Health*; Reference Series in Phytochemistry; Ekiert, H.M., Ramawat, K.G., Arora, J., Eds.; Springer: Cham, Switzerland, 2022; pp. 709–773.
26. Bautista-Silva, J.P.; Seibert, J.B.; Amparo, T.R.; Rodrigues, I.V.; Teixeira, L.F.M.; Souza, G.H.B.; Dos Santos, O.D.H. *Melaleuca leucadendra* essential oil promotes loss of cell membrane and wall integrity and inhibits bacterial growth: An in silico and in vitro approach. *Curr. Microbiol.* **2020**, *77*, 2181–2191. [CrossRef]
27. Monzote, L.; Scherbakov, A.M.; Scull, R.; Satyal, P.; Cos, P.; Shchekotikhin, A.E.; Gille, L.; Setzer, W.N. Essential oil from *Melaleuca leucadendra*: Antimicrobial, antikinetoplastid, antiproliferative and cytotoxic assessment. *Molecules* **2020**, *25*, 5514. [CrossRef]
28. Van, N.T.B.; Vi, O.T.; Yen, N.T.P.; Nhung, N.T.; Cuong, N.V.; Kiet, B.T.; Hoang, N.V.; Hien, V.B.; Thwaites, G.; Campell, J.; et al. Minimum inhibitory concentrations of commercial essential oils against common chicken pathogenic bacteria and their relationship with antibiotic resistance. *J. Appl. Microbiol.* **2022**, *132*, 1025–1035. [CrossRef]
29. Vázquez, A.; Tabanca, N.; Kendra, P.E. HPTLC analysis and chemical composition of selected *Melaleuca* essential oils. *Molecules* **2023**, *28*, 3925. [CrossRef]
30. Hu, W.; Sarengaowa; Guan, Y.; Feng, K. Biosynthe sis of phenolic compounds and antioxidant activity in fresh-cut fruits and vegetables. *Front. Microbiol.* **2022**, *13*, 906069.
31. Saikia, S.; Bordoloi, M. Molecular docking: Challenges, advances and its use in drug discovery perspective. *Curr. Drug Targets* **2019**, *20*, 501–521. [CrossRef]
32. Manzoor, M.F.; Ahmad, N.; Ahmed, Z.; Siddique, R.; Zeng, X.A.; Rahaman, A.; Muhammad Aadil, R.; Wahab, A. Novel extraction techniques and pharmaceutical activities of luteolin and its derivatives. *J. Food Biochem.* **2019**, *43*, e12974. [CrossRef] [PubMed]

33. Moon, N.R.; Kang, S.; Park, S. Consumption of ellagic acid and dihydromyricetin synergistically protects against UV-B induced photoaging, possibly by activating both TGF-beta1 and wnt signaling pathways. *J. Photochem. Photobiol. B* **2018**, *178*, 92–100. [CrossRef] [PubMed]
34. Baradaran Rahimi, V.; Ghadiri, M.; Ramezani, M.; Askari, V.R. Antiinflammatory and anti-cancer activities of pomegranate and its constituent, ellagic acid: Evidence from cellular, animal, and clinical studies. *Phytother. Res.* **2020**, *34*, 685–720. [CrossRef] [PubMed]
35. Pozos-Nonato, S.; Domínguez-Delgado, C.L.; Campos-Santander, K.A.; Benavides, A.A.; Pacheco-Ortin, S.M.; Higuera-Piedrahita, R.I.; Resendiz-González, G.; Molina-Trinidad, E.M. Novel nanotechnological strategies for skin anti-aging. *Curr. Pharm. Biotechnol.* **2023**, *24*, 1397–1419.
36. Qi, W.; Qi, W.; Xiong, D.; Long, M. Quercetin: Its antioxidant mechanism, antibacterial properties and potential application in prevention and control of toxipathy. *Molecules* **2022**, *27*, 6545. [CrossRef]
37. Chagas, M.D.S.S.; Behrens, M.D.; Moragas-Tellis, C.J.; Penedo, G.X.M.; Silva, A.R.; Gonçalves-de-Albuquerque, C.F. Flavonols and flavones as potential anti-inflammatory, antioxidant, and antibacterial compounds. *Oxid. Med. Cell. Longev.* **2022**, *2022*, 9966750. [CrossRef]
38. Tchuente Tchuenmogne, M.A.; Kammalac, T.N.; Gohlke, S.; Kouipou, R.M.T.; Aslan, A.; Kuzu, M.; Comakli, V.; Demirdag, R.; Ngouela, S.A.; Tsamo, E.; et al. Compounds from *Terminalia mantaly* L. (Combretaceae) stem bark exhibit potent inhibition against some pathogenic yeasts and enzymes of metabolic significance. *Medicines* **2017**, *4*, 6. [CrossRef]
39. Kujala, T.S.; Loponen, J.M.; Klika, K.D.; Pihlaja, K. Phenolics and betacyanins in red beetroot (*Beta vulgaris*) root: Distribution and effect of cold storage on the content of total phenolics and three individual compounds. *J. Agric. Food Chem.* **2000**, *48*, 5338–5342. [CrossRef]
40. Pourmorad, F.; Hosseinimehr, S.J.; Shahabimajd, N. Antioxidant activity, phenol and flavonoid contents of some selected Iranian medicinal plants. *Afr. J. Biotechnol.* **2006**, *5*, 1142–1145.
41. Wu, L.; Chen, C.Y.; Cheng, C.Y.; Dai, H.; Ai, Y.; Lin, C.H.; Chung, C.Y. Evaluation of tyrosinase inhibitory, antioxidant, antimicrobial, and antiaging activities of *Magnolia officinalis* extracts after *Aspergillus niger* fermentation. *BioMed Res. Int.* **2018**, *2018*, 5201786. [CrossRef]
42. Merchán Arenas, D.R.; Muñoz Acevedo, A.; Vargas Méndez, L.Y.; Kouznetsov, V.V. Scavenger activity evaluation of the clove bud essential oil (*Eugenia caryophyllus*) and eugenol derivatives employing ABTS$^{+\bullet}$ decolorization. *Sci. Pharm.* **2011**, *79*, 779–792. [CrossRef] [PubMed]
43. Lee, W.C.; Mahmud, R.; Pillai, S.; Perumal, S.; Ismail, S. Antioxidant activities of essential oil of *Psidium guajava* L. leaves. *APCBEE Proc.* **2012**, *2*, 86–91. [CrossRef]
44. Zheng, Z.P.; Tan, H.Y.; Wang, M. Tyrosinase inhibition constituents from the roots of *Morus australis*. *Fitoterapia* **2012**, *83*, 1008–1013. [CrossRef] [PubMed]
45. Sittisart, P.; Chitsomboon, B. Intracellular ROS scavenging activity and downregulation of inflammatory mediators in RAW264.7 macrophage by fresh leaf extracts of *Pseuderanthemum palatiferum*. *Evid. Based Complement. Alternat. Med.* **2014**, *2014*, 309095. [CrossRef]
46. Chen, C.Y.; Hu, C.Y.; Chen, Y.H.; Li, Y.T.; Chung, Y.C. Submerged fermentation with *Lactobacillus brevis* significantly improved the physiological activities of *Citrus aurantium* flower extract. *Heliyon* **2022**, *8*, e10498. [CrossRef]
47. Li, H.; Dasilva, N.A.; Liu, W.; Xu, J.; Dombi, G.W.; Dain, J.A.; Li, D.; Chamcheu, J.C.; Seeram, N.P.; Ma, H. Thymocid®, a standardized black cumin (*Nigella sativa*) seed extract, modulates collagen cross-linking, collagenase and elastase activities, and melanogenesis in murine B16F10 melanoma cells. *Nutrients* **2020**, *12*, 2146. [CrossRef]
48. Sumantran, V.N.; Kulkarni, A.A.; Harsulkar, A.; Wele, A.; Koppikar, S.J.; Chandwaskar, R.; Gaire, V.; Dalvi, M.; Wagh, U.V. Hyaluronidase and collagenase inhibitory activities of the herbal formulation *Triphala guggulu*. *J. Biosci.* **2007**, *32*, 755–761. [CrossRef]
49. Rahman, M.A.; Imran, T.B.; Islam, S. Antioxidative, antimicrobial and cytotoxic effects of the phenolics of *Leea indica* leaf extract. *Saudi J. Biol. Sci.* **2013**, *20*, 213–225. [CrossRef]
50. Trabelsi, N.; Waffo-Teguo, P.; Snoussi, M.; Ksouri, R.; Merillon, J.M.; Smaoui, A.; Abdelly, C. Variability of phenolic composition and biological activities of two Tunisian halophyte species from contrasted regions. *Acta Physiol. Plant* **2013**, *35*, 749–761. [CrossRef]

Disclaimer/Publisher's Note: The statements, opinions and data contained in all publications are solely those of the individual author(s) and contributor(s) and not of MDPI and/or the editor(s). MDPI and/or the editor(s) disclaim responsibility for any injury to people or property resulting from any ideas, methods, instructions or products referred to in the content.

Article

Phytoconstituents, Antioxidant Activity and Cytotoxicity of *Puya chilensis* Mol. Extracts in Colon Cell Lines

Manuel Martínez-Lobos [1], Valentina Silva [1], Joan Villena [2], Carlos Jara-Gutiérrez [3], Waleska E. Vera Quezada [4], Iván Montenegro [5] and Alejandro Madrid [1,*]

[1] Laboratorio de Productos Naturales y Síntesis Orgánica (LPNSO), Departamento de Ciencias y Geografía, Facultad de Ciencias Naturales y Exactas, Universidad de Playa Ancha, Avda. Leopoldo Carvallo 270, Playa Ancha, Valparaíso 2340000, Chile; manuel.martinez@upla.cl (M.M.-L.); silvapedrerosv@gmail.com (V.S.)

[2] Centro Interdisciplinario de Investigación Biomédica e Ingeniería para la Salud (MEDING), Escuela de Medicina, Facultad de Medicina, Universidad de Valparaíso, Valparaíso 2340000, Chile; juan.villena@uv.cl

[3] Centro Interdisciplinario de Investigación Biomédica e Ingeniería para la Salud (MEDING), Escuela de Medicina, Escuela de Kinesiología, Universidad de Valparaíso, Valparaíso 2340000, Chile; carlos.jara@uv.cl

[4] Laboratorio de Química de Metabolitos Bioactivos, Escuela de Química y Farmacia, Facultad de Farmacia, Centro de Investigación Farmacopea Chilena, Universidad de Valparaíso, Valparaíso 2340000, Chile; waleska.vera@uv.cl

[5] Centro Interdisciplinario de Investigación Biomédica e Ingeniería para la Salud (MEDING), Escuela de Obstetricia, Facultad de Medicina, Universidad de Valparaíso, Valparaíso 2340000, Chile; ivan.montenegro@uv.cl

* Correspondence: alejandro.madrid@upla.cl; Tel.: +56-032-250-0526

Citation: Martínez-Lobos, M.; Silva, V.; Villena, J.; Jara-Gutiérrez, C.; Vera Quezada, W.E.; Montenegro, I.; Madrid, A. Phytoconstituents, Antioxidant Activity and Cytotoxicity of *Puya chilensis* Mol. Extracts in Colon Cell Lines. *Plants* 2024, *13*, 2989. https://doi.org/10.3390/plants13212989

Academic Editors: Luis Ricardo Hernández, Eugenio Sánchez-Arreola, Edgar R. López-Mena and Gianluca Caruso

Received: 3 September 2024
Revised: 23 October 2024
Accepted: 24 October 2024
Published: 26 October 2024

Copyright: © 2024 by the authors. Licensee MDPI, Basel, Switzerland. This article is an open access article distributed under the terms and conditions of the Creative Commons Attribution (CC BY) license (https://creativecommons.org/licenses/by/4.0/).

Abstract: *Puya chilensis* Mol. is a plant of the Bromeliaceae family, which has been traditionally used for medicinal applications in various digestive disorders. In this study, the phytoconstituents of six extracts of stems and flowers of *P. chilensis* were evaluated: phenols, flavonoids and total anthraquinones, as well as their antioxidant capacity and cytotoxicity in colon cancer cell lines HT-29. The data demonstrate that the ethyl acetate extract of *P. chilensis* flowers is cytotoxic in HT-29 cell lines (IC_{50} = 41.70 µg/mL) without causing toxic effects on healthy colon cells (IC_{50} > 100 µg/mL); also, this extract concentrated the highest amount of phenols (4.63 µg GAE/g d.e.), flavonoids (31.5 µg QE/g d.e.) and anthraquinones (12.60 µg EE/g d.e.) among all the extracts tested, which also correlated with its highlighted antioxidant capacity (DPPH·IC_{50} = 4.15 mg/mL and FRAP 26.52 mM TEAC) over the other extracts. About thirty-five compounds were identified in this extract—the fatty acid esters present have been shown to have therapeutic effects on several types of cancer and could explain its antiproliferative activity.

Keywords: *Puya chilensis*; antioxidant; cytotoxic; colon cancer; ascorbic acid 2,6 dihexadecanoate

1. Introduction

Cancer is the second leading cause of human death worldwide. Among the different types of human cancer, colorectal cancer is the second deadliest and it is estimated that in 2020 alone there were more than 930,000 deaths from this cause. By 2040, the lethality of colorectal cancer is projected to increase by about 73% [1]. Currently, treatment with surgery, radiation and chemotherapy are limited in terms of tolerance, efficacy and cross-resistance. However, recent research has linked a diet rich in fruit and vegetables to the prevention of colon cancer [2,3]. This benefit is attributed in part to the polyphenols present, which have potent antiproliferative effects on cancer cells [4,5]. In this regard, the Bromeliaceae *Ananas comosus*, known as "pineapple", has been used medicinally by tropical natives for centuries as a digestive aid and wound healer [6]. Studies have indicated that pineapple juice is capable of inhibiting growth of colon cancer cells [7].

Among the species of Bromeliaceae, the genus *Puya* stands out, which is composed of more than 200 species native to the Andes and Central America [8]. These plants are generally characterized by being monocarpic (the plant dies after producing a flower and seeds) [9]. In Chile, a total of nine species of *Puya* are recognized and the most abundant species is the *Puya chilensis* Mol., known as "chagual", a name that comes from the Quechua language word "ch'ahuar" or "ch'auwar" that would mean "tow" or "bristle", which is explained by the ancient extraction of fiber from the leaves to make twine and yarn used for fishing nets; its flowers were also used as ornaments in festivities [10]. *P. chilensis* has been used for centuries as a plant for human food, either fresh or processed [11]. In addition, this plant was used in folk medicine as an astringent and moisturizing agent, as an antipyretic, anti-inflammatory and antidiarrheic [12–14]. There have also been reports that methanol extracts of *P. chilensis* meristems and leaves possess pharmacological properties, such as antioxidant and α-glucosidase inhibitory activities, and extracts of the stems have been reported to exhibit anticancer activity on human hepatocellular carcinoma [15,16]. Despite the few studies conducted on this plant, there is insufficient information on the antioxidant and cytotoxic properties of *P. chilensis* stems and flowers. Therefore, the aim of this study was to estimate the content of phenols, flavonoids and anthraquinones, and to evaluate and compare the in vitro antioxidant and cytotoxic properties of sequential extracts of increasing polarity of *P. chilensis* stems and flowers in colon cancer (HT-29) and colon non-cancer (CCD 841 CoN) cells.

2. Results and Discussion

The plant's constituents' extraction with increased polarity solvents hexane (**H**), ethyl acetate (**EA**) and ethanol (**E**), resulted in six extracts of *P. chilensis*, of which three correspond to stem (**S**) extracts and three to flowers (**F**). The yield of **SH**, **SEA**, **SE**, **FH**, **FEA** and **FE** extraction from *P. chilensis* were 0.33%, 0.81%, 3.44%, 0.23, 0.61 and 1.65% respectively. After the extracts were obtained, the total content of phytoconstituents was measured using colorimetric assays, as summarized in Table 1.

Table 1. Phytoconstituent concentration per extract of *P. chilensis*.

Part of Plant	Extract	Phenols (µg GAE/g d.e.)	Flavonoids (µg QE/g d.e.)	Anthraquinones (µg EE/g d.e.)
Stem	H	2.63 ± 0.24 [a]	0.92 ± 0.17 [a]	1.48 ± 0.12 [a]
	EA	2.66 ± 0.05 [a]	21.06 ± 0.31 [b]	0.48 ± 0.20 [a]
	E	2.50 ± 0.14 [b]	20.29 ± 0.32 [b]	0.93 ± 0.25 [a]
Flower	H	2.84 ± 0.22 [c]	30.07 ± 0.12 [c]	14.71 ± 0.38 [b]
	EA	4.63 ± 0.37 [d]	31.5 ± 0.23 [c]	12.60 ± 0.20 [b]
	E	2.72 ± 0.32 [c]	LOD	LOD

Values expressed as the mean values ± standard deviation (n = 3). Different letters in the same column indicate significant differences; $p < 0.05$; LOD = limit of detection.

The total phenolic content of extracts from *P. chilensis* stems and flowers varied slightly, with the **EA** extract of flowers showing the highest phenolic content of the extracts tested. The presence of flavonoids was recorded in all the extracts except in the **E** extract of the flowers. The non-detection of flavonoids in this extract is probably due to the non-polar nature of flavonoids, such as isoflavones, flavanones, flavones and flavonols, which have an affinity for solvents such as *n*-hexane, chloroform, dichloromethane, diethyl ether and ethyl acetate, which are found in the flower of *P. chilensis* [17,18]. The analysis also showed that the extracts of low and medium polarity from flowers have a higher anthraquinone content, which is to be expected, because an important group of compounds with anthraquinone skeletons are known to be naturally occurring pigments that give a yellow to red coloring to flowers and are often found in extracts of floral origin [19].

The DPPH, FRAP and TRAP assays were used to evaluate the antioxidant activity of *P. chilensis* extracts (see Table 2).

Table 2. Antioxidant activity of *P. chilensis* extracts.

Part of Plant	Sample/Extract	DPPH· (IC$_{50}$ mg/mL)	FRAP (TEAC mM)	TRAP (TEAC μM)
Stem	H	91.33 ± 0.26 [a]	12.53 ± 0.02 [b]	0.02 ± 0.01 [a]
	EA	44.03 ± 0.21 [c]	12.61 ± 0.24 [b]	0.03 ± 0.01 [b]
	E	11.77 ± 0.17 [d]	11.80 ± 0.32 [b]	0.07 ± 0.05 [a]
Flower	H	21.43 ± 0.10 [e]	12.75 ± 0.01 [b]	0.03 ± 0.00 [a]
	EA	4.15 ± 0.10 [d]	26.52 ± 0.01 [a]	0.47 ± 0.01 [b]
	E	5.32 ± 0.01 [d]	11.38 ± 0.01 [b]	0.09 ± 0.01 [a]
	Trolox	0.26 ± 0.02 [f]	n.a	n.a
	Gallic acid	0.06 ± 0.01 [f]	1.72 ± 0.01 [c]	1.14 ± 0.01 [c]
	BHT	n.a	1.52 ± 0.07 [c]	1.06 ± 0.02 [c]

Values expressed as the mean values ± standard deviation of three independent experiments, each performed in triplicate. Different letters in the same column indicate significant differences; $p < 0.05$; n.a = not applicable.

In general, a moderate amount of phytochemicals is correlated with a moderate antioxidant capacity. According to our results presented in Table 2, the highest capacity to neutralize DPPH radicals was found in the **EA** extract of flowers, with an IC$_{50}$ value of 4.15 mg/mL. In turn, the stem **EA** extract and both **H** extracts showed low radical scavenging capacity. On the other hand, both **E** extracts showed low scavenging activity with IC$_{50}$ values of 5.32 and 11.77 mg/mL in flowers and stems, respectively. The FRAP assay represents the electron donating capacity of the samples, thus allowing the determination of their reducing power [20]. The results of the FRAP assay indicated contrasting results for the extracts of both flowers and stems of *P. chilensis* (Table 2). Thus, the highest reducing power among the samples tested was exhibited by the flower **EA** extract with a value of 26.52 mM. However, all samples were more active than the control. Additionally, flower **EA** extract was found to be the most active of all extracts tested in the TRAP assay, however when comparing the active extract activity with the pure controls, it was almost 3 times less potent. This may be explained by the fact that the extract is a mixture of active components, between which antagonism may exist [21].

In the present study, the cytotoxic activity of *P. chilensis* stems and flowers against two human colorectal cell lines was assessed; HT-29 (adenocarcinoma) and CCD 841 CoN (epithelial), using the Sulforhodamine B (SRB) assay using doxorubicin (Doxo) and 5-fluorouracil (5-FU) as control drugs. Results are shown in Table 3.

Table 3. Cytotoxic effect of *P. chilensis* extracts (IC$_{50}$ μg/mL) and selectivity index (SI).

Part of Plant	Extract	Cell Lines		SI
		HT-29	CCD 841 CoN	
Stem	H	>100	>100	I
	EA	>100	>100	I
	E	>100	>100	I
Flower	H	>100	>100	I
	EA	41.70 ± 0.05	>100	2.40
	E	98.6 ± 0.02	>100	1.01
	Doxo	1.75 ± 0.05	5.01 + 0.53	2.86
	5-FU	9.15 ± 0.5	42.41 ± 0.1	4.63

Values expressed as the mean values ± standard deviation of three independent experiments, each performed in triplicate. I= inactive; SI = selectivity index obtained using Equation (3).

Cytotoxic activity can be classified according to the median inhibition values (IC$_{50}$) obtained. A literature review in this regard indicated that activity levels can be labeled as potentially cytotoxic (IC$_{50}$ > 100 μg/mL) and highly selective (SI > 3), or moderately cytotoxic (IC$_{50}$ > 100) but less selective (SI < 3) [22,23]. In this sense, the results shown in

Table 3 show that **EA** from *P. chilensis* flowers presents potentially cytotoxic activity in the HT-29 cell line with an IC_{50} value of 41.70 µg/mL, effectively correlating the phytochemical content and the antioxidant power shown by this extract, as has been validated in other studies of edible or medicinal plants [24]. Furthermore, its selectivity index is close to 3 and according to the above, would prove to be selective. This extract would have a high to moderate selectivity comparable to doxorubicin, which causes approximately 20 times more damage to normal colon epithelial cells CCD 841 CoN than **EA** flower extract.

The antiproliferative activity of **EA** flower extract from *P. chilensis* could be in part due to the action of the flavonoids present in the extract. Some of these, as well as other phenolic constituents, were reported in the leaves of *P. chilensis* and *P. alpestris* [15,25]. The flavonol quercetin, for example, has several biological properties. In fact, quercetin is a unique compound because of its potential to fight cancer-related diseases in a multi-targeted manner [26]. As well, the flavone apigenin has been shown to have broad anticancer effects in several types of cancer, including colorectal [27]. In addition, the methoxylated trisubstituted flavonol laricitrin can inhibit the growth of epithelial colorectal adenocarcinoma cells [28]. It is noteworthy that all extracts presented IC_{50} values above 100 µg/mL in non-cancerous cells, which allows us to conclude that *P. chilensis* extracts have no adverse effects on healthy colon cells and therefore supports its secure consumption for the ancestral therapeutic effects attributed to this species.

Based on the results obtained, the most active extract was characterized by gas chromatography coupled to mass spectrometry (GC-MS). This technique is commonly used to obtain an effective profile of secondary metabolites present in edible plants, flowers and fruits [29,30], and is also a useful way to determine the volatiles present in extracts of medium polarity [31]. The phytochemical profile of the volatile fraction of **EA** flower extract from *P. chilensis* is presented in Table 4.

Table 4. Composition of the volatile fraction of **EA** flower extract from *P. chilensis*.

N°	RT (min)	Components	%A [a]	RI [b]	RI [c]	Match
1	7.66	2,3-butanediol diacetate	0.17	1075	1080	940
2	8.48	Nonanal	0.09	1106	1102	960
3	8.822	1,3-propanediol diacetate	0.10	1121	RINR	870
4	10.30	benzoic acid	0.05	1186	1191	850
5	10.62	ethyl hydrogen succinate	0.25	1201	RINR	920
6	12.03	benzeneacetic acid	0.28	1279	1276	930
7	12.30	nonanoic acid	0.06	1293	1297	890
8	12.91	2-methoxy-4-vinylphenol	0.02	1328	1330	800
9	13.63	γ-nonanolactone	0.05	1367	1363	910
10	14.34	Vanillin	0.06	1408	1409	870
11	16.91	dodecanoic acid	0.09	1578	1576	860
12	17.10	3-hydroxy-4-methoxybenzoic acid	0.40	1591	RINR	860
13	17.95	3-oxo-α-ionol	0.02	1652	1656	810
14	18.81	ethylhexyl benzoate	0.08	1716	RINR	940
15	19.40	methyl vanillate	0.20	1762	RINR	860
16	19.60	myristic acid	0.33	1776	1775	850
17	19.88	Isophorone	0.30	1798	RINR	810
18	20.83	pentadecanoic acid	0.09	1875	1878	850
19	21.09	*trans*-ferulic acid	0.74	1895	1897	880
20	21.56	hexadecanoic acid methyl ester	5.47	1931	1928	890
21	21.86	9-hexadecenoic acid	2.39	1960	1957	920
22	22.17	ascorbic acid 2,6 dihexadecanoate	15.21	1986	RINR	930
23	22.32	palmitic acid	4.88	1999	1996	900

Table 4. Cont.

N°	RT (min)	Components	%A [a]	RI [b]	RI [c]	Match
24	24.17	(Z,Z)-9,12-octadecadienoic acid	10.41	2083	2095	920
26	26.57	Docosane	0.78	2200	2200	900
27	30.21	Tetracosane	0.39	2400	2400	860
28	30.36	bis(ethylhexyl) sebacate	1.11	2408	RINR	940
29	30.93	1-tetracosanol	9.85	2437	RINR	930
30	32.47	1-pentacosanol	12.57	2529	RINR	930
31	33.06	α-tocopherol	1.77	2563	3112	950
32	34.86	β-sitosterol	2.55	2665	3187	860
33	35.29	20b-Dihydroprogesterone	0.92	2690	RINR	800
34	35.73	9,19-cyclolanost-24-en-3-ol	0.69	2715	3465	850
35	36.53	stigmast-4-en-3-one	10.59	2761	3458	880
		Known compounds	82.96			
		Unknown compounds	17.04			

[a] Surface area of GC peak; [b] experimental retention index for RTX-5 capillary column; [c] bibliographic retention index [32]; RINR = retention index not reported on column with similar polarity.

Thirty-five compounds were identified in the **EA** flower extract from *P. chilensis*, which represented 82.96% of the volatiles of the **EA** extract. The major constituents of the volatiles of the **EA** flower extract were ascorbic acid 2,6-dihexadecanoate (15.21%), 1-pentacosanol (12.57%), stigmast-4-en-3-one (10.59%), (Z,Z)-9,12-octadecadienoic acid (10.41%) and 1-tetracosanol (9.85%). Most of the volatile compounds extracted were fatty acid esters, fatty alcohols, fatty acids, steroids and to a lesser extent benzoic acid derivatives; similarly, previous data showed the existence of derivatives of hydroxybenzoic acid in stems of *P. chilensis* [15,16] and other edible plants [33]. In this context, the nature of the metabolites determined can be attributed not only to the polarity of the extraction solvent and genetic diversity of the species [34], but also to the properties of the GC method, characterized to determine volatile compounds [31]. Ascorbic acid esters have demonstrated strong antioxidant, antibacterial and cytotoxic activity [35]; in this sense, the percentage of ascorbic acid 2,6-dihexadecanoate could contribute to the antioxidant and antiproliferative activity of the **EA** flower extract against tumor cells, as has been reported previously [36,37]. Fatty acids and their respective esters constitute a wide range of materials used in the discovery and formulation of active ingredients of pharmaceutical importance, because they have been reported as potential antioxidant and antiproliferative compounds [38,39]. In addition, fatty acid esters have been shown to modulate the anti-inflammatory response of macrophages [40], which would contribute to the antiproliferative capacity of the extract. Likewise, fatty alcohols have shown potent anti-inflammatory activity; among them, 1-tetracosanol has shown antiproliferative effects in human melanoma cell lines [41] and 1-pentacosanol is a potential inhibitor of prostate tumor cell proliferation [42]. Both alcohols could contribute to the activity shown by the **EA** floral extract of *P. chilensis*.

The results obtained in the present study call for further study of *P. chilensis* flowers in ethnopharmacological terms, such as evaluating the antibacterial potential of this plant against pathogens associated with gastric pathologies in order to enhance the therapeutic use of *P. chilensis*, especially as a potent source of active metabolites such as fatty acids, phenolic acids and their respective derivatives, and the possibility for investigators to develop multiple bioactive agents with therapeutic effects against various malignant neoplasms.

3. Materials and Methods

3.1. Plant Material

Stems and flowers of *P. chilensis* were collected in January 2023 from Federico Santa María cliffs (33°03′04″ S 71°39′34″ O) in Valparaíso, Chile. Botanical identification and authentication were verified by Mr. Patricio Novoa, Forest Engineer, Botanical Expert, and Chief of the Horticulture Department, National Botanic Garden of Viña del Mar,

Valparaíso, Chile. Voucher specimen has been deposited in the Herbarium of Natural Products Laboratory of Universidad de Playa Ancha, Valparaíso, Chile (PC-012023).

3.2. Preparation of Plant Extract

At the laboratory, the stems and flowers were rinsed with water and air-dried. The stems and flowers were then dried in a food dehydrator at 60 °C for 14 h and 50 °C for 20 h, respectively. Finally, both plant parts were ground in a kitchen grinder to obtain a fine powder. Later, 300 g of dried powder of *P. chilensis* stems and flowers separately was extracted using maceration containers at room temperature for 72 h using **H** and the residue was extracted successively with **EA** and **E**, using 1 L of each solvent in the extraction. All solutions were evaporated and dried under vacuum (below 40 °C), and then each extract was kept in darkness at room temperature.

3.3. Analysis of Total Polyphenolic, Total Flavonoid and Anthraquinones Content

According to protocols by Jara et al. [43], total phenolic content was established employing the Folin–Ciocalteu method, total flavonoid content was established employing the Dowd method and total anthraquinone content was determined by the Arvouet-Grand method using a UV/VIS spectrometer, UV-2601 (Ray-LEIGH, Beijing, China). It is important to note that the detection limit (LOD) of the equipment is 0.001 nm for absorbance.

3.3.1. Phenols

Five hundred microliters of the extract solution in ethanol (1.0 mg/mL) was mixed with a Folin–Ciocalteu reagent (2.5 mL, 0.2 N) and incubated for 5 min. Then, a 7.5% w/v Na_2CO_3 solution (2.0 mL) was added and the mix was incubated in the dark at room temperature for 2 h. The absorbance of the solution was measured at 700 nm using ethanol as the blank. The obtained absorbance values were interpolated using a gallic acid standard curve (0–100 mg/L) and the total phenolic content was expressed as µg of gallic acid equivalents (GAE) per g of dried extract (d.e.). Values shown are the mean ± standard deviation of three independent experiments performed in triplicate.

3.3.2. Flavonoids

One milliliter of 2% w/v aluminum chloride ($AlCl_3$) in ethanol was mixed with the same volume of the extract solution in ethanol (1.0 mg/mL). The mix was incubated for 10 min at room temperature, and absorbance was measured at 415 nm against a blank sample consisting of a 1.0 mL extract solution with 1.0 mL of ethanol without $AlCl_3$. The absorbance values were interpolated using a quercetin calibrate curve (0–100 mg/L). The total flavonoid content was expressed as µg of quercetin equivalents (QE) per g of dried extract (d.e.). Values shown are the mean ± standard deviation of three independent experiments performed in triplicate.

3.3.3. Anthraquinones

One milliliter of 2% w/v $AlCl_3$ in ethanol was mixed with the same volume of the extract solution in ethanol (1.0 mg/mL). The mix was incubated for 10 min at room temperature, and absorbance was measured at 486 nm against a blank sample consisting of 1.0 mL extract solution with 1.0 mL of ethanol without $AlCl_3$. The absorbance values were interpolated using an emodin calibrate curve (0–100 mg/L). The total anthraquinone content was expressed as µg of emodin equivalents (EE) per g of dried extract (d.e.). Values shown are the mean ± standard deviation of three independent experiments performed in triplicate.

3.4. GC-MS Identification of Compounds

The **EA** flower extract from *P. chilensis* was analyzed by GC-MS for volatiles and semi-volatiles, using Shimadzu GCMS-QP2010 Plus combination (Shimadzu, Kyoto, Japan) coupled with a fused silica RTX-5 capillary column (30 m × 0.25 mm id, 0.25 um film; Restek,

Bellefonte, PA, USA). The protocol and working conditions used were those previously described by Faundes-Gandolfo [44]. The GC was operated in splitless mode (30 s sampling time and 1 µL of sample) with helium as the carrier gas (1 mL min^{-1} flow) and an injector temperature of 200 °C. The oven and the column were programmed from 50 °C (2 min hold) to 280 °C at a rate of 8 °C min^{-1} (15 min hold). The mass spectrum was acquired by electronic impact at 70 eV with a mass range of 35 to 500 m/z in full scan mode (1.56 scan s^{-1}). Compounds in the chromatograms were identified by comparing their mass spectra with those in the NIST20 library. Chromatographic peaks were considered "unknown" when their similarity index (MATCH) and reverse similarity index (RMATCH) were less than 800, and discarded in this identification process [45]. These parameters are referred to by the degree the target spectrum matches the standard spectrum in the NIST Library (the value 1000 indicates a perfect fit), and by comparison of their retention index with data published in other studies for the same type of column. The retention indices were determined under the same operating conditions in relation to a homologous n-alkanes series (C_8–C_{36}) by Equation (1):

$$RI = 100 \times (n + Tr_{(unknown)} - Tr_{(n)} / Tr_{(N)} - Tr_{(n)}) \quad (1)$$

where n = the number of carbon atoms in the smaller n-alkane; N = the number of carbon atoms in the larger n-alkane; and Tr = the retention time. Components' relative concentrations were obtained by peak area normalization.

3.5. DPPH Free Radical Scavenging Assay

P. chilensis extracts were tested in vitro using the DPPH (2,2-diphenyl-1-picrylhydrazyl) assay according to the protocol described previously in [46]. The sample (100 µL, extracts at 0–100 mg/mL) was mixed with a 50 µM DPPH solution (2.9 mL) freshly prepared in ethanol. A 50 µM DPPH solution (2.9 mL) with ethanol (0.1 mL) was used as the control. The sample and control solutions were incubated for 15 min at room temperature, and the absorbance was measured at 517 nm. The inhibition (%) was calculated by Equation (2):

$$I\% = 100\% \, (A_{control} - A_{sample}) / A_{control} \quad (2)$$

From the obtained I% values, the IC_{50} value was determined by linear regression analysis. All the measurements were obtained from three independent experiments, each performed in triplicate.

3.6. Ferric Reducing Power (FRAP) Assay

The experiment was performed on the basis of the protocol described by Mellado et al. [46]. Freshly prepared (10 volumes of 300 mM acetate buffer, pH 3.6, with 1.0 volume of 10 mM TPTZ (2,4,6-tri(2-pyridyl)-s-triazine) in 40 mM hydrochloric acid, and 1.0 volume of 20 mM ferric chloride) FRAP reagent (3.0 mL) was mixed with deionized water (300 µL) and the sample (100 µL, 1.0 mg/mL of each extract). The mix was incubated for 30 min at 37 °C in a water bath and the absorbance was measured at 593 nm using ethanol as the blank solution. The obtained absorbance values were interpolated in a Trolox calibrate curve (0–200 mg/L) and the FRAP values were expressed in mM Trolox equivalent antioxidant capacity (mM TEAC). Values shown are the mean ± standard deviation of three independent experiments performed in triplicate.

3.7. Total Reactive Antioxidant Power (TRAP) Assay

The experiment was performed following the protocol previously published by Mellado et al. [46]. One volume of 10 mM solution of ABAP (2,2′-azo-bis(2-amidino propane) was mixed with the same volume of 150 µM solution of ABTS (2,2′-Azino-bis(3-ethylbenzthiazoline-6-sulfonic acid) using PBS 100 mM at pH of 7.4 (TRAP solution). The mixture was incubated at 45 °C for 30 min and then cooled to room temperature for use. Sample solution (10 µL, 1.0 mg/mL of each extract) was mixed with the TRAP solution

(990 µL), and the absorbance was measured after 50 s at 734 nm against the ABTS solution as the blank. The absorbance values were interpolated in a Trolox standard curve (0–120 mg/L). Values shown are the mean ± standard deviation of three independent experiments performed in triplicate.

3.8. In Vitro Cytotoxicity Assay

3.8.1. Cells and Culture Conditions

The cell lines were purchased from the American Type Culture Collection (Rockville, MD, USA): HT-29 (human colon cancer) and CCD 841 CoN (human colon epithelial cells). The cell lines were maintained in a 1:1 mixture of Dulbecco's modified Eagle's medium (DMEM) and Ham's F12 medium, containing 10% heat-inactivated fetal bovine serum (FBS), penicillin (100 U/mL) and streptomycin (100 µg/mL) in a humidified atmosphere with 5% CO_2 at 37 °C.

3.8.2. In Vitro Growth Inhibition Assay

The assay was performed following the sulforhodamine B (Sigma-Aldrich, St. Louis, MO, USA) method previously published by Villena et al. [47]. Briefly, the cells were set up at 3×10^3 cells per well of a 96-well, flat-bottomed 200 µL microplate. Cells were incubated at 37 °C in a humidified 5% CO_2/95% air atmosphere and treated with the extracts at different concentrations for 72 h. At the end of the crude extract exposure, cells were fixed with 50% trichloroacetic acid at 4 °C (TCA final concentration 10%). After washing with distilled water, cells were stained with 0.1% sulforhodamine B, dissolved in 1% acetic acid (50 µL/well) for 30 min and subsequently washed with 1% acetic acid to remove unbound stain. Protein-bound stain was solubilized with 100 µL of 10 mM unbuffered Tris base. The cell density was determined using a fluorescence plate reader (wavelength 540 nm). Untreated cells were used as the negative control while cells treated with doxorubicin and 5-fluorouracil were used as the positive control. In addition, all of the samples were tested from 0 to 100 µg/mL (concentration of extracts) using ethanol as the carrier solvent. Values shown are the mean ± standard deviation of three independent experiments performed in triplicate. Finally, Sigma Plot software was used to calculate the IC_{50} value.

3.8.3. Selectivity Index

The selectivity index (SI) is the quotient of the IC_{50} value of the *P. chilensis* extracts determined for CCD 841 CoN cells and the value obtained for the cancer cell line, and was calculated following Equation (3):

$$SI = IC_{50\,(CCD\,841\,CoN)} / IC_{50(HT\text{-}29\,cells)} \qquad (3)$$

3.9. Statistical Analysis

Values are averages ± standard deviation of three independent experiments performed in triplicate. Kruskal–Wallis ANOVA with 95% confidence level using STATISTICA 7.0 was used due to non-parametric data.

4. Conclusions

The **EA** extract of the *P. chilensis* flower shows antioxidant activity, outperforming the other extracts in the DPPH, FRAP and TRAP assays, which correlates with the results obtained in this study of its phytoconstituents, since it concentrates the highest amount of phenols, flavonoids and anthraquinones. Among the compounds identified in this extract are mainly fatty acid esters, fatty alcohols, fatty acids and steroids. Some of these compounds, like 2,6-dihexadecanoate ascorbate (15.21%), have antioxidant and antiproliferative activity; 1-tetracosanol (12.57%) has also shown antiproliferative activity, supporting the efficacy of **EA** flower extract in inhibiting the growth of HT-29. In conclusion, the **EA** extract of *P. chilensis* flowers shows a promising profile for antioxidant and antitumor applications, supported by its bioactive compounds. Furthermore, this study demonstrated that its extracts do not cause toxicity in healthy colon cells, which supports its use in traditional medicine. Future research could be carried out to determine the antiproliferative effect of its main compounds separately, to identify the mechanism of action by which they exert their cytotoxic activity.

Author Contributions: Conceptualization, M.M.-L. and A.M.; methodology, A.M., J.V. and M.M.-L.; validation, A.M., I.M. and J.V.; formal analysis, M.M.-L., C.J.-G., W.E.V.Q. and I.M.; investigation, M.M.-L., C.J.-G., I.M., V.S. and W.E.V.Q.; resources, A.M.; data curation, J.V. and C.J.-G.; writing—original draft preparation, M.M.-L.; writing—review and editing, A.M., V.S. and M.M.-L.; funding acquisition, A.M. All authors have read and agreed to the published version of the manuscript.

Funding: This research was funded by Agencia Nacional de Desarrollo y Investigación (ANID), Fondo Nacional de Desarrollo Científico y Tecnológico FONDECYT grant number 1230311.

Institutional Review Board Statement: Not applicable.

Informed Consent Statement: Not applicable.

Data Availability Statement: The original contributions presented in this study are included in this article; further inquiries can be directed to the corresponding authors.

Acknowledgments: The authors thank the Dirección General de Investigación de la Universidad de Playa Ancha for their support in the hiring of the technicians Valentina Silva Pedreros for the "Apoyos técnicos para laboratorios y grupos de investigación UPLA 2024 (SOS Technician)" program. The APC was funded by Universidad de Playa Ancha, Plan de Fortalecimiento Universidades Estatales-Ministerio de Educación, Convenio UPA 1999.

Conflicts of Interest: The authors declare no conflicts of interest.

References

1. WHO. Colorectal Cancer. Available online: https://www.who.int/news-room/fact-sheets/detail/colorectal-cancer (accessed on 27 August 2024).
2. Vallis, J.; Wang, P.P. The Role of Diet and Lifestyle in Colorectal Cancer Incidence and Survival. In *Gastrointestinal Cancers*; Morgado-Diaz, J.A., Ed.; Exon Publications: Brisbane City, QLD, Australia, 2022; Chapter 2. [CrossRef]
3. Alzate-Yepes, T.; Pérez-Palacio, L.; Martínez, E.; Osorio, M. Mechanisms of Action of Fruit and Vegetable Phytochemicals in Colorectal Cancer Prevention. *Molecules* **2023**, *28*, 4322. [CrossRef] [PubMed]
4. Yang, J.; Martinson, T.E.; Liu, R.H. Phytochemical profiles and antioxidant activities of wine grapes. *Food Chem.* **2009**, *116*, 332–339. [CrossRef]
5. Jakobušić Brala, C.; Karković Marković, A.; Kugić, A.; Torić, J.; Barbarić, M. Combination Chemotherapy with Selected Polyphenols in Preclinical and Clinical Studies—An Update Overview. *Molecules* **2023**, *28*, 3746. [CrossRef] [PubMed]
6. Gani, M.B.A.; Nasiri, R.; Hamzehalipour, A.J. In Vitro Antiproliferative Activity of Fresh Pineapple Juices on Ovarian and Colon Cancer Cell Lines. *Int. J. Pept. Res. Ther.* **2015**, *21*, 353–364. [CrossRef]
7. Amini, A.; Ehteda, A.; Masoumi, M.S.; Akhter, J.; Pillai, K.; Morris, D.L. Cytotoxic effects of bromelain in human gastrointestinal carcinoma cell lines (MKN45, KATO-III, HT29-5F12, and HT29-5M21) Lines. *OncoTargets Ther.* **2013**, *6*, 403–409.
8. Jabaily, R.S.; Sytsma, K.J. Phylogenetics of *Puya* (Bromeliaceae): Placement, major lineages, and evolution of Chilean species. *Am. J. Bot.* **2010**, *97*, 337–356. [CrossRef]
9. Metcalf, J.C.; Rose, K.E.; Rees, M. Evolutionary demography of monocarpic perennials. *Trends Ecol. Evol.* **2003**, *18*, 471–480. [CrossRef]
10. Muñoz-Schick, M.; Moreira-Muñoz, A. El género *Schizanthus* (Solanaceae) en Chile. *Rev. Chagual* **2008**, *6*, 21–32. Available online: https://jardinbotanicochagual.cl/wp-content/uploads/2023/10/revista-chagual-1.pd (accessed on 27 August 2024).
11. Zizka, G.; Schmidt, M.; Schulte, K.; Novoa, P.; Pinto, R.; König, K. Chilean Bromeliaceae: Diversity, Distribution and Evaluation of Conservation Status. *Biodivers. Conserv.* **2009**, *18*, 2449–2471. [CrossRef]
12. Muñoz, M.S.; Barrera, E.M.; Meza, I.P. *El uso Medicinal y Alimenticio de Plantas Nativas y Naturalizadas en Chile*; Publicación Ocasional N° 33; Museo Nacional de Historia Natural: Santiago, Chile, 1981; p. 12.
13. Smith, L.B.; Looser, G. Las especies chilenas del género *Puya*. *Rev. Univ. Católica De Chile* **1935**, *20*, 255–295. Available online: https://obtienearchivo.bcn.cl/obtieneimagen?id=documentos/10221.1/53390/1/207614.pdf (accessed on 27 August 2024).
14. Hornung-Leoni, C. Avances sobre Usos Etnobotánicos de las Bromeliaceae en Latinoamérica. *BLACPMA* **2011**, *10*, 297–314. Available online: https://www.redalyc.org/pdf/856/85619300003.pdf (accessed on 28 August 2024).

15. Jiménez-Aspee, F.; Theoduloz, C.; Gómez-Alonso, S.; Hermosín-Gutiérrez, I.; Reyes, M.; Schmeda-Hirschmann, G. Polyphenolic Profile and Antioxidant Activity of Meristem and Leaves from "Chagual" (*Puya chilensis* Mol.), a Salad from Central Chile. *Food Res. Int.* **2018**, *114*, 90–96. [CrossRef] [PubMed]
16. Echeverria-Echeverria, C.; Valderrama-Villarroel, A.; Ortega, M.; Contreras, R.A.; Zúñiga, G.E.; Alvarado-Soto, L.; Ramírez-Tagle, R. Cytotoxic Effect of *Puya chilensis* Collected in Central Chile. *Nat. Prod. Commun.* **2022**, *17*, 1–5. [CrossRef]
17. Kumar, S.; Pandey, A.K. Chemistry and Biological Activities of Flavonoids: An Overview. *Sci. World J.* **2013**, *2013*, 162750. [CrossRef] [PubMed]
18. Rodríguez, S.L.; Ramírez-Garza, R.E.; Serna, S.O. Environmentally Friendly Methods for Flavonoid Extraction from Plant Material: Impact of Their Operating Conditions on Yield and Antioxidant Properties. *Sci. World J.* **2020**, *2020*, 6792069. [CrossRef]
19. Zhang, X.; Thuong, P.T.; Jin, W.; Su, N.D.; Sok, D.E.; Bae, K.; Kang, S.S. Antioxidant Activity of Anthraquinones and Flavonoids from Flower of *Reynoutria sachalinensis*. *Arch. Pharm. Res.* **2005**, *28*, 22–27. [CrossRef]
20. Ben Mrid, R.; Bouchmaa, N.; Bouargalne, Y.; Ramdan, B.; Karrouchi, K.; Kabach, I.; El Karbane, M.; Idir, A.; Zyad, A.; Nhiri, M. Phytochemical Characterization, Antioxidant and In Vitro Cytotoxic Activity Evaluation of *Juniperus oxycedrus* Subsp. *oxycedrus* Needles and Berries. *Molecules* **2019**, *24*, 502. [CrossRef]
21. Stankovic, M.S.; Niciforovic, N.; Topuzovic, M.; Solujic, S. Total Phenolic Content, Flavonoid Concentrations and Antioxidant Activity, of The Whole Plant and Plant Parts Extracts from *Teucrium montanum* L. Var. *Montanum*, F. *supinum* (L.) Reichenb. *Biotechnol. Biotechnol. Equip.* **2011**, *25*, 2222–2227. [CrossRef]
22. Prayong, P.; Barusrux, S.; Weerapreeyakul, N. Cytotoxic Activity Screening of Some Indigenous Thai Plants. *Fitoterapia* **2008**, *79*, 598–601. [CrossRef]
23. Chothiphirat, A.; Nittayaboon, K.; Kanokwiroon, K.; Srisawat, T.; Navakanitworakul, R. Anticancer Potential of Fruit Extracts from *Vatica diospyroides* Symington Type SS and Their Effect on Program Cell Death of Cervical Cancer Cell Lines. *Sci. World J.* **2019**, *2019*, 5491904. [CrossRef]
24. Jaradat, N.A.; Al-Ramahi, R.; Zaid, A.N.; Ayesh, O.I.; Eid, A.M. Ethnopharmacological Survey of Herbal Remedies Used for Treatment of Various Types of Cancer and Their Methods of Preparations in the West Bank-Palestine. *BMC Complement. Altern. Med.* **2016**, *16*, 93. [CrossRef] [PubMed]
25. Mizuno, T.; Sugahara, K.; Tsutsumi, C.; Iino, M.; Koi, S.; Noda, N.; Iwashina, T. Identification of Anthocyanin and Other Flavonoids from the Green–Blue Petals of *Puya alpestris* (Bromeliaceae) and a Clarification of Their Coloration Mechanism. *Phytochem.* **2021**, *181*, 112581. [CrossRef]
26. Hashemzaei, M.; Far, A.D.; Yari, A.; Heravi, R.E.; Tabrizian, K.; Taghdisi, S.M.; Sadegh, S.E.; Tsarouhas, K.; Kouretas, D.; Tzanakakis, G.; et al. Anticancer and Apoptosis-Inducing Effects of Quercetin in Vitro and in Vivo. *Oncol. Rep.* **2017**, *38*, 819–828. [CrossRef] [PubMed]
27. Yan, X.; Qi, M.; Li, P.; Zhan, Y.; Shao, H. Apigenin in Cancer Therapy: Anti-Cancer Effects and Mechanisms of Action. *Cell Biosci.* **2017**, *7*, 50. [CrossRef]
28. Gómez-Alonso, S.; Collins, V.J.; Vauzour, D.; Rodríguez-Mateos, A.; Corona, G.; Spencer, J.P.E. Inhibition of Colon Adenocarcinoma Cell Proliferation by Flavonols Is Linked to a G2/M Cell Cycle Block and Reduction in Cyclin D1 Expression. *Food Chem.* **2012**, *130*, 493–500. [CrossRef]
29. Giannetti, V.; Biancolillo, A.; Marini, F.; Boccacci Mariani, M.; Livi, G. Characterization of the Aroma Profile of Edible Flowers Using HS-SPME/GC–MS and Chemometrics. *Food Res. Int.* **2024**, *178*, 114001. [CrossRef]
30. Abed, S.S.; Kiranmayi, P.; Imran, K.; Lateef, S.S. Gas Chromatography-Mass Spectrometry (GC-MS) Metabolite Profiling of *Citrus limon* (L.) Osbeck Juice Extract Evaluated for Its Antimicrobial Activity Against *Streptococcus mutans*. *Cureus* **2023**, *15*, e33585. [CrossRef] [PubMed]
31. Ajilogba, C.F.; Babalola, O.O. GC–MS Analysis of Volatile Organic Compounds from Bambara Groundnut Rhizobacteria and Their Antibacterial Properties. *World J. Microbiol. Biotechnol.* **2019**, *35*, 83. [CrossRef]
32. Adams, P.R. *Identification of Essential Oil Components by Gas Chromatography/Mass Spectrometry*, 4th ed.; Allured Publishing Corp.: Carol Stream, IL, USA, 2007.
33. Baeshen, N.A.; Almulaiky, Y.Q.; Afifi, M.; Al-Farga, A.; Ali, H.A.; Baeshen, N.N.; Abomughaid, M.M.; Abdelazim, A.M.; Baeshen, M.N. GC-MS Analysis of Bioactive Compounds Extracted from Plant *Rhazya stricta* Using Various Solvents. *Plants* **2023**, *12*, 960. [CrossRef]
34. Pais, A.L.; Li, X.; Xiang, Q. Discovering Variation of Secondary Metabolite Diversity and Its Relationship with Disease Resistance in *Cornus florida* L. *Ecol. Evol.* **2018**, *8*, 5619–5636. [CrossRef]
35. Mohy El-Din, S.M.; El-Ahwany, A.M.D. Bioactivity and Phytochemical Constituents of Marine Red Seaweeds (*Jania rubens*, *Corallina mediterranea* and *Pterocladia capillacea*). *J. Taibah Univ. Sci.* **2016**, *10*, 471–484. [CrossRef]
36. Akinmoladun, A.C.; Ibukun, E.O.; Afor, E.; Akinsinlola, B.L.; Onibon, T.R.; Akinboye, A.O. Chemical constituents and antioxidant activity of *Boonei alstonia*. *Afr. J. Biotechnol.* **2007**, *6*, 1197.
37. Begum, S.M.F.M.; Priya, S.; Sundararajan, R.; Hemalatha, S. Novel anticancerous compounds from *Sargassum wightii*: In silico and in vitro approaches to test the antiproliferative efficacy. *J. Adv. Pharm. Edu Res.* **2017**, *7*, 272–277.
38. Rodríguez, J.P.; Guijas, C.; Astudillo, A.M.; Rubio, J.M.; Balboa, M.A.; Balsinde, J. Sequestration of 9-Hydroxystearic Acid in FAHFA (Fatty Acid Esters of Hydroxy Fatty Acids) as a Protective Mechanism for Colon Carcinoma Cells to Avoid Apoptotic Cell Death. *Cancers* **2019**, *11*, 524. [CrossRef]

39. Jóźwiak, M.; Filipowska, A.; Fiorino, F.; Struga, M. Anticancer Activities of Fatty Acids and Their Heterocyclic Derivatives. *Eur. J. Pharmacol.* **2020**, *871*, 172937. [CrossRef]
40. Korbecki, J.; Bajdak-Rusinek, K. The Effect of Palmitic Acid on Inflammatory Response in Macrophages: An Overview of Molecular Mechanisms. *Inflamm. Res.* **2019**, *68*, 915–932. [CrossRef] [PubMed]
41. Vergara, M.; Olivares, A.; Altamirano, C. Antiproliferative Evaluation of Tall-Oil Docosanol and Tetracosanol over CHO-K1 and Human Melanoma Cells. *Electro J. Biotecnol.* **2015**, *18*, 291–294. [CrossRef]
42. Sowmya, S.; Perumal, P.C.; Ravi, S.; Anusooriya, P.; Shanmughavel, P.; Murugesh, E.; Chaithany, K.K.; Gopalakrishnan, V.K. 1-Pentacosanol Isolated from Stem Ethanolic Extract of *Cayratia trifolia* (L.) Is a Potential Target for Prostate Cancer-In silico Approach. *JJBS* **2021**, *14*, 359–365. [CrossRef]
43. Jara, C.; Leyton, M.; Osorio, M.; Silva, V.; Fleming, F.; Paz, M.; Madrid, A.; Mellado, M. Antioxidant, Phenolic and Antifungal Profiles of *Acanthus mollis* (Acanthaceae). *Nat. Prod. Res.* **2017**, *31*, 2325–2328. [CrossRef]
44. Faundes-Gandolfo, N.; Jara-Gutiérrez, C.; Párraga, M.; Montenegro, I.; Vera, W.; Escobar, M.; Madrid, A.; Valenzuela, M.; Villena, J. *Kalanchoe pinnata* (Lam.) Pers. leaf ethanolic extract exerts selective anticancer activity through ROS-induced apoptotic cell death in human cancer cell lines. *BMC Complement. Med. Ther.* **2024**, *24*, 269. [CrossRef]
45. NIST. *NIST Mass Spectral Database for NIST/EPA/NIH and Mass Spectral Search Program (Version 2.3)*; National Institute of Standards and Technology NIST: Gaithersburg, MD, USA, 2017; Volume 6, pp. 1–73.
46. Mellado, M.; Soto, M.; Madrid, A.; Montenegro, I.; Jara-Gutiérrez, C.; Villena, J.; Werner, E.; Godoy, P.; Aguilar, L.F. In Vitro Antioxidant and Antiproliferative Effect of the Extracts of *Ephedra chilensis* K Presl Aerial Parts. *BMC Complement. Altern. Med.* **2019**, *19*, 53. [CrossRef] [PubMed]
47. Villena, J.; Montenegro, I.; Said, B.; Werner, E.; Flores, S.; Madrid, A. Ultrasound Assisted Synthesis and Cytotoxicity Evaluation of Known 2′,4′-Dihydroxychalcone Derivatives against Cancer Cell Lines. *Food Chem. Toxicol.* **2021**, *148*, 111969. [CrossRef] [PubMed]

Disclaimer/Publisher's Note: The statements, opinions and data contained in all publications are solely those of the individual author(s) and contributor(s) and not of MDPI and/or the editor(s). MDPI and/or the editor(s) disclaim responsibility for any injury to people or property resulting from any ideas, methods, instructions or products referred to in the content.

Article

Antimicrobial and Antibiofilm Potential of *Flourensia retinophylla* against *Staphylococcus aureus*

Minerva Edith Beltrán-Martínez [1], Melvin Roberto Tapia-Rodríguez [2], Jesús Fernando Ayala-Zavala [1], Agustín Gómez-Álvarez [3], Ramon Enrique Robles-Zepeda [4], Heriberto Torres-Moreno [5], Diana Jasso de Rodríguez [6,*] and Julio César López-Romero [5,*]

[1] Coordinación de Tecnología de Alimentos de Origen Vegetal, Centro de Investigación en Alimentación y Desarrollo, A.C. Carretera Gustavo Astiazarán Rosas No. 46, Colonia la Victoria, Hermosillo 83304, Mexico; mbeltran122@estudiantes.ciad.mx (M.E.B.-M.); jayala@ciad.mx (J.F.A.-Z.)
[2] Departamento de Biotecnología y Ciencias Alimentarias, Instituto Tecnológico de Sonora, 5 de Febrero 818 sur, Col. Centro, Ciudad Obregón 85000, Mexico; melvin.tapia14987@potros.itson.edu.mx
[3] Departamento de Ingeniería Química y Metalurgia, Universidad de Sonora, Hermosillo 83000, Mexico; agustin.gomez@unison.mx
[4] Departamento de Ciencias Químico Biológicas, Universidad de Sonora, Hermosillo 83000, Mexico; robles.zepeda@unison.mx
[5] Departamento de Ciencias Químico Biológicas y Agropecuarias, Universidad de Sonora, Caborca 83600, Mexico; heriberto.torres@unison.mx
[6] Universidad Autónoma Agraria Antonio Narro, Saltillo 25315, Mexico
* Correspondence: dianajassocantu@yahoo.com.mx (D.J.d.R.); julio.lopez@unison.mx (J.C.L.-R.)

Citation: Beltrán-Martínez, M.E.; Tapia-Rodríguez, M.R.; Ayala-Zavala, J.F.; Gómez-Álvarez, A.; Robles-Zepeda, R.E.; Torres-Moreno, H.; de Rodríguez, D.J.; López-Romero, J.C. Antimicrobial and Antibiofilm Potential of *Flourensia retinophylla* against *Staphylococcus aureus*. *Plants* **2024**, *13*, 1671. https://doi.org/10.3390/plants13121671

Academic Editors: Luis Ricardo Hernández, Eugenio Sánchez-Arreola and Edgar López-Mena

Received: 28 May 2024
Revised: 13 June 2024
Accepted: 14 June 2024
Published: 17 June 2024

Copyright: © 2024 by the authors. Licensee MDPI, Basel, Switzerland. This article is an open access article distributed under the terms and conditions of the Creative Commons Attribution (CC BY) license (https://creativecommons.org/licenses/by/4.0/).

Abstract: *Staphylococcus aureus* is a Gram-positive bacteria with the greatest impact in the clinical area, due to the high rate of infections and deaths reaching every year. A previous scenario is associated with the bacteria's ability to develop resistance against conventional antibiotic therapies as well as biofilm formation. The above situation exhibits the necessity to reach new effective strategies against this pathogen. *Flourensia retinophylla* is a medicinal plant commonly used for bacterial infections treatments and has demonstrated antimicrobial effect, although its effect against *S. aureus* and bacterial biofilms has not been investigated. The purpose of this work was to analyze the antimicrobial and antibiofilm potential of *F. retinophylla* against *S. aureus*. The antimicrobial effect was determined using an ethanolic extract of *F. retinophylla*. The surface charge of the bacterial membrane, the K^+ leakage and the effect on motility were determined. The ability to prevent and remove bacterial biofilms was analyzed in terms of bacterial biomass, metabolic activity and viability. The results showed that *F. retinophylla* presents inhibitory (MIC: 250 µg/mL) and bactericidal (MBC: 500 µg/mL) activity against *S. aureus*. The MIC extract increased the bacterial surface charge by 1.4 times and the K^+ concentration in the extracellular medium by 60%. The MIC extract inhibited the motility process by 100%, 61% and 40% after 24, 48 and 72 h, respectively. The MIC extract prevented the formation of biofilms by more than 80% in terms of biomass production and metabolic activity. An extract at $10 \times$ MIC reduced the metabolic activity by 82% and the viability by ≈50% in preformed biofilms. The results suggest that *F. retinophylla* affects *S. areus* membrane and the process of biofilm formation and removal. This effect could set a precedent to use this plant as alternative for antimicrobial and disinfectant therapies to control infections caused by this pathogen. In addition, this shrub could be considered for carrying out a purification process in order to identify the compounds responsible for the antimicrobial and antibiofilm effect.

Keywords: natural products; *F. retinophylla*; *S. aureus*; planktonic cells; biofilms

1. Introduction

Bacterial infections constitute a public health problem worldwide due to the substantial increase in the number of cases reported in recent years [1]. This has been associated with the ability of microorganisms to resist conventional antibiotic therapies used in clinical practice, which has resulted in increased morbidity and mortality rates [2]. In this regard, the World Health Organization (WHO) estimates that bacterial infections will be the leading cause of death worldwide by 2050, causing more than 10 million deaths per year [3].

One of the most relevant Gram-positive bacteria in public health is *Staphylococcus aureus*, considered by the WHO as a high priority microorganism due to its antibiotic resistance [4]. Furthermore, *S. aureus* has been classified within the ESKAPE bacteria because of its high virulence and resistance to antibiotics [5]. In addition, this microorganism is highly associated with community and hospital infections, with more than 110,500 deaths estimated in 2019 [6]. Some pathologies associated with this bacteria include skin infections, soft tissue infections of the lower respiratory tract, bacteremia, osteomyelitis and endocarditis, which can become chronic, persistent and a cause of death [7]. Additionally, this bacteria is one of the main sources of contamination of medical devices and instrumentation [8]. This is mainly associated with the ability of *S. aureus* to adhere to different surfaces and form biofilms [9].

Biofilms are defined as communities of microorganisms that grow embedded in a layer of exopolysaccharides, which are mainly composed of polysaccharides, proteins, genetic material and lipids [10]. These communities are formed in a four-stage process: adhesion, synthesis of extracellular matrix, formation and maturation of the biofilm, and detachment of bacterial cells [11]. These structures confer to the bacterial community resistance against antibiotics and the immune system, which makes treatment difficult and causes persistent infections, since they act as a continuous focus of infection [12].

As shown above, the current strategies for the control of *S. aureus* are not effective, demonstrating the claim to develop new effective alternatives for the control of *S. aureus* infections and their biofilms to reduce their impact on health. In this sense, plants could represent a feasible strategy, used in traditional medicine against different health conditions, including bacterial infections, where it is estimated around 80% of the world population uses plants as primary treatment for different health conditions [13]. *Flourensia retinophylla* S.F. blake, a plant known as "yerba de mula", is widely distributed in Coahuila, Mexico and it is used in traditional medicine against infections [14]. Recent research has provided information about different biological activities such as antimicrobial; however, its antimicrobial effect against *S. aureus* has not been reported and its antibiofilm effect has not been studied [14–16]. The biological potential of this plant is associated with the presence of bioactive compounds, especially phenolic compounds and terpenes [14,15].

Based on the above, the objective of this work was to determine the effect of *F. retinophylla* on planktonic cells of *S. aureus* and on the prevention and removal of bacterial biofilms of such pathogens.

2. Results and Discussions

In recent years, there has been a global increase in infections caused by bacteria. Those caused by antibiotic-resistant bacteria are of the greatest concern since the existing pharmacological treatments have decreased or even lost their efficiency [17]. Given this scenario, the need arises to explore new therapies capable of combating or reducing the incidence of infectious diseases associated with resistant pathogens. In this context, traditional medicine becomes more relevant since currently around 80% of the population uses it as primary treatment [13]. In addition, a considerable number of drugs have been derived from natural sources and plants are a notable source due to the presence of various secondary metabolites that have been reported to confer antimicrobial activity [18]. Among the plants used by ethnic groups, *F. retinophylla* stands out. Despite limited previous research on this plant, there has been interest in examining its antimicrobial and antibiofilm effects. This study reports for the first time the antimicrobial and antibiofilm activity of *F. retinophylla* against *S. aureus*, one of the main microorganisms causing nosocomial infections.

Antimicrobial assessment showed that the ethanolic extract of *F. retinophylla* effectively inhibited the growth of *S. aureus* with a minimum inhibitory concentration (MIC) of

250 µg/mL and a minimum bactericidal concentration (MBC) of 500 µg/mL. According to Simoes et al. [19], an MIC equal to or lower than 1000 µg/mL suggests that the natural source is a potential antimicrobial agent, thus revealing a promising role of this plant as bactericidal. Additionally, regarding edible plant extracts or their parts, it is estimated that they are very active if they show MIC values <100 µg/mL, significantly active if $100 \leq \text{MIC} \leq 512$ µg/mL, moderately active if $512 \leq \text{MIC} \leq 2048$ µg/mL and not very active if the MIC > 2048 µg/mL [20]. The biological potential of this plant could be associated with the nature of the chemical compounds present in the plant. Previously, our work group demonstrated the presence of bioactive compounds such as flavonoids, phenolic acids, and terpenes in this extract [14,15], which have been shown to exhibit antimicrobial effects against *S. aureus*, such as apigenin (MIC: 31.25 µg/mL), quercetin (MIC: 300 µg/mL), ellagic acid (MIC: 128 µg/mL) and phytol (MIC: >1000 µg/mL) [21–24].

On the other hand, the antimicrobial mode of action of *F. retinophylla* has not been previously reported. Thus, in order to know the mode of action associated with the antimicrobial effect of *F. retinophylla* extract, we evaluated its effect on the surface charge of *S. aureus* (Figure 1). We observed an increase of 1.4 times the surface charge of the bacteria treated with *F. retinophylla* MIC extract compared to the vehicle control ($p < 0.05$).

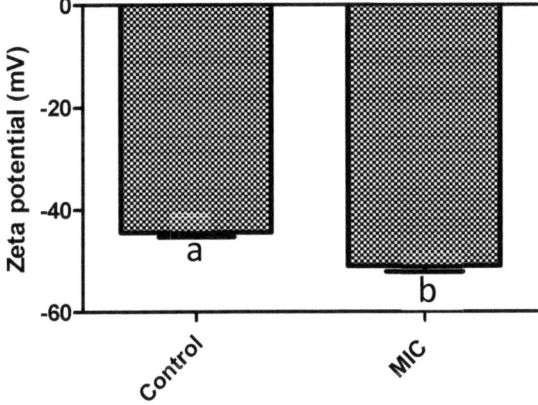

Figure 1. Zeta potential values (mV) of *S. aureus* after 1 h of exposure to *F. retinophylla* MIC ethanolic extract. Data are presented as mean ± standard deviation. $^{a-b}$ Mean with different letter are different ($p < 0.05$).

It is well known that the bacterial cell membrane has a negative cell surface charge due to its constituents [25]. Hence, modifications in the surface charge may indicate modifications in the bacterial membrane integrity. The *F. retinophylla* extract increased the surface charge of *S. aureus*, suggesting that the bioactive compounds present in the extract interact with the bacterial cell membrane. This charge change could be related to the ability of the plant compounds to modulate membrane potential, similar to other studies that have shown that phenolic compounds and terpenes have the ability to interact with the bacterial cell wall and membrane of *S. aureus*, altering the charge of the bacterial cell surface [26–28].

In the study of the mode of action of *F. retinophylla* against *S. aureus*, K^+ leak was assessed, which identifies changes in the permeability of the bacterial cell membrane. After 1 h of treatment with *F. retinophylla*, MIC extract produced a 60% increase in the K^+ concentration in the extracellular medium compared to the control group ($p < 0.05$) (Figure 2).

The cytoplasmic membrane is crucial to maintain the homeostasis of bacterial cells, regulating the entry and exit of intracellular components [29]. K^+ is one of the main constituents of the bacterial cytoplasm because it plays an important role in the metabolic processes [30]. The high concentration of K^+ in the extracellular medium found in this study suggests an alteration of the cytoplasmic membrane which may cause membrane disruption and subsequent leakage of intracellular components leading to bacterial cell death [31]. This effect may be associated with the bioactive compounds present in the *F. retinophylla* extract, since phenolic compounds and terpenes were shown to induce an alteration in the cytoplasmic membrane of *S. aureus*, inducing leakage of intracellular components including K^+ [24,27,28].

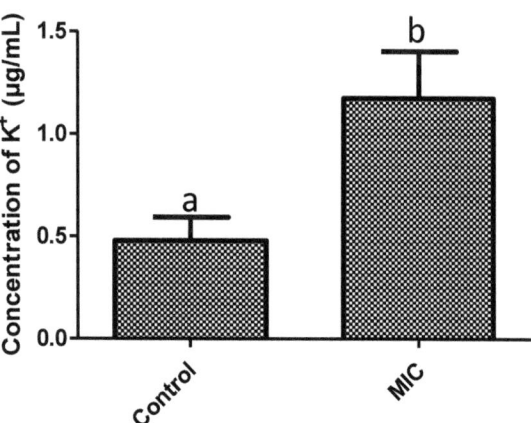

Figure 2. Concentration of K^+ (µg/mL) in solution of *S. aureus* after 1 h of exposure to *F. retinophylla* MIC ethanolic extract. Data are presented as mean ± standard deviation. [a-b] Mean with different letter are different ($p < 0.05$).

On the other hand, bacterial motility plays an essential role in the virulence and pathogenicity of bacteria. We analyzed the effect of the *F. retinophylla* extract on the motility of *S. aureus* (Figure 3). Our results showed that the extract inhibited the motility of *S. aureus* at all concentrations tested. MIC was the most active concentration ($p < 0.05$), showing a dose–response effect and inhibiting motility by 100%, 61% and 40% after 24 h, 48 h and 72 h, respectively. Furthermore, ½ MIC (between 41–15% at 24–72 h, respectively) and ¼ MIC (between 24–15% at 24–72 h, respectively) also decreased pathogen motility ($p < 0.05$).

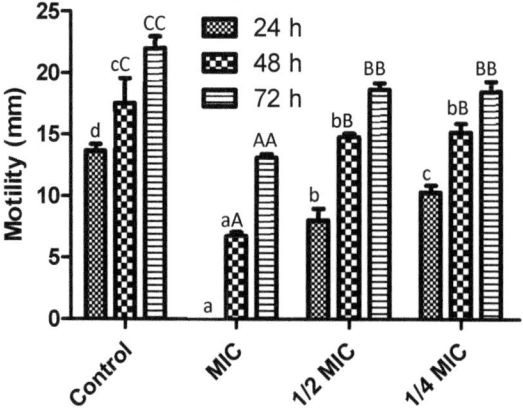

Figure 3. Motility (mm) of *S. aureus* after 72 h of exposure to different *F. retinophylla* ethanolic extract concentration. Data are presented as mean ± standard deviation. [a–d: 24 h; aA–cC: 48 h; AA–CC: 72 h] Mean with different letter between the different time of incubation (24, 48 and 72 h) are different ($p < 0.05$).

Results reveals that MIC concentration inhibits the motility process and subsequently increases over time, similar with the other evaluated concentrations. In this sense, it was demonstrated that MIC or lower concentrations of plant extracts could be able to inhibit the *S. aureus* motility process after 24 h [32]. Also, Vazquez-Armenta et al. [33] observed that grape stem extract inhibited the motility process in *L. monocytogenes* after 24 h; however, this process gradually increased after 48 and 72 h. The bacterial motility mechanism is not elucidated at all; however, it is important to highlight that motility correlates with the cellular state of the microorganism [34]. In this sense, MIC did not inhibit the entire bacterial viability; therefore, it could be suggested the presence of a minimum count of viable cells that subsequently managed to adapt and develop under stress conditions, such as antimicrobial treatments. Otherwise, the motility is vital for the adaptation, survival, colonization, biofilm development and virulence of bacteria [35]. In particular, *S. aureus* displays a motility process called spreading [36]. This process is related to two factors: the agr quorum sensing system and the production of phenol-soluble modulin surfactants. These molecules have lytic activity against leukocytes and erythrocytes, cause pro-inflammatory effects and interfere with the development of biofilms [36,37]. The results of this study showed that the *F. retinophylla* extract inhibited the motility process of *S. aureus* which could be related to the bioactive compounds present in the extract. For example, it was previously shown that phenolic compounds (luteolin, -hydroxyemodin and 3-hydroxybenzoic acid) and terpenes (eugenol and (+)-nootkatone) affect the *quorum sensing* system of *S. aureus* and its agr system [38–42], inhibiting the transcriptional units RNAII and RNAIII, which are involved in the production of virulence factors such as phenol-soluble modulin [38,43,44]. Furthermore, the agr *quorum sensing* system of *S. aureus* is associated with the bacterial membrane [45]. In this study, we showed that the *F. microphylla* extract alters the membrane integrity of *S. aureus*. Based on this, we hypothesize that: (i) the *F. retinophylla* extract could affect the agr *quorum sensing* system of *S. aureus*, inhibiting the production of phenol-soluble modulin, and (ii) the alteration of the bacterial membrane could affect the detection of the central *quorum sensing* system in *S. aureus*.

This study is the first one to report the antimicrobial effect of *F. retinophylla* against *S. aureus*, one of the most relevant Gram-positive pathogens in the clinic. We demonstrated that the effect against planktonic cells of *S. aureus* is related to the ability of the extract to induce damage in the bacterial cell wall and membrane, causing irreversible damage and subsequent cell death. Furthermore, we revealed that *F. retinophylla* significantly reduces motility on *S. aureus*, which may diminish bacterial pathogenicity.

Another important factor in the pathogenesis of *S. aureus* is its ability to form biofilms. In this sense, we assessed the ability of *F. retinophylla* to inhibit the formation of *S. aureus* biofilms. The extract significantly inhibited biomass formation ($p < 0.05$) by 80% at MIC, by 67% at 1/2 MIC and by 32% at 1/4 MIC (Figure 4). The extract also reduced the metabolic activity of the bacterial cells of the biofilms ($p < 0.05$), being most effective at MIC with a reduction of 82% (Figure 5). Concentrations of 1/2 MIC and 1/4 MIC also reduced ($p < 0.05$) the metabolic activity of biofilms by 66% and 43%, respectively.

The first step for the formation of biofilms is the reversible attachment of cells to the surface to be colonized [46]. This process is associated with motility and with the proteins and genes that regulate adhesion [47]. Here we showed that *F. retinophylla* significantly reduces biofilm formation of *S. aureus*. This effect could be initially associated with damage to the planktonic cells, on top of damage to the membrane where the proteins that regulate adhesion are located. Furthermore, the extract demonstrated to reduce the motility process of *S. aureus* and possibly affect the quorum sensing system. The bioactive compounds present in the extracts, such as flavonoids, demonstrated interaction with proteins that regulate adhesion and biofilm formation such as AtlE, Bap, IcaA, SarA, SasG [48,49]. Other studies have shown that phenolic compounds and terpenes exhibit the ability to affect the expression of genes associated with biofilm formation such as *sarA*, *agrA*, *icaA*, *spa*, *sdrD*, *hld*, *cap5B* and *cap5C* [50,51]. These studies also demonstrated that flavonoids and terpenes exhibit the ability to inhibit biofilm formation of *S. aureus*. Based on these findings, we suggest that the previously reported bioactive compounds present in the *F. retinophylla* extract could interact with proteins involved in the regulation of genes that control adhesion and biofilm formation of *S. aureus*, resulting in a high inhibition of biofilm formation of this pathogen.

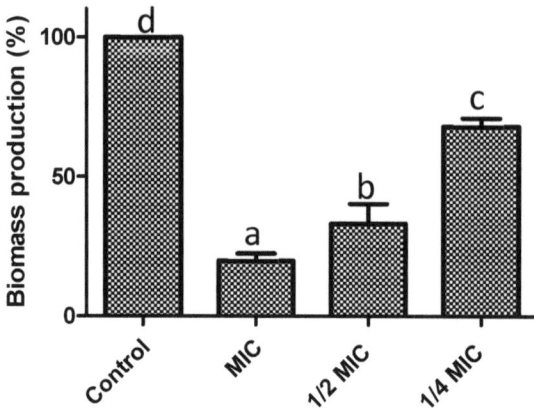

Figure 4. Preventive effect on biofilm formation of *F. retinophylla* ethanolic extract at MIC, ½ MIC and ¼ MIC on biomass production of *S. aureus*. Data are presented as mean ± standard deviation. [a–d] Mean with different letter are different ($p < 0.05$).

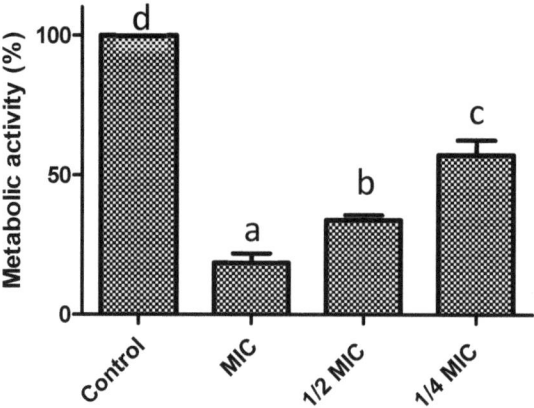

Figure 5. Preventive effect on biofilm formation of *F. retinophylla* ethanolic extract at MIC, ½ MIC and ¼ MIC on metabolic activity of *S. aureus*. Data are presented as mean ± standard deviation. [a–d] Mean with different letter are different ($p < 0.05$).

Another crucial factor is biofilms maturation, as it becomes highly complicated to control or eradicate it at this stage. In this sense, we assessed the ability of *F. retinophylla* to remove 24-h-preformed biofilms of *S. aureus* after being exposed to the extract for 1 h (Figure 6). Initially, the capacity of the extract to affect the metabolic activity of the preformed biofilms was evaluated. A concentration-dependent effect was observed ($p < 0.05$) since the highest evaluated concentration (10 MIC) showed the greatest reduction (83%), followed by the concentration of 5 MIC (50%) and 2 MIC (11%) after 1 h of contact. Then, the ability of *F. retinophylla* to affect the viability of the cells contained in the 24-h-preformed biofilms was determined by exposed them in contact for 1 h (Figure 7). Again, a similar effect was observed, where the concentration of 10 MIC decreased the viability of the biofilm cells by ~50% ($p < 0.05$), followed by the extract at 5 MIC with a 15% reduction ($p < 0.05$); however, the extract at 2 MIC did not affect bacterial viability.

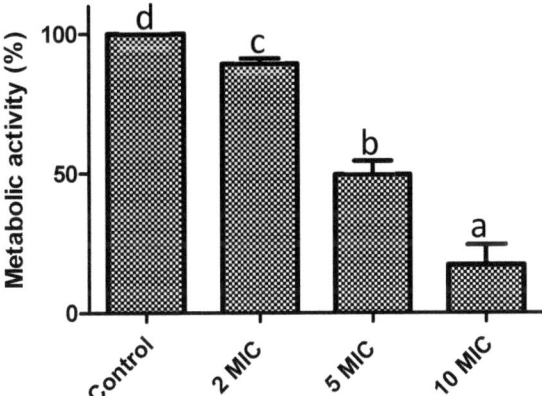

Figure 6. Effect of *F. retinophylla* ethanolic extract against pre-established biofilms (24 h) of *S. aureus* in terms of metabolic activity after 1 h of exposure to different concentration of *F. retinophylla* ethanolic extract. Data are presented as mean ± standard deviation. [a–d] Mean with different letter are different ($p < 0.05$).

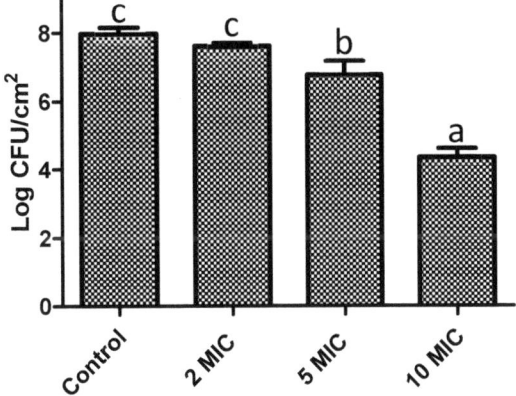

Figure 7. Effect of *F. retinophylla* ethanolic extract against pre-established biofilms (24 h) of *S. aureus* in terms of remaining viable biofilm cells after 1 h of exposure to different concentration of *F. retinophylla* ethanolic extract. Data are presented as mean ± standard deviation. [a–c] Mean with different letter are different ($p < 0.05$).

These results demonstrate that *F. retinophylla* has a high capacity to remove preformed biofilms of *S. aureus* after 1 h of exposition. This effect may be associated with the compounds previously reported to be present in the extract. Phenolic compounds and terpenes have been shown to alter bacterial biofilms by modifying the exopolysaccharide structure, i.e., reducing the concentration of polysaccharides, proteins and DNA of these structures [52–55]. Furthermore, microscopy studies have demonstrated that these bioactive compounds can degrade bacterial biofilms of *S. aureus* by being in contact [56,57]. This suggests that the bioactive compounds of the *F. retinophylla* extract may interact with the exopolysaccharide structure of *S. aureus* and damage its architecture, possibly allowing the passage of bioactive compounds that may interact with the cells and modify their metabolism and reduce their viability, resulting in bacterial cell death. In turn, it could be suggested that the bioactive compounds could cross the biofilms by passive diffusion and interact with the bacterial cells on the inside, producing a loss of their viability.

This work may be of interest in the clinic since a large number of infections caused by *S. aureus* and more than 80% of nosocomial infections are associated with bacterial biofilms [58,59]. In addition, these structures represent a serious challenge for the public health, as there are currently no effective strategies that can eradicate them after they form in the human organism [60]. It is worth mentioning that the exopolysaccharide structures provide the bacterial community with the ability to evade the effect of immune system cells, such as neutrophils, macrophages and antibodies [61]. It has also been observed that biofilms confer resistance against antimicrobials used in clinical practice and that it can be between 10 to 100 times higher compared to planktonic cells [62]. Moreover, once biofilms reach maturity, bacterial cells begin to detach and may colonize other biological surfaces, generating what is known as microbial metastasis [63]. These behaviors cause that the development of biofilms in the organisms gives rise to chronic, persistent infections and possible spread that in severe cases can lead to the death of the patient [64].

The results found in this study report unpublished and original data. To our knowledge, this is the first study to report the antibiofilm effect of *F. retinophylla*, specifically against *S. aureus* which constitutes a challenge in public health due to the antibiotic resistance that it has gained in recent years. Based on this work, we consider that *F. retinophylla* represents a possible source for the development of antimicrobial and/or disinfectant agents, which may be applied in humans for the treatment of bacterial infections caused by *S. aureus*. However, it is necessary to research and guarantee the safety of the use of this natural source.

3. Materials and Methods

3.1. Plant Material Obtention

F. retinophylla S.F. Blake was collected in Sierra Paila, Coahuila, Mexico during September 2021. The plant was identified at the Herbarium of the Universidad Autónoma Agraria Antonio Narro by Dr. José Ángel Villarreal Quintanilla (voucher 82956). The plants were transported to the Phytochemistry Laboratory of the Universidad Autónoma Agraria Antonio Narro, where they were dried in an oven (60 °C for 48 h) and grounded (2 mm sieve).

3.2. Preparation of the Extract

The extraction process was carried out using a Soxhlet method with ethanol as the extraction solvent. A total of 14 g of ground *F. retinophylla* leaves was extracted with 200 mL of ethanol for 72 h. After, the solvent was removed using a rotary evaporator. The residual solvent was further eliminated in an oven (50 °C for 24 h). The obtained dry extract was stored frozen at −20 °C until use.

All methods used in this study were performed in triplicate.

3.3. Bacterial Strain

The microorganism used in this research was *Staphylococcus aureus* ATCC 25923. The bacterial strain was preserved at −80 °C in cryovials containing Mueller–Hinton broth and glycerol (30% v/v). Before use, the bacterial strain was activated (37 °C for 24 h) on Mueller–Hinton broth.

3.4. Antimicrobial Evaluation

The antimicrobial effect of *F. retinophylla* extract was carried out using a reported method by Velazquez et al. [65]. Fresh overnight growth bacteria (16–18 h at 37 °C) in Mueller–Hinton broth was adjusted at 0.5 McFarland (1×10^8 colony forming units (CFU)/mL). Afterward, 15 µL of the adjusted bacteria were inoculated into 96-well polystyrene microplate (Costar, Corning, NY, USA) containing 200 µL of different extract concentrations (62.5–1000 µg/mL). *F. retinophylla* extract was dissolved in dimethyl sulfoxide (DMSO) and diluted in Mueller–Hinton broth. The concentration of DMSO was less than 2% (weight/volume: w/v). DMSO

(2%, w/v) and gentamicin (12 µg/mL) were utilized as controls. The microplates were incubated at 37 °C for 24 h. Afterward, the absorbance was read at 620 nm in a microplate reader (Multiskan Fc, Thermo Scientific, Bohemia, NY, USA). The minimal inhibitory concentration (MIC) was defined as the lowest concentration that inhibited bacterial growth. After, 10 µL of equal or lower concentration than MIC was inoculated onto plate count agar and incubated at 37 °C for 24 h. The concentration that showed no growth was defined as the minimal bactericidal concentration (MBC).

3.5. Surface Charge Determination

Fresh overnight culture (16–18 h) was adjusted to a cell density of 1×10^6 CFU/mL. The adjusted bacteria were brought into contact with the MIC extract and incubated for 1 h at 37 °C. The mixture was centrifugated twice ($6000 \times g$ for 10 min) and resuspended in sterile water. Finally, bacteria surface charge was calculated by zeta potential using Zeta-sizer Nano-ZS90 (Malvern Instruments Ltd., Worcestershire, UK), with deionized water utilized as a diluent [27].

3.6. Potassium (K^+) Leakage

The potassium leakage was analyzed using flame emission and atomic absorption spectroscopy for K+ tritation in *S. aureus* suspension incorporated with MIC extract. The mixture was kept in contact for 1 h and was filtered (sterile membrane filter size of 0.22 µM). The sample was analyzed by atomic absorption spectroscopy, using a Perking-Elmer atomic absorption equipment AAnalyst 400 (Perkin Elmer, Shelton, CT, USA) [27].

3.7. Motility Assay

Bacterial motility was evaluated following the method described by Abreu et al. [66] with some modifications. Fresh overnight culture (16–18 h) was adjusted at 1×10^8 CFU/mL. After that, 10 µL of the bacterial inoculum was transferred in the center of a Petri dish with 0.3% agar in the presence of MIC, ½ MIC, and ¼ MIC of *F. retinophylla* extract. The extract was incorporated into the medium at 45 °C. DMSO was used as a control. The inoculated Petri dish was stored at 30 °C and measured at 24, 48 and 72 h.

3.8. Biofilm Inhibition Effect—Inhibitory Effect of Initial Bacterial Cells Attachment

The effect of *F. retinophylla* extract to prevent the biofilm formation of *S. aureus* was performed using the method described by Bazargani and Rohloff [67]. A total of 100 µL of MIC, ½ MIC and ¼ MIC of *F. retinophylla* extract were incorporated into each well of a 96-well polystyrene microplate (Costar, Corning, NY, USA). DMSO was used as a control. Subsequently, 100 µL of inoculum (1×10^6 CFU/mL) was added to the wells. The microplates were incubated for 8 h at 37 °C. Subsequently, the biomass production and metabolic activity were quantified.

Biomass production: the content of each well was washed with 200 µL of saline solution (0.85% w/v) and fixed with 200 µL of methanol for 15 min. Subsequently, the microplates were air-dried and stained with 200 µL of crystal violet (CV, 1%, volume/volume, v/v). The CV was removed, and 200 µL of glacial acetic acid (33%, v/v) was added to the wells. Finally, a microplate reader (Multiskan Fc, Thermo Scientific) was used to measure the absorbance at 590 nm. Results were expressed as a percentage of biomass production [62].

Metabolic activity: the content of each well was washed with saline solution (0.85% w/v). XTT was dissolved in saline solution (0.85%, w/v) to obtain a 1 mg/mL concentration. Subsequently, the solution was filter-sterilized and stored at −80 °C. Menadione was dissolved in acetone to obtain a 1 mM concentration and sterilized. Each well was then filled with 200 µL of saline solution (0.85%, w/v), and 27 µL of an XTT (sodium 3′-[1-(phenylaminocarbonyl)-3,4-tetrazolium]-bis(4-methoxy6-nitro) benzene sulfonic acid hydrate)/menadione mixture (relation 12.5:1) was added. The microplates were incubated in darkness at 37 °C for 2–3 h. Absorbance was measured at 490 nm in a microplate reader

(Multiskan Fc, Thermo Scientific). Results were represented as a percentage of metabolic activity [67].

3.9. Biofilm Control

The effect of *F. retinophylla* extract to remove preformed biofilm of *S. aureus* was evaluated based on the method previously described by Borges et al. [62]. Fresh overnight culture (16–18 h) was adjusted to a cell density of 1×10^8 CFU/mL. Subsequently, 200 µL of inoculum was added to a 96-well polystyrene microplate (Costar, Corning, NY, USA) and incubated for 24 h at 37 °C. After, the content of each well was washed with 200 µL of saline solution (0.85% w/v). *F. retinophylla* extract was performed at different concentrations ($2 \times$ MIC, $5 \times$ MIC, $10 \times$ MIC) and incorporated in each well of a 96-well microplate and incubated for 1 h at 37 °C. DMSO was used as a control. After, the metabolic activity and biofilm viable cells were quantified.

Metabolic activity was analyzed using the method previously described.

Biofilm viable cells: the content of each well was washed with saline solution (0.85% w/v). Each well was then filled with 200 µL of saline solution (0.85%, w/v) and scrapped from the microtiter plate using a pipette tip for 1 min. This process was replicated three times. After, serial dilutions (1:10) were performed in saline solution (0.85%, w/v). Subsequently, 10 µL of each dilution was cultured on plate count agar. The inoculated plates were incubated for 24 h at 37 °C. Finally, the bacterial colonies were quantified. Results were represented as log CFU/cm^2 [62].

3.10. Statistical Analysis

The data were analyzed through ANOVA using the software NCSS, 2007. The analyzed variables in this study were surface charge, potassium leakage, motility, biomass production, metabolic activity and viable count. In all cases, data is presented as mean ± standard deviation. When significance differences were observed between treatments means, a Tukey–Kramer test were carried out. The level of significance in the error was $p < 0.05$.

4. Conclusions

This study is the first to showcase the potential antimicrobial and antibiofilm effects of *F. retinophylla* against *S. aureus*. Further research is required to conduct toxicological tests to evaluate the safety of *F. retinophylla*. Future studies should focus on identifying the specific compounds in *F. retinophylla* responsible for its antimicrobial and antibiofilm effects. Additionally, the obtained results open the door for the analysis of the antimicrobial and antibiofilm effect of this plant source against other microorganisms of relevance in the health sector. Finally, the results give additional scientific support to the medicinal use of the analyzed species to treat infections.

Author Contributions: Conceptualization, J.C.L.-R., H.T.-M. and D.J.d.R.; methodology, J.C.L.-R., H.T.-M., D.J.d.R., M.E.B.-M., M.R.T.-R. and A.G.-Á.; validation, J.C.L.-R.; formal analysis, J.C.L.-R. and M.E.B.-M.; investigation, J.C.L.-R., D.J.d.R. and M.E.B.-M.; resources, J.C.L.-R., D.J.d.R., J.F.A.-Z. and A.G.-Á.; writing—original draft preparation, J.C.L.-R., D.J.d.R. and M.E.B.-M.; writing—review and editing, J.C.L.-R., D.J.d.R., H.T.-M., R.E.R.-Z., M.R.T.-R. and A.G.-Á.; supervision, J.C.L.-R.; project administration, J.C.L.-R. All authors have read and agreed to the published version of the manuscript.

Funding: This research received no funding.

Data Availability Statement: The original contributions presented in the study are included in the article, further inquiries can be directed to the corresponding authors.

Acknowledgments: We extend our sincere gratitude to QB. Rosa Idalia Armenta Corral for her active support in the analysis of the surface charge of *Staphylococcus aureus* exposed to antimicrobial treatments. QB. Armenta provided invaluable guidance in developing the technique and training in properly using the Zeta-sizer equipment.

Conflicts of Interest: The authors declare no conflicts of interest.

References

1. Salam, M.A.; Al-Amin, M.Y.; Salam, M.T.; Pawar, J.S.; Akhter, N.; Rabaan, A.A.; Alqumber, M.A. Antimicrobial resistance: A growing serious threat for global public health. *Healthcare* **2023**, *11*, 1946. [CrossRef] [PubMed]
2. Hou, J.; Long, X.; Wang, X.; Li, L.; Mao, D.; Luo, Y.; Ren, H. Global trend of antimicrobial resistance in common bacterial pathogens in response to antibiotic consumption. *J. Hazard. Mater.* **2023**, *442*, 130042. [CrossRef] [PubMed]
3. Kim, C.; Holm, M.; Frost, I.; Hasso-Agopsowicz, M.; Abbas, K. Global and regional burden of attributable and associated bacterial antimicrobial resistance avertable by vaccination: Modelling study. *BMJ Glob. Health* **2023**, *8*, e011341. [CrossRef]
4. WHO. WHO Publishes List of Bacteria for Which New Antibiotics Are Urgently Needed. Available online: https://www.who.int/news/item/27-02-2017-who-publishes-list-of-bacteria-for-which-new-antibiotics-are-urgently-needed (accessed on 2 February 2024).
5. Roch, M.; Sierra, R.; Andrey, D.O. Antibiotic heteroresistance in ESKAPE pathogens, from bench to bedside. *Clin. Microbiol. Infect.* **2023**, *29*, 320–325. [CrossRef] [PubMed]
6. Linz, M.S.; Mattappallil, A.; Finkel, D.; Parker, D. Clinical impact of *Staphylococcus aureus* skin and soft tissue infections. *Antibiotics* **2023**, *12*, 557. [CrossRef] [PubMed]
7. Gherardi, G. *Staphylococcus aureus* infection: Pathogenesis and antimicrobial resistance. *Int. J. Mol. Sci.* **2023**, *24*, 8182. [CrossRef] [PubMed]
8. Ciandrini, E.; Morroni, G.; Cirioni, O.; Kamysz, W.; Kamysz, E.; Brescini, L.; Baffone, W.; Campana, R. Synergistic combinations of antimicrobial peptides against biofilms of methicillin-resistant *Staphylococcus aureus* (MRSA) on polystyrene and medical devices. *J. Glob. Antimicrob. Resist.* **2020**, *21*, 203–210. [CrossRef] [PubMed]
9. Tuon, F.F.; Suss, P.H.; Telles, J.P.; Dantas, L.R.; Borges, N.H.; Ribeiro, V.S.T. Antimicrobial treatment of *Staphylococcus aureus* biofilms. *Antibiotics* **2023**, *12*, 87. [CrossRef] [PubMed]
10. Zhao, A.; Sun, J.; Liu, Y. Understanding bacterial biofilms: From definition to treatment strategies. *Front. Cell. Infect. Microbiol.* **2023**, *13*, 1137947. [CrossRef]
11. Vani, S.; Vadakkan, K.; Mani, B. A narrative review on bacterial biofilm: Its formation, clinical aspects and inhibition strategies. *Future J. Pharm. Sci.* **2023**, *9*, 50. [CrossRef]
12. Mohamad, F.; Alzahrani, R.R.; Alsaadi, A.; Alrfaei, B.M.; Yassin, A.E.B.; Alkhulaifi, M.M.; Halwani, M. An explorative review on advanced approaches to overcome bacterial resistance by curbing bacterial biofilm formation. *Infect. Drug Resist.* **2023**, *16*, 19–49. [CrossRef] [PubMed]
13. Haq, A.; Badshah, L.; Hussain, W.; Ullah, I. Quantitative ethnobotanical exploration of wild medicinal plants of Arang Valley, District Bajaur, Khyber Pakhtunkhwa, Pakistan: A mountainous region of the Hindu Kush Range. *Ethnobot. Res. Appl.* **2023**, *25*, 1–29. [CrossRef]
14. de Rodríguez, D.J.; Victorino-Jasso, M.; Rocha-Guzmán, N.; Moreno-Jiménez, M.; Díaz-Jiménez, L.; Rodríguez-García, R.; Villarreal-Quintanilla, J.; Peña-Ramos, F.; Carrillo-Lomelí, D.; Genisheva, Z. *Flourensia retinophylla*: An outstanding plant from northern Mexico with antibacterial activity. *Ind. Crops Prod.* **2022**, *185*, 115120. [CrossRef]
15. de Rodríguez, D.J.; Torres-Moreno, H.; López-Romero, J.C.; Vidal-Gutiérrez, M.; Villarreal-Quintanilla, J.Á.; Carrillo-Lomelí, D.A.; Robles-Zepeda, R.E.; Vilegas, W. Antioxidant, anti-inflammatory, and antiproliferative activities of *Flourensia* spp. *Biocatal. Agric. Biotechnol.* **2023**, *47*, 102552. [CrossRef]
16. de Rodríguez, D.J.; Hernández-Castillo, D.; Angulo-Sánchez, J.; Rodríguez-García, R.; Villarreal Quintanilla, J.; Lira-Saldivar, R. Antifungal activity in vitro of *Flourensia* spp. extracts on Alternaria sp., Rhizoctonia solani, and *Fusarium oxysporum*. *Ind. Crop. Prod.* **2007**, *25*, 111–116. [CrossRef]
17. Bai, H.-J.; Geng, Q.-F.; Jin, F.; Yang, Y.-L. Epidemiologic analysis of antimicrobial resistance in hospital departments in China from 2022 to 2023. *J. Health Popul. Nutr.* **2024**, *43*, 39. [CrossRef] [PubMed]
18. Zhou, H.; Eun, H.; Lee, S.Y. Systems metabolic engineering for the production of pharmaceutical natural products. *Curr. Opin. Syst. Biol.* **2023**, *37*, 100491. [CrossRef]
19. Simoes, M.; Bennett, R.N.; Rosa, E.A. Understanding antimicrobial activities of phytochemicals against multidrug resistant bacteria and biofilms. *Nat. Prod. Rep.* **2009**, *26*, 746–757. [CrossRef] [PubMed]
20. Tamokou, J.; Mbaveng, A.; Kuete, V. Antimicrobial activities of African medicinal spices and vegetables. In *Medicinal Spices and Vegetables from Africa*; Elsevier: Amsterdam, The Netherlands, 2017; pp. 207–237.
21. de Moraes Alves, M.M.; Brito, L.M.; Souza, A.C.; de Carvalho, T.P.; Viana, F.J.C.; de Alcântara Oliveira, F.A.; Barreto, H.M.; Oliveira, J.S.d.S.M.; Chaves, M.H.; Arcanjo, D.D.R. Antimicrobial activity and cytotoxic assessment of gallic and ellagic acids. *J. Interdiscip. De Biociências* **2018**, *3*, 17.
22. Liu, R.; Zhang, H.; Yuan, M.; Zhou, J.; Tu, Q.; Liu, J.-J.; Wang, J. Synthesis and biological evaluation of apigenin derivatives as antibacterial and antiproliferative agents. *Molecules* **2013**, *18*, 11496–11511. [CrossRef]
23. Ghaneian, M.T.; Ehrampoush, M.H.; Jebali, A.; Hekmatimoghaddam, S.; Mahmoudi, M. Antimicrobial activity, toxicity and stability of phytol as a novel surface disinfectant. *Environ. Health Eng. Manag. J.* **2015**, *2*, 13–16.
24. Amin, M.U.; Khurram, M.; Khattak, B.; Khan, J. Antibiotic additive and synergistic action of rutin, morin and quercetin against methicillin resistant *Staphylococcus aureus*. *BMC Complement. Altern. Med.* **2015**, *15*, 1–12. [CrossRef] [PubMed]
25. Zhang, J.; Su, P.; Chen, H.; Qiao, M.; Yang, B.; Zhao, X. Impact of reactive oxygen species on cell activity and structural integrity of Gram-positive and Gram-negative bacteria in electrochemical disinfection system. *Chem. Eng. J.* **2023**, *451*, 138879. [CrossRef]

26. Kurinčič, M.; Jeršek, B.; Klančnik, A.; Možina, S.S.; Fink, R.; Dražić, G.; Raspor, P.; Bohinc, K. Effects of natural antimicrobials on bacterial cell hydrophobicity, adhesion, and zeta potential. *Arh. Za Hig. Rada I Toksikol.* **2016**, *67*, 39–45. [CrossRef] [PubMed]
27. Borges, A.; Ferreira, C.; Saavedra, M.J.; Simões, M. Antibacterial activity and mode of action of ferulic and gallic acids against pathogenic bacteria. *Microb. Drug Resist.* **2013**, *19*, 256–265. [CrossRef] [PubMed]
28. Lopez-Romero, J.C.; González-Ríos, H.; Borges, A.; Simões, M. Antibacterial effects and mode of action of selected essential oils components against *Escherichia coli* and *Staphylococcus aureus*. *Evid.-Based Complement. Altern. Med.* **2015**, *2015*, 795435.
29. Wei, H.; Shan, X.; Wu, L.; Zhang, J.; Saleem, M.; Yang, J.; Liu, Z.; Chen, X. Microbial cell membrane properties and intracellular metabolism regulate individual level microbial responses to acid stress. *Soil Biol. Biochem.* **2023**, *177*, 108883. [CrossRef]
30. Beagle, S.D.; Lockless, S.W. Unappreciated roles for K+ channels in bacterial physiology. *Trends Microbiol.* **2021**, *29*, 942–950. [CrossRef]
31. Sinlapapanya, P.; Sumpavapol, P.; Nirmal, N.; Zhang, B.; Hong, H.; Benjakul, S. Ethanolic cashew leaf extract: Antimicrobial activity, mode of action, and retardation of spoilage bacteria in refrigerated Nile tilapia slices. *Foods* **2022**, *11*, 3461. [CrossRef]
32. Hayat, S.; Sabri, A.N.; McHugh, T.D. Chloroform extract of turmeric inhibits biofilm formation, EPS production and motility in antibiotic resistant bacteria. *J. Gen. Appl. Microbiol.* **2017**, *63*, 325–338. [CrossRef]
33. Vazquez-Armenta, F.; Bernal-Mercado, A.; Lizardi-Mendoza, J.; Silva-Espinoza, B.; Cruz-Valenzuela, M.; Gonzalez-Aguilar, G.; Nazzaro, F.; Fratianni, F.; Ayala-Zavala, J. Phenolic extracts from grape stems inhibit *Listeria monocytogenes* motility and adhesion to food contact surfaces. *J. Adhes. Sci. Technol.* **2018**, *32*, 889–907. [CrossRef]
34. Zheng, S.; Bawazir, M.; Dhall, A.; Kim, H.-E.; He, L.; Heo, J.; Hwang, G. Implication of surface properties, bacterial motility, and hydrodynamic conditions on bacterial surface sensing and their initial adhesion. *Front. Bioeng. Biotechnol.* **2021**, *9*, 643722. [CrossRef] [PubMed]
35. Duan, Q.; Zhou, M.; Zhu, L.; Zhu, G. Flagella and bacterial pathogenicity. *J. Basic Microbiol.* **2013**, *53*, 1–8. [CrossRef] [PubMed]
36. Pollitt, E.J.; Crusz, S.A.; Diggle, S.P. *Staphylococcus aureus* forms spreading dendrites that have characteristics of active motility. *Sci. Rep.* **2015**, *5*, 17698. [CrossRef] [PubMed]
37. Pollitt, E.J.; Diggle, S.P. Defining motility in the Staphylococci. *Cell. Mol. Life Sci.* **2017**, *74*, 2943–2958. [CrossRef] [PubMed]
38. Yuan, Q.; Feng, W.; Wang, Y.; Wang, Q.; Mou, N.; Xiong, L.; Wang, X.; Xia, P.; Sun, F. Luteolin attenuates the pathogenesis of *Staphylococcus aureus* by interfering with the agr system. *Microb. Pathog.* **2022**, *165*, 105496. [CrossRef] [PubMed]
39. Daly, S.M.; Elmore, B.O.; Kavanaugh, J.S.; Triplett, K.D.; Figueroa, M.; Raja, H.A.; El-Elimat, T.; Crosby, H.A.; Femling, J.K.; Cech, N.B. ω-Hydroxyemodin limits *Staphylococcus aureus* quorum sensing-mediated pathogenesis and inflammation. *Antimicrob. Agents Chemother.* **2015**, *59*, 2223–2235. [CrossRef] [PubMed]
40. Ganesh, P.S.; Veena, K.; Senthil, R.; Iswamy, K.; Ponmalar, E.M.; Mariappan, V.; Girija, A.S.; Vadivelu, J.; Nagarajan, S.; Challabathula, D. Biofilm-associated Agr and Sar *quorum sensing* systems of *Staphylococcus aureus* are inhibited by 3-hydroxybenzoic acid derived from *Illicium verum*. *ACS Omega* **2022**, *7*, 14653–14665. [CrossRef] [PubMed]
41. Li, H.; Li, C.; Shi, C.; Alharbi, M.; Cui, H.; Lin, L. Phosphoproteomics analysis reveals the anti-bacterial and anti-virulence mechanism of eugenol against *Staphylococcus aureus* and its application in meat products. *Int. J. Food Microbiol.* **2024**, *414*, 110621. [CrossRef]
42. Farha, A.K.; Yang, Q.-Q.; Kim, G.; Zhang, D.; Mavumengwana, V.; Habimana, O.; Li, H.-B.; Corke, H.; Gan, R.-Y. Inhibition of multidrug-resistant foodborne *Staphylococcus aureus* biofilms by a natural terpenoid (+)-nootkatone and related molecular mechanism. *Food Control* **2020**, *112*, 107154. [CrossRef]
43. Coelho, P.; Oliveira, J.; Fernandes, I.; Araújo, P.; Pereira, A.R.; Gameiro, P.; Bessa, L.J. Pyranoanthocyanins Interfering with the Quorum Sensing of *Pseudomonas aeruginosa* and *Staphylococcus aureus*. *Int. J. Mol. Sci.* **2021**, *22*, 8559. [CrossRef] [PubMed]
44. Nakagawa, S.; Hillebrand, G.G.; Nunez, G. *Rosmarinus officinalis* L. (rosemary) extracts containing carnosic acid and carnosol are potent quorum sensing inhibitors of *Staphylococcus aureus* virulence. *Antibiotics* **2020**, *9*, 149. [CrossRef] [PubMed]
45. Bojer, M.S.; Lindemose, S.; Vestergaard, M.; Ingmer, H. Quorum sensing-regulated phenol-soluble modulins limit persister cell populations in *Staphylococcus aureus*. *Front. Microbiol.* **2018**, *9*, 336511. [CrossRef]
46. Kilic, T.; Bali, E.B. Biofilm control strategies in the light of biofilm-forming microorganisms. *World J. Microbiol. Biotechnol.* **2023**, *39*, 131. [CrossRef] [PubMed]
47. Moormeier, D.E.; Bayles, K.W. *Staphylococcus aureus* biofilm: A complex developmental organism. *Mol. Microbiol.* **2017**, *104*, 365–376. [CrossRef]
48. Parai, D.; Banerjee, M.; Dey, P.; Mukherjee, S.K. Reserpine attenuates biofilm formation and virulence of *Staphylococcus aureus*. *Microb. Pathog.* **2020**, *138*, 103790. [CrossRef]
49. Matilla-Cuenca, L.; Gil, C.; Cuesta, S.; Rapún-Araiz, B.; Žiemytė, M.; Mira, A.; Lasa, I.; Valle, J. Antibiofilm activity of flavonoids on staphylococcal biofilms through targeting BAP amyloids. *Sci. Rep.* **2020**, *10*, 18968. [CrossRef]
50. Salinas, C.; Florentín, G.; Rodríguez, F.; Alvarenga, N.; Guillén, R. Terpenes combinations inhibit biofilm formation in *Staphylococcus aureus* by interfering with initial adhesion. *Microorganisms* **2022**, *10*, 1527. [CrossRef]
51. Wu, X.; Wang, H.; Xiong, J.; Yang, G.-X.; Hu, J.-F.; Zhu, Q.; Chen, Z. *Staphylococcus aureus* biofilm: Formulation, regulatory, and emerging natural products-derived therapeutics. *Biofilm* **2024**, *7*, 100175. [CrossRef]
52. Vazquez-Armenta, F.; Bernal-Mercado, A.; Tapia-Rodriguez, M.; Gonzalez-Aguilar, G.; Lopez-Zavala, A.; Martinez-Tellez, M.; Hernandez-Oñate, M.; Ayala-Zavala, J. Quercetin reduces adhesion and inhibits biofilm development by *Listeria monocytogenes* by reducing the amount of extracellular proteins. *Food Control* **2018**, *90*, 266–273. [CrossRef]

53. Ivanov, M.; Nović, K.; Malešević, M.; Dinić, M.; Stojković, D.; Jovčić, B.; Soković, M. Polyphenols as inhibitors of antibiotic resistant bacteria—Mechanisms underlying rutin interference with bacterial virulence. *Pharmaceuticals* **2022**, *15*, 385. [CrossRef]
54. Peng, Q.; Tang, X.; Dong, W.; Zhi, Z.; Zhong, T.; Lin, S.; Ye, J.; Qian, X.; Chen, F.; Yuan, W. Carvacrol inhibits bacterial polysaccharide intracellular adhesin synthesis and biofilm formation of mucoid *Staphylococcus aureus*: An in vitro and in vivo study. *RSC Adv.* **2023**, *13*, 28743–28752. [CrossRef]
55. Zhang, C.; Li, C.; Abdel-Samie, M.A.; Cui, H.; Lin, L. Unraveling the inhibitory mechanism of clove essential oil against *Listeria monocytogenes* biofilm and applying it to vegetable surfaces. *LWT* **2020**, *134*, 110210. [CrossRef]
56. Gu, K.; Ouyang, P.; Hong, Y.; Dai, Y.; Tang, T.; He, C.; Shu, G.; Liang, X.; Tang, H.; Zhu, L. Geraniol inhibits biofilm formation of methicillin-resistant *Staphylococcus aureus* and increase the therapeutic effect of vancomycin in vivo. *Front. Microbiol.* **2022**, *13*, 960728. [CrossRef] [PubMed]
57. Cui, S.; Ma, X.; Wang, X.; Zhang, T.-A.; Hu, J.; Tsang, Y.F.; Gao, M.-T. Phenolic acids derived from rice straw generate peroxides which reduce the viability of *Staphylococcus aureus* cells in biofilm. *Ind. Crops Prod.* **2019**, *140*, 111561. [CrossRef]
58. Jamal, M.; Ahmad, W.; Andleeb, S.; Jalil, F.; Imran, M.; Nawaz, M.A.; Hussain, T.; Ali, M.; Rafiq, M.; Kamil, M.A. Bacterial biofilm and associated infections. *J. Chin. Med. Assoc.* **2018**, *81*, 7–11. [CrossRef] [PubMed]
59. Shakibaie, M.R. Bacterial biofilm and its clinical implications. *Ann. Microbiol. Res.* **2018**, *2*, 45–50.
60. Ali, A.; Zahra, A.; Kamthan, M.; Husain, F.M.; Albalawi, T.; Zubair, M.; Alatawy, R.; Abid, M.; Noorani, M.S. Microbial biofilms: Applications, clinical consequences, and alternative therapies. *Microorganisms* **2023**, *11*, 1934. [CrossRef]
61. Ramírez-Larrota, J.S.; Eckhard, U. An introduction to bacterial biofilms and their proteases, and their roles in host infection and immune evasion. *Biomolecules* **2022**, *12*, 306. [CrossRef]
62. Borges, A.; Lopez-Romero, J.; Oliveira, D.; Giaouris, E.; Simões, M. Prevention, removal and inactivation of *Escherichia coli* and *Staphylococcus aureus* biofilms using selected monoterpenes of essential oils. *J. Appl. Microbiol.* **2017**, *123*, 104–115. [CrossRef]
63. Chauhan, A.; Lebeaux, D.; Decante, B.; Kriegel, I.; Escande, M.-C.; Ghigo, J.-M.; Beloin, C. A rat model of central venous catheter to study establishment of long-term bacterial biofilm and related acute and chronic infections. *PLoS ONE* **2012**, *7*, e37281. [CrossRef] [PubMed]
64. Nesse, L.L.; Osland, A.M.; Vestby, L.K. The role of biofilms in the pathogenesis of animal bacterial infections. *Microorganisms* **2023**, *11*, 608. [CrossRef] [PubMed]
65. Velazquez, C.; Navarro, M.; Acosta, A.; Angulo, A.; Dominguez, Z.; Robles, R.; Robles-Zepeda, R.; Lugo, E.; Goycoolea, F.; Velazquez, E. Antibacterial and free-radical scavenging activities of Sonoran propolis. *J. Appl. Microbiol.* **2007**, *103*, 1747–1756. [CrossRef] [PubMed]
66. Abreu, A.C.; Borges, A.; Mergulhão, F.; Simões, M. Use of phenyl isothiocyanate for biofilm prevention and control. *Int. Biodeterior. Biodegrad.* **2014**, *86*, 34–41. [CrossRef]
67. Bazargani, M.M.; Rohloff, J. Antibiofilm activity of essential oils and plant extracts against *Staphylococcus aureus* and *Escherichia coli* biofilms. *Food Control* **2016**, *61*, 156–164. [CrossRef]

Disclaimer/Publisher's Note: The statements, opinions and data contained in all publications are solely those of the individual author(s) and contributor(s) and not of MDPI and/or the editor(s). MDPI and/or the editor(s) disclaim responsibility for any injury to people or property resulting from any ideas, methods, instructions or products referred to in the content.

Article

Antimicrobial, Cytotoxic, and Anti-Inflammatory Activities of *Tigridia vanhouttei* Extracts

Jorge L. Mejía-Méndez [1,*,†], Ana C. Lorenzo-Leal [2,†], Horacio Bach [2,*,†], Edgar R. López-Mena [3], Diego E. Navarro-López [3], Luis R. Hernández [1], Zaida N. Juárez [4] and Eugenio Sánchez-Arreola [1,*]

[1] Laboratory of Phytochemistry Research, Chemical Biological Sciences Department, Universidad de las Américas Puebla, Ex Hacienda Sta. Catarina Mártir S/N, San Andrés Cholula 72810, Mexico; luisr.hernandez@udlap.mx
[2] Division of Infectious Diseases, Faculty of Medicine, University of British Columbia, Vancouver, BC V6H 3Z6, Canada; anacecylole@gmail.com
[3] Tecnologico de Monterrey, Escuela de Ingeniería y Ciencias, Campus Guadalajara, Av. Gral. Ramón Corona No 2514, Colonia Nuevo México, Zapopan 45121, Mexico; edgarl@tec.mx (E.R.L.-M.); diegonl@tec.mx (D.E.N.-L.)
[4] Chemistry Area, Biological Sciences, Universidad Popular Autónoma del Estado de Puebla, 21 Sur 1103 Barrio Santiago, Puebla 72410, Mexico; zaidanelly.juarez@upaep.mx
* Correspondence: jorge.mejiamz@udlap.mx (J.L.M.-M.); hbach@mail.ubc.ca (H.B.); eugenio.sanchez@udlap.mx (E.S.-A.)
† These authors contributed equally to this work.

Citation: Mejía-Méndez, J.L.; Lorenzo-Leal, A.C.; Bach, H.; López-Mena, E.R.; Navarro-López, D.E.; Hernández, L.R.; Juárez, Z.N.; Sánchez-Arreola, E. Antimicrobial, Cytotoxic, and Anti-Inflammatory Activities of *Tigridia vanhouttei* Extracts. *Plants* **2023**, *12*, 3136. https://doi.org/10.3390/plants12173136

Academic Editor: Ain Raal

Received: 21 August 2023
Revised: 30 August 2023
Accepted: 30 August 2023
Published: 31 August 2023

Copyright: © 2023 by the authors. Licensee MDPI, Basel, Switzerland. This article is an open access article distributed under the terms and conditions of the Creative Commons Attribution (CC BY) license (https://creativecommons.org/licenses/by/4.0/).

Abstract: In this work, bulb extracts of *Tigridia vanhouttei* were obtained by maceration with solvents of increasing polarity. The extracts were evaluated against a panel of pathogenic bacterial and fungal strains using the minimal inhibitory concentration (MIC) assay. The cytotoxicity of the extracts was tested against two cell lines (THP-1 and A549) using the MTT assay. The anti-inflammatory activity of the extracts was evaluated in THP-1 cells by measuring the secretion of pro-inflammatory (IL-6 and TNF-α) and anti-inflammatory (IL-10) cytokines by ELISA. The chemical composition of the extracts was recorded by FTIR spectroscopy, and their chemical profiles were evaluated using GC-MS. The results revealed that only hexane extract inhibited the growth of the clinical isolate of *Pseudomonas aeruginosa* at 200 μg/mL. Against THP-1 cells, hexane and chloroform extracts were moderately cytotoxic, as they exhibited LC_{50} values of 90.16, and 46.42 μg/mL, respectively. Treatment with methanol extract was weakly cytotoxic at LC_{50} 443.12 μg/mL against the same cell line. Against the A549 cell line, hexane, chloroform, and methanol extracts were weakly cytotoxic because of their LC_{50} values: 294.77, 1472.37, and 843.12 μg/mL. The FTIR analysis suggested the presence of natural products were confirmed by carboxylic acids, ketones, hydroxyl groups, or esters. The GC-MS profile of extracts revealed the presence of phytosterols, tetracyclic triterpenes, multiple fatty acids, and sugars. This report confirms the antimicrobial, cytotoxic, and anti-inflammatory activities of *T. vanhouttei*.

Keywords: traditional medicine; Iridaceae; *Iris*; *Tigridia vanhouttei*; biological activities

1. Introduction

Currently, human health is being threatened by infections caused by drug-resistant microorganisms, such as bacteria and fungi, and high-incident types of cancer [1,2].

According to the Centers for Disease Control and Prevention (CDC), it has been documented that resistance to current antimicrobials has resulted in more than 2.8 million infections and 35,000 deaths in the United States of America (USA) [3]. Multi-drug resistant bacteria related to these events include members of the ESKAPE (*Enterococcus faecium*, *Staphylococcus aureus*, *Klebsiella pneumoniae*, *Acinetobacter baumannii*, *Pseudomonas aeruginosa*, and *Enterobacter* species) pathogens.

In comparison to other bacteria, ESKAPE pathogens are the main cause of nosocomial infections since they can evade the activity of antibacterial agents due to the development of several drug-resistance mechanisms, for example, drug-binding site alteration, changes in the permeability of drugs, and aberrations in drug efflux transporters [4]. Similar effects are observed for drug-resistant fungi strains that belong to the *Aspergillus*, *Candida*, *Cryptococcus*, and *Pneumocystis* genera [5], which can lead to invasive fungal infections in the bloodstream, lungs, brain, and skin [6].

Cancer arises from the uncontrolled proliferation and growth of cells, and it is classified in view of its tissue or cell of origin. Lung cancer (LC) can originate from the central area and peripheric regions of the lungs [7]. In contrast to other types of cancer, LC is characterized by its invasiveness, aggressiveness, and high prevalence worldwide [8]. Epidemiologically, LC is the most common type of cancer diagnosed, and the leading cause of cancer mortality among men and women [9]. Another highly prevalent type of cancer is acute myeloid leukemia (AML). AML constitutes a heterogeneous disorder that arises from the clonal expansion of myeloid progenitors in peripheral blood and bone marrow. This results in bone marrow failure and hampers erythropoiesis [10]. Among leukemias, it is estimated that AML accounts for 80% of all cases in the adult population and more than 10,000 deaths over the last years [11]. The latter represented 1.8% of all cancer deaths in the United States [12].

Despite their molecular and cellular differences, LC and AML are treated with chemotherapy, radiotherapy, immune therapy, and targeted therapy regimens. Despite their possible efficacy, their use is prone to failure due to their limited specificity, poor solubility, the possibility of relapse, and numerous toxicities that can affect the cardiovascular, pulmonary, and musculoskeletal systems [13]. Since the registered numbers for infections caused by pathogenic bacteria and fungi, and patients diagnosed with LC or AML are expected to increase in the next decades, it is imperative to continue exploring, considering, and evaluating sources with potential biological activity.

Traditional medicine combines knowledge, practices, and experiences from indigenous cultures. In this discipline, compounds isolated from animals, microorganisms, or plants are used to treat, prevent, maintain, or improve human health [14]. Nowadays, natural products isolated from sources used in traditional medicine are broadly investigated through integrative and interdisciplinary approaches to continue developing relevant information regarding their use in the pharmaceutical, food, and healthcare industries. This has been reviewed in an innovative platform known as the International Natural Product Sciences Taskforce (INPST), which considers medicinal plants as important sources of bioactive molecules for modern medicine [15]. Medicinal plants are representative sources of secondary metabolites that can exert multiple therapeutic properties such as antimicrobial, antioxidant, antidiabetic, anti-inflammatory, etc. The family Iridaceae includes many flowering plants distributed in South Africa, the Eastern Mediterranean, and Central America [16].

In the past, species that belong to this family have been used to prepare decoctions, pastes, syrups, and extracts to treat muscle pains, respiratory syndromes, neurological disorders, gastrointestinal diseases, and cancers [17]. The genus *Iris* is the largest genus of the Iridaceae family, and its species are differentiated because of their violet-like scent and broad presence in North America, Europe, and Asia [18]. In traditional medicine, species from the genus *Iris* have been considered to treat infections caused by bacteria, viruses, cancer, and inflammatory disorders [18]. The documented biological activities are attributed to the bioactive natural products that they contain such as spiroiridals, flavonoids, triterpenoids, and xanthones [16].

Taxonomically, the genus *Iris* is subdivided into the following four subfamilies: Ixoideae, Isophysidoideae, Nivenioideae, and Iridoideae [19]. The latter is divided into the following tribes: Mariceae, Irideae, Sisyrinchieae, and Tigridieae [19]. The tribe Tigridieae constitutes a monophyletic group that is organized into thirteen genera, sixty-six species, and seven subspecies and is widely distributed in North America [20]. Over the last decades, the phytochemical composition of some *Tigridieae* species has been reported. For example, iridals such as spiroiridal, belamcandal, and 16-hydroxyiridal have been identified in the essential oil of *T. pavonia* [21]. Moreover, glucosyl xanthones, such as mangiferin, have been identified in methanol extracts from *T. alpestris* [22]. However, to our knowledge, for other *Tigridieae* species, such as *T. vanhouttei*, no biological activities or chemical composition have been reported.

Continuing with our research program about scientifically validating the medicinal use of plants, this study aimed to investigate the antimicrobial activity of extracts from *T. vanhouttei* against a panel of human bacterial and fungal pathogens. The cytotoxicity of extracts was tested against human-derived macrophage THP-1 cells and A549 cells. To evaluate the chemical composition of extracts, FTIR spectroscopy was used, whereas GC-MS analyses were considered to assess their chemical profile.

2. Results and Discussion

2.1. FTIR Analysis

In contrast to other spectroscopy techniques, FTIR spectroscopy is based on the absorption of infrared light by proteins, lipids, carbohydrates, and fatty acids. For plant extracts, FTIR analyses are utilized to preliminary characterize their chemical composition by determining the presence of multiple functional groups (e.g., ketones, esters, and carbonyl groups) that can be related to their phytoconstituents. As presented in Figure 1, hexane, chloroform, and methanol extracts exhibit similar bands within the 4000 to 400 cm^{-1} range. Initially, it can be noted that the three extracts present two sharp bands at 2920 and 2835 cm^{-1}, which are related to the symmetrical and asymmetrical stretching of C-H bonds. As expected, the FTIR spectrum of methanol extract displays a broad band at 3300 cm^{-1}, which corresponds to the stretching of the O-H bond possibly from the methanol used to prepare this extract or phenolic compounds. Moreover, the three extracts share a series of peaks located from 1700 to 900 cm^{-1} that can be related to the presence of carboxylic acids, aromatic amines, and alkenes. To record the chemical profile of extracts from *T. vanhouttei*, GC/MS was used.

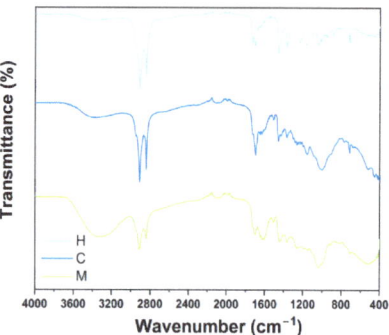

Figure 1. FTIR analysis of hexane (H), chloroform (C), and methanol (M) extract from *T. vanhouttei*.

2.2. GC-MS Analysis

Among chromatography techniques, GC-MS represents a robust and sensitive approach for analyzing chemical and biological samples. In the study of plant extracts, using GC-MS predominantly enables the determination of volatile compounds with low molecular weight [23]. Using this technique, we assessed the phytochemical content of extracts from *T. vanhouttei*; see Table 1. Chromatograms are presented in the Supplementary Material (Figures S1–S3).

Table 1. GC/MS analysis of extracts from *T. vanhouttei*.

Extract	Rt (min)	R Match	Match	%	Compound
Hexane	42.493	851	869	3.62	Palmitic acid
	46.386	864	811	0.04	Myristic acid
	46.717	791	783	0.54	Ascorbic acid
	54.295	939	895	13.29	Linolelaidic acid
	54.664	920	908	12.27	Oleic acid
	101.676	875	814	1.34	Octacosane
	115.255	873	844	2.58	Nonacosane
	120.443	779	757	0.77	Isopsoralen
	122.133	730	721	0.53	β-Stigmasterol
	126.025	813	799	1.17	ε-Sitosterol
	127.875	866	811	2.22	Untriacontane
	133.475	773	703	0.42	Sitostenone
Chloroform	67.704	878	869	5.26	Hexadecanoic acid
	74.340	862	809	8.19	Tetradecanoic acid
	88.372	872	820	5.01	Linolelaidic acid
	90.104	831	800	1.73	Oleic acid
	95.842	790	764	6.11	Stearic acid
	129.154	771	739	2.16	Arachidic acid
	168.336	829	794	1.21	Methyl lignocerate
	180.281	730	707	1.11	Campesterol
	181.254	747	742	1.58	Stigmasterol
Methanol	43.475	927	843	27.51	Tridecanoic acid methyl ester
	46.164	827	794	33.26	Pentadecanoic acid
	55.765	871	854	9.93	Linoleic acidmethyl ester
	56.264	805	787	2.86	Myristoleic acid
	57.959	877	801	2.18	Methyl tetradecanoate
	58.412	853	801	7.35	Stearolic acid
	58.728	749	703	5.03	Oleic acid
	100.201	782	756	1.93	β-Sitosterol

2.3. Antimicrobial Activity

Plant extracts constitute an attractive alternative to evaluate and obtain innovative antimicrobial agents. The advantages of plant extracts over current antimicrobials rely on their intrinsic biological activity, synergistic effects, limited toxicity, and capacity to suppress drug-resistance mechanisms [24]. However, their activity can be limited due to possible antagonism between their phytoconstituents [25].

As indicated in Table 2, results revealed that only hexane extract inhibited growth of the clinical isolate of *P. aeruginosa* at 200 µg/mL (Table 2). No antibacterial activity was observed during treatment with chloroform and methanol extract. In addition, extracts did not exert antifungal activity against the tested strains at the proposed concentrations (50, 100, 150, and 200 µg/mL).

Table 2. Antimicrobial activity of extracts from *T. vanhouttei* expressed as the minimal inhibitory concentration (µg/mL).

Extract	Bacteria								Fungi		
	AB	EC	MRSA	PA	LM	SA	AB_c	PA_c	CA	CN	TM
Hexane	R	R	R	R	R	R	R	200	R	R	R
Chloroform	R	R	R	R	R	R	R	R	R	R	R
Methanol	R	R	R	R	R	R	R	R	R	R	R

Abbreviations: AB, *Acinetobacter baumannii*; EC, *Escherichia coli*; MRSA, methicillin-resistant *Staphylococcus aureus*; PA, *Pseudomonas aeruginosa*; LM, *Listeria monocytogenes*; SA, *Staphylococcus aureus*; AB_c, *Acinetobacter baumannii* clinical isolate; PA_c, *Pseudomonas aeruginosa* clinical isolate; CA, *Candida albicans*; CN, *Cryptococcus neoformans*; TM, *Trichophyton mentagrophytes*; R, resistant.

In the healthcare system, *P. aeruginosa* is considered an opportunistic pathogen that can infect patients with burn wounds, immunodeficiency, or cystic fibrosis [26]. In contrast to other gram-negative strains, *P. aeruginosa* has represented a serious source of mortality and morbidity among long-term care hospitals and intensive care units over the last few years [27].

The activity of hexane extract against this strain can be attributed to the presence of palmitic and oleic acids, which can display antibacterial activity due to their ability to inhibit the activity of essential components for bacterial fatty acid biosynthesis, such as the enoyl-acyl carrier protein reductase component [28]. In addition, phytosterols such as stigmasterol have been reported to exhibit bactericidal activity against gram-negative bacteria due to their ability to modify bacterial membrane composition [29]. These findings suggest the potential use of *T. vanhouttei* extracts to treat infections caused by clinical isolates of *P. aeruginosa*.

2.4. Cytotoxic Activity

Cytotoxicity assays are required to determine the potential toxicity of bioactive substances before their consideration in the development of pharmaceutical formulations [30]. This is often applied to plant extracts or isolated natural products against several models of cancer cell lines.

Here, we tested the cytotoxicity of extracts from *T. vanhouttei* against the THP-1 and the A549 cell lines. The former is a human leukemia monocytic cell line that is widely cultured to investigate the molecular and cellular functionality of monocytes or macrophages [31] and screen the toxicity of candidate molecules against them [32]. The latter comprehends human alveolar basal epithelial cells that are broadly used to assess the functionality of alveolar cells [33], and the potential use of plant extracts or synthetic molecules against lung cancer [34,35].

As represented in Figure 2A, treatment with hexane and chloroform extracts decreased the viability of THP-1 cells in a dose-dependent manner (50, 100, 150, and 200 µg/mL). Initially, it can be observed that treatment with 50 µg/mL of hexane extract resulted in 22.07% of THP-1 cell death, whereas treatment with 100 µg/mL resulted in 71.22% cell death ($p < 0.005$). Against treatment with 150 and 200 µg/mL, 88.14 and 89.61% of cell death were registered ($p < 0.0005$), respectively. This phenomenon was more evident with chloroform extract treatment.

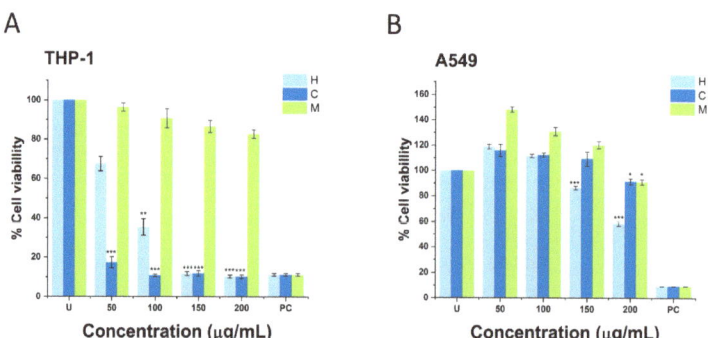

Figure 2. % of cell viability of human-derived macrophage (**A**) THP-1 and (**B**) A549 cells against treatment with 50, 100, 150, and 200 µg/mL of *T. vanhouttei* extracts. U, untreated cells; H, hexane extract; C, chloroform extract; M, methanol extract. PC represents SDS, which was used as a positive control. Shown is the mean ± S.D. of three independent experiments. * p values < 0.05.

In Figure 2A, it can be noted that treatment with 50 µg/mL of chloroform extract resulted in 82.56% cell death ($p < 0.005$). During treatment with 100, 150, and 200 µg/mL of chloroform extract, 88.06, 89.09, and 89.69% of THP-1 cell deaths were recorded, respectively

($p < 0.0005$). In the same figure, it can be observed that the cytotoxicity of methanol extract against the THP-1 cell line was weak since the cells continued to proliferate at 50 µg/mL. However, at 100, 150, and 200 µg/mL, THP-1 cells presented a 9.15, 20.39, and 27.01% in cell death. Despite the importance of these results, the cytotoxicity of plant extracts at different concentrations against other cell lines can vary, which in this case was observed against the A549 cell line.

It can be noted in Figure 2B, that treatment with 50 or 100 µg/mL of hexane extract was not cytotoxic against A549 cells. In fact, cells continued to proliferate after 24 h of exposure to treatment. For the same extract, treatment with 150 and 200 µg/mL resulted significantly in the death of 13.58 and 41.61% A549 cells, respectively ($p < 0.005$). Even though treatment with chloroform extract at 50, 100, and 150 µg/mL did not induce the death of A549 cells, treatment with 200 µg/mL occurred in 8.55% of A549 cell deaths. At the same concentration, treatment with 200 µg/mL of methanol extract resulted in 8.96% cell death. Only statistical differences were registered for treatment with hexane extract against the A549 cell line.

In view of the results obtained in both cytotoxic assays, we assessed the LC_{50} for extracts against each cell line. Against THP-1 cells, the calculated LC_{50} values of the hexane, chloroform, and methanol extract were 90.16, 46.42, and 443.12 µg/mL, respectively. Against the A549 cell line, hexane, chloroform, and methanol extract presented the following LC_{50} values: 294.77, 1472.37, and 843.12 µg/mL. Following the National Cancer Institute (NCI) of the United States, extracts can be considered moderately cytotoxic against the THP cell line, whereas against the A549 cell line, their cytotoxicity is considered weak [36]. The LC_{50} values calculated for each cell line are compiled in Table 3.

Table 3. LC_{50} values of hexane, chloroform, and methanol extract from *T. vanhouttei* against THP-1 and A549 cell lines. Concentrations are expressed in µg/mL.

Extract	THP-1	A549
Hexane	90.16	294.77
Chloroform	46.42	1472.37
Methanol	443.12	843.12

Plant extracts can display distinct biological activities through bioactive secondary metabolites. Even though the biological activities of *Tigridia* species are unknown, recent studies about medicinal plants, such as *Polygonum hydropiper* L., have demonstrated that sterols, such as β-sitosterol, can decrease the viability of breast and cervical cancer cell lines at 1 mg/mL [37]. For the same sterol, its capacity to interfere with the apoptosis, cell cycle, cell signaling pathways, invasion, and survival of lung, stomach, colon, and leukemia cells has been reviewed recently [38].

In other studies, where the cytotoxicity of natural products, such as furanocoumarins, has been investigated, it was reported that isopsoralen could exert cytotoxic activity against human hepatoma cells at the micromolar range (10–200 mM) in a time-dependent manner [39]. Regarding the cytotoxicity of fatty acids, it has been unveiled that palmitic acid can enhance the generation of reactive oxygen species (ROS), induce apoptosis by promoting the activity of caspase 3, decrease mitochondrial membrane potential, and cause cell damage among hepatocyte cell cultures in the millimolar range (0.125–2 mmol/L). In the same study, the activity of oleic acid was tested; however, no significant changes were observed [40]. The presence of these compounds among extracts from *T. vanhouttei* can explain the observed cytotoxicity.

2.5. Anti-Inflammatory Activity

Inflammation constitutes a complex biological response induced by pathogens, toxic compounds, or damaged cells [41]. Depending on its progression, inflammation can lead to tissue damage or disease development. Common inflammation-related diseases include atherosclerosis, rheumatoid arthritis, diabetes, and cancer [42]. It is well-known that plant extracts constitute sources of promising bioactive compounds that can interfere with inflammatory processes. Therefore, there is a need to continue exploring them.

The capacity of extracts from *T. vanhouttei* to elicit an inflammatory response in THP-1-derived macrophages is presented in Figure 3. Results revealed that THP-1 cells treated with LPS or methanol extract decreased IL-6 levels to 50.33 ± 5.49 and 50.99 ± 1.98 pg/mL, respectively. Conversely, cells treated with 50 µg/mL of hexane and chloroform extracts significantly exhibited 81.94 ± 6.68 and 75.95 ± 0.41 IL-6 levels, respectively ($p < 0.001$). IL-6 is a pleiotropic cytokine that is commonly associated with inflammatory processes when dysregulated. However, it is also involved in the hematopoiesis process, acute phase responses against infections and tissue injuries, immune cell functionalities, and immune reactions [43]. The obtained results suggest the potential use of hexane and chloroform extracts from *T. vanhouttei* as anti-inflammatory agents that modulate the secretion of IL-6 to treat bone destruction disorders [44] or regulate metabolic and cardiovascular events [45]. The observed phenomenon can be different against other pro-inflammatory cytokines, such as TNF-α.

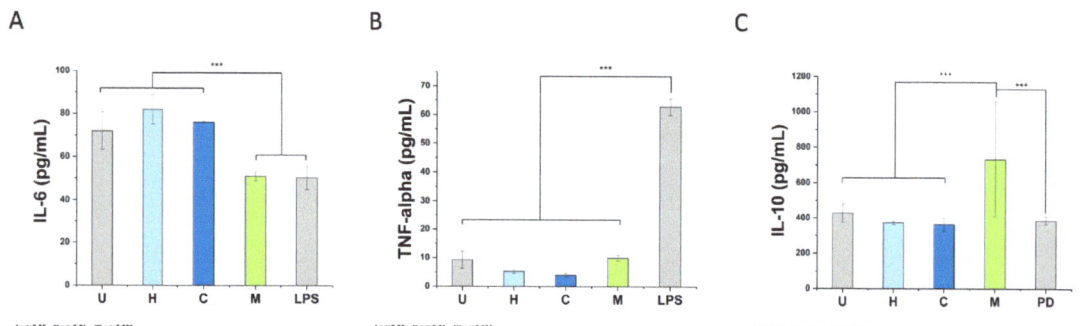

Figure 3. Immunological response of *T. vanhouttei* extracts: hexane (H), chloroform (C), and methanol (M) on human-derived THP-1 cells using ELISA for (**A**) IL-6, (**B**) TNF-α, and (**C**) IL-10. Untreated cells (U); PD, prednisone (positive control for anti-inflammatory analysis); LPS, lipopolysaccharide (positive control for inflammatory analysis). Shown is the mean ± S.D. of three independent experiments.

TNF-α is also a pleiotropic cytokine that can regulate inflammatory responses but is commonly associated with the progression of both autoimmune and inflammatory diseases [46]. According to Figure 3B, treatment with LPS significantly promoted the secretion of TNF-α (62.71 ± 2.88 pg/mL), whereas treatment with 50 µg/mL of hexane, chloroform, or methanol extract exhibited the following TNF-α levels: 5.28 ± 0.70, 3.97 ± 0.68, and 9.93 ± 0.97 pg/mL, respectively. Since the levels of TNF-α were not enhanced during treatment with extracts from *T. vanhouttei*, these results suggest their anti-inflammatory activity. To continue evaluating the anti-inflammatory activity of extracts, the secretion of IL-10 was investigated.

Among anti-inflammatory cytokines, IL-10 mediates the host's anti-inflammatory response against external stimuli, stimulates immune cells' activation, differentiation, and proliferation, and inhibits non-specific immunological responses [47]. As depicted in Figure 3C, cells treated with 50 µg/mL of hexane extract presented no significant levels of IL-10 (373.28 ± 10.09 pg/mL) in contrast to prednisone, which was used as a positive control. Comparably, no significant differences were observed during treatment with 50 µg/mL chloroform extract, as it exhibited 362.03 ± 36.58 pg/mL IL-10 levels. This effect varied with methanol extract treatment as it showed a significant IL-10 level: 732.30 ± 324.87 pg/mL. The observed anti-inflammatory activities of extracts from *T. vanhouttei* can be due to their phytochemical content.

Plant extracts possess anti-inflammatory activities due to their ability to interfere with oxidation-reduction reactions, modulate cell signaling pathways involved in inflammatory processes, and interfere with reactive species generation [48].

In the case of hexane and chloroform extracts, their anti-inflammatory activity can be due to the presence of stigmasterol, which has been observed to exert this effect in murine models [49] and prevent the generation of pro-inflammatory cytokines [50].

In addition, the potential anti-inflammatory activity of chloroform extract can be attributed to the existence of campesterol, which has been reported to interfere with the release of TNF-α in a dose-dependent manner (25–200 mM) [51]. Compounds such as β-sitosterol and stigmasterol can also inhibit the production of TNF-α at the same concentration range [51] and promote the secretion of IL-10 [52]. The presence of β-sitosterol can be related to the capacity of methanol extract to induce the secretion of IL-10.

3. Materials and Methods

3.1. Plant Material and Extract Preparation

Bulbs from *T. vanhouttei* were collected in Texcoco, Estado de México (19°29′ N 98°53′ W). Specimens were identified by the biologist Lilián López-Chávez and deposited with voucher number 36230 at the herbarium of Universidad Autónoma de Chapingo (Carr. Federal México-Texcoco, 56230, Texcoco, Estado de Mexico). For extract preparation, 500 g of bulbs were dried at room temperature, powdered using a mechanical blender, and progressively macerated with *n*-hexane, chloroform, and methanol for 72 h at room temperature. Mixtures were manually mixed at the beginning of this process. The time of extraction was selected since it is the time recommended to efficiently extract bioactive components from plants [53]. The ratio of this process was 0.33 g of plant per milliliter of solvent. Solvents were evaporated under reduced pressure to dryness using a rotary evaporator (Heidolph Laborota 4000; Schwabach, Germany). Extracts were collected and maintained under refrigeration for further evaluation.

3.2. FTIR Evaluation

FTIR spectroscopy has been widely used to amplify the knowledge regarding the identification and differentiation between the chemical composition of extracts from foods, fruits, nanomaterials, and plant extracts by providing a characteristic fingerprint [54]. To evaluate the chemical composition of extracts from *T. vanhouttei*, a Cary 630 Fourier-transform infrared (FTIR) spectrometer (Agilent Technologies, Santa Clara, CA, USA) was used. An ethanol solution (100% v/v) was added to clear the detection diamond, and background spectra were determined without samples at 25 °C. To perform sample analysis, 20 mg of each extract was placed, and ethanol was used again to clean the detection diamond after each measurement. Spectra were recorded within the 4000 to 400 cm^{-1} wavenumber region. Experiments were performed in triplicate.

3.3. Analysis of Phytoconstituents by GC-MS

The extracts' chemical profiles of *T. vanhouttei* were established using a Varian CP-3800 gas chromatograph coupled to a Varian 1200 quadrupole mass spectrometer. The extracts were analyzed according to published protocols [55]. Briefly, 1 μL of samples prepared at 1% (w/v) chloroform were injected into a Factor Four capillary column: VF-5MS (5% phenyl-methyl polysiloxane–95% polydimethylsiloxane; Agilent Technologies), 30 m × 0.25 mm, and 0.25 μm thickness. The separation of phytoconstituents was achieved by using helium as the carrier gas (1 mL/min flow rate) at the following gradient temperature: 60 °C for 2 min, 120 °C for 16 min, 160 °C for 15 min, 180 °C for 15 min, 200 °C for 10 min, 230 °C for 15 min, 290 °C for 20 min, and 300 °C for 30 min. The extracts' components were determined according to their fragmentation patterns and retention times by consulting the National Institute of Standards and Technology Mass Spectral (NIST-MS) database. The relative percentage of phytoconstituents was registered based on the total area of the peaks.

3.4. Strains and Culture Media

In this work, a panel of gram-positive and gram-negative bacteria was used. Gram-positive bacteria included *Staphylococcus aureus* (ATCC 25923) and methicillin-resistant *Staphylococcus aureus* (MRSA) (ATCC 700698) strains. Gram-negative bacteria included *Acinetobacter baumannii* (ATCC BAA-747), *Escherichia coli* (ATCC 25922), and *Pseudomonas aeruginosa* (ATCC 14210) strains. Clinical isolates of *A. baumannii* and *P. aeruginosa* were also tested. In addition, this study evaluated the *Trichophyton mentagrophytes* (ATCC 9533) strain. The pathogenic fungi *Cryptococcus neoformans* var. *grubii* (provided by Dr. Karen Bartlett, University of British Columbia, BC, Canada) and *Candida albicans* (ATCC 10231) strains were also evaluated. Bacterial strains were cultured in Mueller–Hinton broth (Becton and Dickinson (B&D)) at 37 °C in a shaker, whereas Sabouraud broth (B&D) was used for fungal strains at 28 °C.

3.5. Minimal Inhibitory Concentration (MIC) Assay

Following previous protocols [56], 50, 100, 150, and 200 μg/mL of *T. vanhouttei* extracts dissolved in DMSO were tested in a 96-well plate at a final volume of 100 μL/well of Mueller–Hinton or Sabouraud broth. Microbial strains were prepared to have a final optical density of 0.05 at 600 nm. MICs were defined as the concentration of the extracts at which no microbial growth was observed. For bacteria, treatment with amikacin or gentamicin was considered the positive control, whereas treatment with DMSO was used as a negative control. For fungi, amphotericin and terbinafine were used as positive controls, and DMSO remained as a negative control. All experiments were performed in triplicate.

3.6. Cell Culture

The cytotoxicity of *T. vanhouttei* extracts was analyzed using human-derived THP-1 monocytic (ATCC TIB-202) and pulmonary A549 (CCL-185) cells. The THP-1 cell line was cultured using an RPMI 1640 (Hyclone, GE Healthcare, Logan, UT, USA) medium supplemented with 2 mmol L-glutamine (Stem cell Technologies, Vancouver, BC, Canada) and 5% fetal bovine serum (FBS) (Hyclone) and differentiated using 20 ng/mL of phorbol 12-myristate 13-acetate (PMA, Sigma). The A549 (CCL-185) cell was cultured using Dulbecco's Modified Eagle's Medium (DMEM; Gibco, Carlsbad, CA, USA) containing 10% FBS and 100 μg/mL streptomycin. Both cell lines were maintained in a humidified atmosphere supplemented with 5% CO_2 at 37 °C.

3.7. Cytotoxicity Assay

The cytotoxicity of *T. vanhouttei* extracts was analyzed using human-derived THP-1 monocytic (ATCC TIB-202) and pulmonary A549 (CCL-185) cells [57]. The THP-1 cell line was cultured using an RPMI 1640 (Hyclone, GE Healthcare, Logan, UT, USA) medium supplemented with 2 mmol L-glutamine (Stem cell Technologies, Vancouver, BC, Canada) and 5% fetal bovine serum (FBS) (Hyclone) and differentiated using 20 ng/mL of phorbol 12-myristate 13-acetate (PMA, Sigma). The A549 cell line was cultured using Dulbecco's Modified Eagle's Medium (Gibco, Carlsbad, CA, USA) containing 10% FBS, and 100 μg/mL streptomycin. To perform the cytotoxicity assay, 1×10^5 of THP-1 or A549 cells were dispensed (per well) in a 96-well plate in a final volume of 100 μL, respectively. The plate was incubated at 37 °C and supplemented with an atmosphere of 5% CO_2. The next day, the medium was changed, and the cells were treated with extracts at concentrations ranging from 50 to 200 μg/mL. The plate was incubated under the same conditions, and the next day, 25 μL of an MTT (3-(4,5-dimethylthiazol-2-yl)-2,5-diphenyltetrazolium bromide, Sigma) solution (5 mg/mL) was added per well, and the plate was incubated for 4 h at 37 °C under a supplemented atmosphere with 5% CO_2. Formazan crystals were dissolved with 100 μL of extraction buffer, which was prepared with 20% (w/v) of sodium dodecyl sulfate (SDS) in a warm solution of dimethyl formamide at 50% containing 2.5% HCl and 2.5% acetic acid. The plate was placed in an incubator overnight at 37 °C. The next day, the absorbance was measured at 570 nm utilizing a plate reader (Epoch, BioTek). Untreated cells were considered negative controls, whereas cells treated with SDS (2%) were considered positive controls. The half-maximal lethal concentration (LC_{50}) was calculated by plotting the extract concentrations against the percentage of damaged cells. The percentage of cell death for both cell lines was calculated by normalizing the absorbance of untreated cells to 100% and considering published reports where the cytotoxicity of extracts has been assessed [58].

3.8. Inflammatory Assay

The inflammatory assay was performed following published protocols [57]. THP-1 cells, differentiated with PMA, were dispensed at a final concentration of 1×10^5 cells/well using a 96-well plate. Considering instructions from the manufacturer, the secretion of the anti-inflammatory cytokine IL-10 and the pro-inflammatory cytokine IL-6 and TNF-α was measured using commercial kits (B&D). As controls, cells treated with DMSO were used as a negative control, whereas cells treated with 1 μg/mL of lipopolysaccharide (LPS) from *Escherichia coli* (Sigma-Aldrich) were considered the positive control. For the measurement of the secretion of IL-10, prednisone (PD) was used as a positive control. Untreated cells were used as a negative control. Readings were determined utilizing a microplate reader at 450 nm. Experiments were performed in triplicate.

3.9. Statistical Analysis

To assess significant statistical differences between data obtained from cell viability, a two-way analysis of variance (ANOVA) followed by Tukey's mean separation test was applied using OriginPro 2023 data processing software (OriginLab, Northampton, MA, USA).

4. Conclusions

This work reported, for the first time, the antimicrobial, cytotoxic, and anti-inflammatory activities of extracts from *T. vanhouttei*. Among the tested extracts, only hexane extract inhibited the growth of the clinical isolate of *P. aeruginosa*. The cytotoxicity of the obtained extracts was tested against representative cancer cell lines from leukemia and lung cancer. Even though the obtained extracts exhibited moderate cytotoxic activity against THP-1 cells and weak cytotoxic activity against A549 cells, these results suggest their potential use against cancer. Regarding their anti-inflammatory activity, methanol extract promoted the secretion of IL-10, whereas hexane and chloroform extract did not elicit the secretion of TNF-α, suggesting their potential anti-inflammatory effect. The recorded biological activities of extracts can be attributed to the various fatty acids and sterols that they contain, which were preliminary identified by FTIR spectroscopy and confirmed by GC/MS analysis. Taking together these results, the novelty of this research relies on demonstrating, for the first time, the therapeutic application of Tigridieae species using in vitro models. The findings of this work can have several applications in pharmacognosy and pharmacotherapy.

Supplementary Materials: The following supporting information can be downloaded at https://www.mdpi.com/article/10.3390/plants12173136/s1, Figure S1: Chromatogram of hexane extract from *T. vanhouttei*. Figure S2: Chromatogram of chloroform extract from *T. vanhouttei*. Figure S3: Chromatogram of methanol extract from *T. vanhouttei*.

Author Contributions: Conceptualization, H.B., L.R.H., Z.N.J. and E.S.-A.; data curation, J.L.M.-M., A.C.L.-L., H.B., E.R.L.-M. and E.S.-A.; formal analysis, J.L.M.-M., H.B. and E.S.-A.; investigation, J.L.M.-M., A.C.L.-L., H.B. and E.S.-A.; methodology, J.L.M.-M., A.C.L.-L., H.B. and E.S.-A.; project administration, J.L.M.-M., A.C.L.-L., H.B., L.R.H., Z.N.J. and E.S.-A.; resources, H.B., E.R.L.-M., D.E.N.-L. and E.S.-A.; supervision, H.B., L.R.H. and E.S.-A.; validation, H.B., L.R.H., Z.N.J. and E.S.-A.; visualization, J.L.M.-M., E.R.L.-M. and E.S.-A.; writing—original draft, J.L.M.-M., H.B., L.R.H. and E.S.-A.; writing—review and editing, A.C.L.-L., E.R.L.-M., D.E.N.-L. and Z.N.J. All authors have read and agreed to the published version of the manuscript.

Funding: This research received no external funding.

Data Availability Statement: Experimental data generated in this work can be requested from the authors for correspondence.

Acknowledgments: J.L.M.-M. thanks Consejo Nacional de Humanidades, Ciencias y Tecnologías (CONAHCyT), for his doctoral fellowship. This work was partially funded by Tecnologico de Monterrey through the Nanodevices Research Group (E.R.L.-M. and D.E.N.-L.). The authors acknowledge the support of Edith Salomé-Castañeda for collecting the bulbs. J.L.M.-M. dedicates this work to Enrique Mejía-Sáenz; he is dearly missed.

Conflicts of Interest: The authors declare no conflict of interest.

References

1. Deen, J.; Von Seidlein, L.; Clemens, J.D. Issues and Challenges of Public-Health Research in Developing Countries. *Mansons Trop. Infect. Dis.* **2014**, 40–48.e1. [CrossRef]
2. Cooper, L.A.; Purnell, T.S.; Showell, N.N.; Ibe, C.A.; Crews, D.C.; Gaskin, D.J.; Foti, K.; Thornton, R.L.J. Progress on Major Public Health Challenges: The Importance of Equity. *Public Health Rep.* **2018**, *133*, 15S–19S. [CrossRef]
3. What Exactly is Antibiotic Resistance? Available online: https://www.cdc.gov/drugresistance/about.html (accessed on 1 July 2023).
4. Aloke, C.; Achilonu, I. Coping with the ESKAPE pathogens: Evolving strategies, challenges and future prospects. *Microb. Pathog.* **2023**, *175*, 105963. [CrossRef]
5. Arastehfar, A.; Gabaldón, T.; Garcia-Rubio, R.; Jenks, J.D.; Hoenigl, M.; Salzer, H.J.F.; Ilkit, M.; Lass-Flörl, C.; Perlin, D.S. Drug-Resistant Fungi: An Emerging Challenge Threatening Our Limited Antifungal Armamentarium. *Antibiotics* **2020**, *9*, 877. [CrossRef]
6. Garvey, M.; Rowan, N.J. Pathogenic Drug Resistant Fungi: A Review of Mitigation Strategies. *Int. J. Mol. Sci.* **2023**, *24*, 1584. [CrossRef]
7. Xie, X.; Li, X.; Tang, W.; Xie, P.; Tan, X. Primary tumor location in lung cancer: The evaluation and administration. *Chin. Med. J.* **2022**, *135*, 127–136. [CrossRef]
8. Mejía-Méndez, J.L.; López-Mena, E.R.; Sánchez-Arreola, E. Activities against Lung Cancer of Biosynthesized Silver Nanoparticles: A Review. *Biomedicines* **2023**, *11*, 389. [CrossRef]
9. Thandra, K.C.; Barsouk, A.; Saginala, K.; Aluru, J.S.; Barsouk, A. Epidemiology of lung cancer. *Contemp. Oncol.* **2021**, *25*, 45–52. [CrossRef]
10. Saultz, J.N.; Garzon, R. Acute Myeloid Leukemia: A Concise Review. *J. Clin. Med.* **2016**, *5*, 33. [CrossRef]
11. De Kouchkovsky, I.; Abdul-Hay, M. 'Acute myeloid leukemia: A comprehensive review and 2016 update'. *Blood Cancer J.* **2016**, *6*, e441. [CrossRef]
12. Nix, N.M.; Price, A. Acute Myeloid Leukemia: An Ever-Changing Disease. *J. Adv. Pract. Oncol.* **2019**, *10*, 4–8. [CrossRef]
13. Nekhlyudov, L.; Campbell, G.B.; Schmitz, K.H.; Brooks, G.A.; Kumar, A.J.; Ganz, P.A.; Von Ah, D. Cancer-related impairments and functional limitations among long-term cancer survivors: Gaps and opportunities for clinical practice. *Cancer* **2022**, *128*, 222–229. [CrossRef]
14. Yuan, H.; Ma, Q.; Ye, L.; Piao, G. The Traditional Medicine and Modern Medicine from Natural Products. *Molecules* **2016**, *21*, 559. [CrossRef]
15. Singla, R.K.; De, R.; Efferth, T.; Mezzetti, B.; Uddin, M.S.; Sanusi; Ntie-Kang, F.; Wang, D.; Schultz, F.; Kharat, K.R.; et al. The International Natural Product Sciences Taskforce (INPST) and the power of Twitter networking exemplified through #INPST hashtag analysis. *Phytomedicine* **2023**, *108*, 154520. [CrossRef]
16. Singab, A.N.B.; Ayoub, I.M.; El-Shazly, M.; Korinek, M.; Wu, T.-Y.; Cheng, Y.-B.; Chang, F.-R.; Wu, Y.-C. Shedding the light on Iridaceae: Ethnobotany, phytochemistry and biological activity. *Ind. Crops Prod.* **2016**, *92*, 308–335. [CrossRef]
17. Khatib, S.; Faraloni, C.; Bouissane, L. Exploring the Use of Iris Species: Antioxidant Properties, Phytochemistry, Medicinal and Industrial Applications. *Antioxidants* **2022**, *11*, 526. [CrossRef] [PubMed]
18. Amin, H.I.M.; Hussain, F.H.S.; Najmaldin, S.K.; Thu, Z.M.; Ibrahim, M.F.; Gilardoni, G.; Vidari, G. Phytochemistry and Biological Activities of Iris Species Growing in Iraqi Kurdistan and Phenolic Constituents of the Traditional Plant Iris postii. *Molecules* **2021**, *26*, 264. [CrossRef] [PubMed]
19. Goldblatt, P. Phylogeny and Classification of the Iridaceae and the Relationships of Iris. *Ann. Bot.* **2000**, *58*.
20. Munguía-Lino, G.; Vargas-Ponce, O.; Rodríguez, A.; Munguía-Lino, G.; Vargas-Ponce, O.; Rodríguez, A. Tigridieae (Iridaceae) in North America: Floral diversity, flower preservation methods and keys for the identification of genera and species. *Bot. Sci.* **2017**, *95*, 473–502. [CrossRef]
21. Effers, K.; Scholz, B.; Nickel, C.; Hanisch, B.; Marner, F.-J. Structure Determination of Tigridial, an Iridopentaene from Tigridia pavonia (Iridaceae). *Eur. J. Org. Chem.* **1999**, *1999*, 2793–2797. [CrossRef]
22. Williams, C.A.; Harborne, J.B. Biflavonoids, Quinones and Xanthones as Rare Chemical Markers in the Family Iridaceae. *Z. Für Naturforschung C* **1985**, *40*, 325–330. [CrossRef]
23. Rockwood, A.L.; Kushnir, M.M.; Clarke, N.J. 2—Mass Spectrometry. In *Principles and Applications of Clinical Mass Spectrometry*; Rifai, N., Horvath, A.R., Wittwer, C.T., Eds.; Elsevier: Amsterdam, The Netherlands, 2018; pp. 33–65. ISBN 978-0-12-816063-3.

24. Vaou, N.; Stavropoulou, E.; Voidarou, C.; Tsigalou, C.; Bezirtzoglou, E. Towards Advances in Medicinal Plant Antimicrobial Activity: A Review Study on Challenges and Future Perspectives. *Microorganisms* **2021**, *9*, 2041. [CrossRef] [PubMed]
25. Caesar, L.K.; Cech, N.B. Synergy and antagonism in natural product extracts: When 1 + 1 does not equal 2. *Nat. Prod. Rep.* **2019**, *36*, 869–888. [CrossRef]
26. Qin, S.; Xiao, W.; Zhou, C.; Pu, Q.; Deng, X.; Lan, L.; Liang, H.; Song, X.; Wu, M. Pseudomonas aeruginosa: Pathogenesis, virulence factors, antibiotic resistance, interaction with host, technology advances and emerging therapeutics. *Signal Transduct. Target. Ther.* **2022**, *7*, 199. [CrossRef]
27. Reynolds, D.; Kollef, M. The Epidemiology and Pathogenesis and Treatment of Pseudomonas aeruginosa Infections: An Update. *Drugs* **2021**, *81*, 2117–2131. [CrossRef]
28. Zheng, C.J.; Yoo, J.-S.; Lee, T.-G.; Cho, H.-Y.; Kim, Y.-H.; Kim, W.-G. Fatty acid synthesis is a target for antibacterial activity of unsaturated fatty acids. *FEBS Lett.* **2005**, *579*, 5157–5162. [CrossRef] [PubMed]
29. Bakrim, S.; Benkhaira, N.; Bourais, I.; Benali, T.; Lee, L.-H.; El Omari, N.; Sheikh, R.A.; Goh, K.W.; Ming, L.C.; Bouyahya, A. Health Benefits and Pharmacological Properties of Stigmasterol. *Antioxidants* **2022**, *11*, 1912. [CrossRef]
30. McGaw, L.J.; Elgorashi, E.E.; Eloff, J.N. 8—Cytotoxicity of African Medicinal Plants Against Normal Animal and Human Cells. In *Toxicological Survey of African Medicinal Plants*; Kuete, V., Ed.; Elsevier: Amsterdam, The Netherlands, 2014; pp. 181–233. ISBN 978-0-12-800018-2.
31. Chanput, W.; Mes, J.J.; Wichers, H.J. THP-1 cell line: An in vitro cell model for immune modulation approach. *Int. Immunopharmacol.* **2014**, *23*, 37–45. [CrossRef]
32. Pick, N.; Cameron, S.; Arad, D.; Av-Gay, Y. Screening of compounds toxicity against human Monocytic cell line-THP-1 by flow cytometry. *Biol. Proced. Online* **2004**, *6*, 220–225. [CrossRef]
33. Smith, B.T. Cell line A549: A model system for the study of alveolar type II cell function. *Am. Rev. Respir. Dis.* **1977**, *115*, 285–293.
34. Liu, M.; Chen, Y.-L.; Kuo, Y.-H.; Lu, M.-K.; Liao, C.-C. Aqueous extract of Sapindus mukorossi induced cell death of A549 cells and exhibited antitumor property in vivo. *Sci. Rep.* **2018**, *8*, 4831. [CrossRef]
35. Olivito, F.; Amodio, N.; Di Gioia, M.L.; Nardi, M.; Oliverio, M.; Juli, G.; Tassone, P.; Procopio, A. Synthesis and preliminary evaluation of the anti-cancer activity on A549 lung cancer cells of a series of unsaturated disulfides †Electronic supplementary information (ESI) available: Experimental details and compound characterization. *MedChemComm* **2018**, *10*, 116–119. [CrossRef] [PubMed]
36. Abdel-Hameed, E.-S.S.; Bazaid, S.A.; Shohayeb, M.M.; El-Sayed, M.M.; El-Wakil, E.A. Phytochemical Studies and Evaluation of Antioxidant, Anticancer and Antimicrobial Properties of Conocarpus erectus L. Growing in Taif, Saudi Arabia. *Eur. J. Med. Plants* **2012**, *2*, 93–112. [CrossRef]
37. Ayaz, M.; Sadiq, A.; Wadood, A.; Junaid, M.; Ullah, F.; Zaman Khan, N. Cytotoxicity and molecular docking studies on phytosterols isolated from Polygonum hydropiper L. *Steroids* **2019**, *141*, 30–35. [CrossRef] [PubMed]
38. Khan, Z.; Nath, N.; Rauf, A.; Emran, T.B.; Mitra, S.; Islam, F.; Chandran, D.; Barua, J.; Khandaker, M.U.; Idris, A.M.; et al. Multifunctional roles and pharmacological potential of β-sitosterol: Emerging evidence toward clinical applications. *Chem. Biol. Interact.* **2022**, *365*, 110117. [CrossRef]
39. Zhang, C.; Zhao, J.-Q.; Sun, J.-X.; Li, H.-J. Psoralen and isopsoralen from Psoraleae Fructus aroused hepatotoxicity via induction of aryl hydrocarbon receptor-mediated CYP1A2 expression. *J. Ethnopharmacol.* **2022**, *297*, 115577. [CrossRef] [PubMed]
40. Moravcová, A.; Červinková, Z.; Kučera, O.; Mezera, V.; Rychtrmoc, D.; Lotková, H. The effect of oleic and palmitic acid on induction of steatosis and cytotoxicity on rat hepatocytes in primary culture. *Physiol. Res.* **2015**, *64*, S627–S636. [CrossRef] [PubMed]
41. Chen, L.; Deng, H.; Cui, H.; Fang, J.; Zuo, Z.; Deng, J.; Li, Y.; Wang, X.; Zhao, L. Inflammatory responses and inflammation-associated diseases in organs. *Oncotarget* **2017**, *9*, 7204–7218. [CrossRef] [PubMed]
42. Okin, D.; Medzhitov, R. Evolution of Inflammatory Diseases. *Curr. Biol. CB* **2012**, *22*, R733–R740. [CrossRef] [PubMed]
43. Tanaka, T.; Narazaki, M.; Kishimoto, T. IL-6 in Inflammation, Immunity, and Disease. *Cold Spring Harb. Perspect. Biol.* **2014**, *6*, a016295. [CrossRef]
44. Balto, K.; Sasaki, H.; Stashenko, P. Interleukin-6 Deficiency Increases Inflammatory Bone Destruction. *Infect. Immun.* **2001**, *69*, 744–750. [CrossRef] [PubMed]
45. Xu, Y.; Zhang, Y.; Ye, J. IL-6: A Potential Role in Cardiac Metabolic Homeostasis. *Int. J. Mol. Sci.* **2018**, *19*, 2474. [CrossRef] [PubMed]
46. Parameswaran, N.; Patial, S. Tumor Necrosis Factor-α Signaling in Macrophages. *Crit. Rev. Eukaryot. Gene Expr.* **2010**, *20*, 87. [CrossRef] [PubMed]
47. Iyer, S.S.; Cheng, G. Role of Interleukin 10 Transcriptional Regulation in Inflammation and Autoimmune Disease. *Crit. Rev. Immunol.* **2012**, *32*, 23–63. [CrossRef]
48. Rodríguez-Yoldi, M.J. Anti-Inflammatory and Antioxidant Properties of Plant Extracts. *Antioxidants* **2021**, *10*, 921. [CrossRef]
49. Morgan, L.V.; Petry, F.; Scatolin, M.; de Oliveira, P.V.; Alves, B.O.; Zilli, G.A.L.; Volfe, C.R.B.; Oltramari, A.R.; de Oliveira, D.; Scapinello, J.; et al. Investigation of the anti-inflammatory effects of stigmasterol in mice: Insight into its mechanism of action. *Behav. Pharmacol.* **2021**, *32*, 640–651. [CrossRef]
50. Gabay, O.; Sanchez, C.; Salvat, C.; Chevy, F.; Breton, M.; Nourissat, G.; Wolf, C.; Jacques, C.; Berenbaum, F. Stigmasterol: A phytosterol with potential anti-osteoarthritic properties. *Osteoarthr. Cartil.* **2010**, *18*, 106–116. [CrossRef]

51. Yuan, L.; Zhang, F.; Shen, M.; Jia, S.; Xie, J. Phytosterols Suppress Phagocytosis and Inhibit Inflammatory Mediators via ERK Pathway on LPS-Triggered Inflammatory Responses in RAW264.7 Macrophages and the Correlation with Their Structure. *Foods* **2019**, *8*, 582. [CrossRef]
52. Kasirzadeh, S.; Ghahremani, M.H.; Setayesh, N.; Jeivad, F.; Shadboorestan, A.; Taheri, A.; Beh-Pajooh, A.; Azadkhah Shalmani, A.; Ebadollahi-Natanzi, A.; Khan, A.; et al. β-Sitosterol Alters the Inflammatory Response in CLP Rat Model of Sepsis by Modulation of NFκB Signaling. *BioMed Res. Int.* **2021**, *2021*, 5535562. [CrossRef] [PubMed]
53. Zhang, Q.-W.; Lin, L.-G.; Ye, W.-C. Techniques for extraction and isolation of natural products: A comprehensive review. *Chin. Med.* **2018**, *13*, 20. [CrossRef]
54. Durazzo, A.; Kiefer, J.; Lucarini, M.; Camilli, E.; Marconi, S.; Gabrielli, P.; Aguzzi, A.; Gambelli, L.; Lisciani, S.; Marletta, L. Qualitative Analysis of Traditional Italian Dishes: FTIR Approach. *Sustainability* **2018**, *10*, 4112. [CrossRef]
55. Cuevas-Cianca, S.I.; Leal, A.C.L.; Hernández, L.R.; Arreola, E.S.; Bach, H. Antimicrobial, toxicity, and anti-inflammatory activities of Buddleja perfoliata Kunth. *Phytomedicine Plus* **2022**, *2*, 100357. [CrossRef]
56. Cruz Paredes, C.; Bolívar Balbás, P.; Gómez-Velasco, A.; Juárez, Z.N.; Sánchez Arreola, E.; Hernández, L.R.; Bach, H. Antimicrobial, Antiparasitic, Anti-Inflammatory, and Cytotoxic Activities of Lopezia racemosa. *Sci. World J.* **2013**, *2013*, e237438. [CrossRef] [PubMed]
57. Juárez, Z.N.; Bach, H.; Sánchez-Arreola, E.; Bach, H.; Hernández, L.R. Protective antifungal activity of essential oils extracted from Buddleja perfoliata and Pelargonium graveolens against fungi isolated from stored grains. *J. Appl. Microbiol.* **2016**, *120*, 1264–1270. [CrossRef] [PubMed]
58. Domínguez, F.; Maycotte, P.; Acosta-Casique, A.; Rodríguez-Rodríguez, S.; Moreno, D.A.; Ferreres, F.; Flores-Alonso, J.C.; Delgado-López, M.G.; Pérez-Santos, M.; Anaya-Ruiz, M. Bursera copallifera Extracts Have Cytotoxic and Migration-Inhibitory Effects in Breast Cancer Cell Lines. *Integr. Cancer Ther.* **2018**, *17*, 654–664. [CrossRef]

Disclaimer/Publisher's Note: The statements, opinions and data contained in all publications are solely those of the individual author(s) and contributor(s) and not of MDPI and/or the editor(s). MDPI and/or the editor(s) disclaim responsibility for any injury to people or property resulting from any ideas, methods, instructions or products referred to in the content.

Article

Biological Activities and Chemical Profiles of *Kalanchoe fedtschenkoi* Extracts

Jorge L. Mejía-Méndez [1,*], Horacio Bach [2], Ana C. Lorenzo-Leal [2], Diego E. Navarro-López [3], Edgar R. López-Mena [3], Luis Ricardo Hernández [1] and Eugenio Sánchez-Arreola [1,*]

1. Laboratory in Phytochemistry Research, Chemical Biological Sciences Department, Universidad de las Américas Puebla, Ex Hacienda Sta. Catarina Mártir S/N, San Andres Cholula 72810, Mexico; luisr.hernandez@udlap.mx
2. Division of Infectious Diseases, Faculty of Medicine, University of British Columbia, Vancouver, BC V6G 3Z6, Canada; hbach@mail.ubc.ca (H.B.); anacecylole@gmail.com (A.C.L.-L.)
3. Tecnologico de Monterrey, Escuela de Ingeniería y Ciencias, Campus Guadalajara, Av. Gral. Ramón Corona No 2514, Colonia Nuevo México, Zapopan 45121, Mexico; diegonl@tec.mx (D.E.N.-L.); edgarl@tec.mx (E.R.L.-M.)
* Correspondence: jorge.mejiamz@udlap.mx (J.L.M.-M.); eugenio.sanchez@udlap.mx (E.S.-A.)

Citation: Mejía-Méndez, J.L.; Bach, H.; Lorenzo-Leal, A.C.; Navarro-López, D.E.; López-Mena, E.R.; Hernández, L.R.; Sánchez-Arreola, E. Biological Activities and Chemical Profiles of *Kalanchoe fedtschenkoi* Extracts. *Plants* **2023**, *12*, 1943. https://doi.org/10.3390/plants12101943

Academic Editor: Corina Danciu

Received: 28 March 2023
Revised: 3 May 2023
Accepted: 8 May 2023
Published: 10 May 2023

Copyright: © 2023 by the authors. Licensee MDPI, Basel, Switzerland. This article is an open access article distributed under the terms and conditions of the Creative Commons Attribution (CC BY) license (https://creativecommons.org/licenses/by/4.0/).

Abstract: In this study, the leaves of *Kalanchoe fedtschenkoi* were consecutively macerated with hexane, chloroform, and methanol. These extracts were used to assess the bioactivities of the plant. The antimicrobial activity was tested against a panel of Gram-positive and -negative pathogenic bacterial and fungal strains using the microdilution method. The cytotoxicity of *K. fedtschenkoi* extracts was investigated using human-derived macrophage THP-1 cells through the MTT assay. Finally, the anti-inflammatory activity of extracts was studied using the same cell line by measuring the secretion of IL-10 and IL-6. The phytoconstituents of hexane and chloroform extracts were evaluated using gas chromatography–mass spectrometry (GC/MS). In addition, high-performance liquid chromatography (HPLC) was used to study the phytochemical content of methanol extract. The total flavonoid content (TFC) of methanol extract is also reported. The chemical composition of *K. fedtschenkoi* extracts was evaluated using Fourier-transform infrared spectroscopy (FTIR). Results revealed that the chloroform extract inhibited the growth of *Pseudomonas aeruginosa* at 150 µg/mL. At the same concentration, methanol extract inhibited the growth of methicillin-resistant *Staphylococcus aureus* (MRSA). Regarding their cytotoxicity, the three extracts were highly cytotoxic against the tested cell line at $IC_{50} < 3$ µg/mL. In addition, the chloroform extract significantly stimulated the secretion of IL-10 at 50 µg/mL ($p < 0.01$). GC/MS analyses revealed that hexane and chloroform extracts contain fatty acids, sterols, vitamin E, and triterpenes. The HPLC analysis demonstrated that methanol extract was constituted by quercetin and kaempferol derivatives. This is the first report in which the bioactivities and chemical profiles of *K. fedtschenkoi* are assessed for non-polar and polar extracts.

Keywords: traditional medicine; Crassulaceae; *Kalanchoe fedtschenkoi*; bioactivities; antimicrobial; cytotoxicity; anti-inflammatory properties

1. Introduction

Traditional medicine comprises knowledge, practices, and beliefs that, when integrated, are useful to treat or avert multiple diseases. Given its accessibility and affordability, traditional medicine is utilized as a first line of response against medical emergencies in African countries [1]. In traditional medicine, syrups, decoctions, infusions, and extracts from herbs or medicinal plants are used as antimicrobial, anticancer, anti-inflammatory, antidepressant, or antiaggregant agents [2].

Medicinal plants and their parts are widely used to prepare therapeutic extracts [3]. In traditional medicine, medicinal plants are predominantly utilized to treat diseases, as they are a cost-effective alternative that exerts lower side effects than current treatment modalities [4]. However, over the last decades, there has been an increasing interest in screening the bioactivities of plant extracts against human health concerns such as infectious diseases, inflammatory processes, and distinct types of cancer.

Communicable diseases (CD) are caused by pathogenic microorganisms (e.g., bacteria, fungi, and viruses) or their products [5]. There are various mechanisms by which they can be transmitted, for example, through contact with contaminated objects or blood products, insect bites, and contact with bodily fluids (e.g., saliva). Common examples of communicable diseases include infections caused by hepatitis A and B viruses, Rift Valley fever, influenza, salmonella, and tuberculosis [5,6]. According to the World Health Organization (WHO), infections caused by critical (e.g., *Acinetobacter baumannii* and *Pseudomonas aeruginosa*) and high (e.g., *Staphylococcus aureus*) priority multidrug-resistant bacteria constitute a human health concern, as their incidence has been correlated to the 4.95 million deaths estimated in 2019 [7,8]. However, these numbers are expected to increase in the next decades [9]. Multidrug bacteria infections are challenging to treat due to their limited effectiveness and resistance to current antibiotics. Another major threat to human health is non-communicable diseases (NCDs).

NCDs are also known as chronic diseases. They include a series of clinical conditions characterized by their long duration and gradual progress [10]. Examples of NCDs include cancer, chronic respiratory diseases, diabetes, and cardiovascular diseases. In addition, in view of their clinical features, gastrointestinal diseases, endocrine, neurological, and genetic disorders are also included in this category [11]. Compared to CDs, NCDs represent an increasing concern due to their high mortality rate and impact on the global economy [10]. Epidemiologically, it has been estimated that NCDs account for 71% of all deaths worldwide and predominantly impact low-income and middle-income countries [12].

The Crassulaceae family belongs to the order Saxifragales, or the orpine family or Stonecrop family [13]. Regarding its distribution, the presence of members of the Crassulaceae family is documented in the Mediterranean region, Southern Africa, and the Southwestern United States [14], which includes 35 genera and ~1410 species [15]. The genus *Kalanchoe* is widely recognized for its ornamental use, mainly attributed to its adaptability to drought, exquisite flowers, easy cultivation, clone growth, and asexual reproduction [16]. In addition, for therapeutic purposes, the biological activities and bioactive compounds from *Kalanchoe* species have been broadly studied.

Over the last years, it has been revealed that *Kalanchoe* species are of great importance in traditional medicine, since they possess different bioactive molecules (e.g., quercetin, afzelin, bryophyllin A, bersaldegenin-3-acetate) that can exert strong antitumor, antimicrobial, anti-inflammatory, antileishmanial, antioxidant, and anti-urolithiasis properties [17,18]. Traditionally, preparations from species of the *Kalanchoe* genus are utilized in different countries such as Brazil, India, and China. In addition, some of them (i.e., *K. pinnata*) belong to the list of medicinal plants to be used in national public health systems, such as the Sistema Único de Saúde (SUS) [19].

The importance of *Kalanchoe* species relies on their capacity to exert different biological activities. Therefore, they have been proposed to treat rheumatic disorders, abscesses, wounds, and burns [20]. Comparably, it has been reported that extracts from their leaves can execute hepatoprotective, hypocholesterolemic, nephroprotective, and nematicide activities [21]. In addition, species from the genus are used in traditional medicine to induce smooth muscle relaxation [22] and prevent premature labor [23]. Moreover, the increasing interest in expanding the therapeutic knowledge about *Kalanchoe* species has resulted in their use as transgenic plants to produce peptides (cecropin P1) with fungicide and wound-healing activity in Wistar rats [24].

K. fedtschenkoi, also known as *Bryophyllum fedtschenkoi*, is a native species from Madagascar that has been poorly studied. For instance, only one study reported that the ethanol extract from this specimen exhibited antibacterial and cytotoxic properties against the group of ESKAPE (*Enterococcus faecium*, *S. aureus*, *Klebsiella pneumoniae*, *A. baumannii*, *P. aeruginosa*, and *Enterobacter cloacae*) pathogens and human keratinocytes, respectively [25].

Furthermore, the same study demonstrated that ethanol extract contained different bioactive nature compounds such as quercetin and caffeic acid [25]. On the other hand, recent studies have revealed that aqueous extracts from its leaves can exert antioxidant activity, since they contain distinct flavonoids such as quercetin di-O-hexoside, methylquercetin-O-hexoside-O-deoxyhexoside, kaempferol 3-O-glucopyranoside 7-O-rhamnopyranoside, and kaempferol-O-hexoside-O-deoxyhexoside-O-pentoside, among others [26].

Continuing with our research program of studying the bioactivities of traditional medicinal plants, we examined the antimicrobial and cytotoxic activities of hexane, chloroform, and methanol extracts from K. fedtschenkoi leaves. The antimicrobial activity of K. fedtschenkoi extracts was evaluated against a panel of pathogenic Gram-positive and -negative bacteria and yeast strains. In addition, these extracts' cytotoxicity and inflammatory response were studied using a human-derived monocyte cell line as an ex vivo model.

2. Results and Discussion

2.1. GC/MS Analysis

Hexane and chloroform extracts contained fatty acids, vitamins, sterols, and triterpenoids commonly found in the Kalanchoe genus (Table 1). For example, among Kalanchoe species, fatty acids such as heptacosane and stearic acid have been unveiled from K. beharensis and K. pinnata, respectively [27,28]. Similarly, K. pinnata leaves have been reported as abundant sources of sterols such as stigmasterol [29] and triterpenes such as friedelin [30]. These compounds and diterpenes, such as phytol, have been identified in K. tomentosa extracts [31]. On the other hand, fat-soluble phenolic molecules such as vitamin E, also known as tocopherol, have been documented among K. daigremontiana [32] and K. crenata extracts [33]. The chromatograms of hexane and chloroform extract are presented as Supplementary Figures (Figures S1 and S2). Given the polarity of methanol extract, its phytochemical content was studied using HPLC, and it was compared with the literature reporting on polar extracts from K. fedtschenkoi.

Table 1. Chemical composition of hexane and chloroform extracts from K. fedtschenkoi.

Extract	Match	R match	Rt (min)	%	Name
Hexane	749	776	45.11	0.59	n-hexadecanoic acid
	888	902	54.86	1.74	Phytol
	735	750	58.22	0.58	Stearic acid
	885	916	94.96	1.23	Squalene
	878	883	103.47	3.83	δ-Tocopherol
	770	809	109.55	0.84	ε-Tocopherol
	883	887	116.01	4.35	α-Tocopherol
	767	771	125.10	0.88	Stigmasterol
	919	925	128.74	15.34	Heptacosane
	813	884	131.90	1.44	Simiarenol
	854	856	138.60	1.90	Friedelin
	759	804	148.48	0.62	Octadecanal
	704	723	151.54	0.305	2-Hexadecanol
Chloroform	756	785	37.82	4.34	Phytol
	726	756	45.09	1.86	Hexadecanoic acid
	757	772	58.28	2.97	Stearic acid
	678	750	95.54	0.44	1-Hexadecanol
	680	683	97.30	0.24	1-Pentatriacontanol
	853	862	103.79	4.85	δ-Tocopherol
	747	850	109.43	5.48	1-Docosene
	812	850	110.75	3.83	ε-Tocopherol
	856	869	115.85	8.13	α-Tocopherol
	724	730	122.14	4.48	β-Stigmasterol
	895	913	128.09	24.22	Heptacosane
	750	836	131.18	2.34	β-Simiarenol
	748	799	136.77	0.25	Octadecanal
	742	861	148.59	0.62	Hexadecanal

Abbreviations: Rt, retention time; min, minutes.

2.2. HPLC Analysis

HPLC is a proper technique in which analytes dissolved in a mobile phase are pumped through a stationary phase. Depending upon the chemical features of the sample, solvent, and stationary phase, analytes exhibit different retention times (Rt). Here, we used HPLC to study the chemical composition of methanol extract from *K. fedtschenkoi*. As presented in the Supplementary Materials (Figure S3), the chromatogram of methanol extract exhibits characteristic peaks corresponding to flavonoids identified in other reports in which *K. fedtschenkoi* has also been studied [25,26]. In accordance with their results, we show that methanol extract also contains quercetin di-*O*-hexoside (Rt: 14.98), methylquercetin-*O*-hexoside-*O*-deoxyhexoside (Rt: 17.37), kaempferol *O*-hexoside-di-*O*-deoxyhexoside (Rt: 19.34 min), and kaempferol *O*-hexoside-di-*O*-deoxyhexoside (Rt: 20.13 min). On the other hand, the same chromatogram also presents a small peak at 38.54 min and a sharp peak at 47.23 min, which, comparably to HPLC analyses of extracts from *K. brasiliensis*, might suggest the presence of patuletin-*O*-deoxy-hexoside-*O*-acetyl-deoxy-hexoside [34]. To estimate the amount of flavonoids in the methanol extract, we performed the TFC assay.

2.3. TFC of Methanol Extract from K. fedtschenkoi

The TFC assay is a widely performed colorimetric method required to assess the presence of flavonoids in plant extracts. This technique is based on aluminum chloride ($AlCl_3$)'s capacity to form complexes with hydroxyl and carbonyl groups from various flavonoids.

Flavonoids constitute a broad category of secondary metabolites. Structurally, flavonoids are formed by two benzene rings (A and B) joined by a three-carbon-based pyran ring (C). According to their substitution pattern and the number of functional groups, flavonoids are categorized into anthocyanidins, flavonols, flavones, and isoflavones. It is known that *Kalanchoe* species can contain multiple flavonoids such as quercetin (Qu), kaempferol, myricetin, luteolin, eupafolin, or their derivatives [35]. These compounds have been reported among several species, for example, *K. pinnata*, *K. gracilis*, *K. blossfeldiana*, *K. tomentosa*, and *K. pathulate* [19,20,36].

A calibration curve was constructed considering various concentrations of Qu to estimate the TFC of methanol extract from *K. fedtschenkoi*. To perform the TFC assay, bioactive nature products such as catechin and rutin are commonly used. However, Qu is also preferred, as it is a flavonol that reacts with $AlCl_3$ due to its keto group at C4 and hydroxyl groups at C3 or C5. Therefore, using the regression equation ($y = 0.0007x + 0.3518$, $R^2 = 0.9995$) presented in the Supplementary Materials (Figure S4), the TFC of this extract was estimated and represented in milligrams of quercetin equivalents per gram of the plant extract (mg Qu/g). In this regard, y was considered as the absorbance of the test sample, whereas x was appraised as the concentration from the calibration curve. Following our calculations, the TFC of methanol extract is 384.54 ± 2.25 mg Qu/g. This result can be comparable to the TFC of methanol extracts prepared from other *Kalanchoe* species, such as *K. pinnata* (106 mg Qu/g) and *K. integra* (178 mg Qu/g) [37].

2.4. Antimicrobial Activity

In this study, the chloroform and methanol extract from *K. fedtschenkoi* exhibited weak antimicrobial activity against the panel tested (see Table 2). For instance, treatment with 150 µg/mL of chloroform extract only inhibited the growth of *P. aeruginosa*. At the same concentration, methanol extract inhibited the growth of MRSA. No inhibition of yeast strains was recorded.

Medicinal plants can exhibit antimicrobial properties against pathogenic bacteria, fungi, protozoa, and viruses through bioactive secondary metabolites such as alkaloids, flavonoids, terpenes, and polysaccharides. Generally, extracts from medicinal plants or herbs can inhibit the growth of pathogenic bacteria and fungi by damaging cell membranes or walls, interfering with protein synthesis, and increasing intracellular osmotic pressure [38]. These mechanisms are due to the phytochemical content of plant extracts; for example, given the existence of hydroxyl groups and delocalized electrons among polyphenols' architecture, they can increase the permeability of the bacterial membrane, alter its potential, and cause structural changes [39]. In contrast, the acyl chains, numerous hydroxyl groups, and glycosylated moieties of flavonoids enable their capacity to reduce nucleic acid synthesis, disrupt energy metabolisms, and suppress cytoplasmic bacterial membrane functionality [40].

Table 2. Antimicrobial activity of K. fedtschenkoi extracts expressed as the minimal inhibitory concentration (µg/mL).

Extract	Bacteria								Fungi	
	MRSA	SA	AB	PA	EC	LM	ABc	PAc	CA	CN
Hexane	R	R	R	R	R	R	R	R	R	R
Chloroform	R	R	R	150	R	R	R	R	R	R
Methanol	150	R	R	R	R	R	R	R	R	R

Abbreviations: MRSA, methicillin-resistant *Staphylococcus aureus*; SA, *Staphylococcus aureus*; AB, *Acinetobacter baumannii*, PA, *Pseudomonas aeruginosa*; EC, *Escherichia coli*; LM, *Listeria monocytogenes*; ABc, *Acinetobacter baumannii* clinical isolate; PAc, *Pseudomonas aeruginosa* clinical isolate; CA, *Candida albicans*; CN, *Cryptococcus neoformans*; R, resistant.

Among *Kalanchoe* species, many antibacterial and antifungal compounds have been identified over the last decades. For example, *K. pinnata* and *K. daigremontiana* are well-known medicinal plants that contain flavonols (e.g., quercetin and kaempferol), flavones (e.g., luteolin), and bufadienolides [41] that can reduce the formation of biofilms, inhibit the growth of pathogenic bacteria strains, decrease protein synthesis, or inhibit the expression of genes related to antimicrobial resistance [20,42–44]. According to published reports [45], these mechanisms of action might be related to the glycosyl derivatives of kaempferol identified on the methanol extract of *K. fedtschenkoi*. In the same regard, extracts from other species, such as *K. blossfeldiana*, contain palmitic acid, gallic acid, methyl gallate, and carbohydrates that can be correlated to their antimicrobial properties [46].

Among multidrug-resistant bacteria, *P. aeruginosa* is a challenging pathogen in human health care. It can cause acute or chronic infections in patients diagnosed with cystic fibrosis, cancer, and coronavirus disease-19 (COVID-19) [47]. Only one study has reported the inhibition of *P. aeruginosa*, using ethanol extracts from *K. fedtschenkoi* at concentrations ranging from IC_{50} 128 to 256 µg/mL; similar concentrations were reported against *A. baumannii* and *S. aureus* [25]. These findings can be attributed to differences in harvesting places, climate, soil characteristics, extract polarity, implemented methodology, and strain culture conditions.

MRSA is a global human health threat characterized by its prevalence in the community, ease of spread, and capacity to cause endocarditis, bacteremia, osteomyelitis, pneumonia, and purulent infections [48]. These results can be compared with the activity of hydroethanolic extracts from *K. brasiliensis* that have inhibited the growth of MRSA strains at MIC > 5000 µg/mL. The antibacterials eupafolin and patuletin or their glycosylated derivatives can explain this activity [49].

2.5. Cytotoxicity Activity

The evaluation of the toxicity of plant extracts or bioactive nature products is necessary to determine their possible application in the development of pharmaceutical formulations or use against other diseases, such as cancer [50]. In this study, we assessed the cytotoxicity using the MTT ((3-(4,5-dimethylthiazol-2-yl)-2,5-diphenyltetrazolium bromide) assay [51]. Results demonstrated the three extracts exhibited a significant cytotoxic effect ($p < 0.0001$) at the tested concentrations (Figure 1). In this regard, it should be noted that even though treatment with hexane and chloroform extracts was cytotoxic to THP-1 cells, treatment with the methanol extract exhibited the highest cytotoxicity towards the tested cell line ($p < 0.0001$).

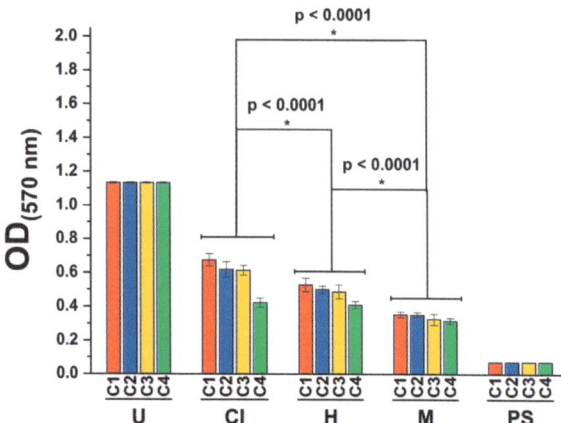

Figure 1. Viability of human-derived THP-1 cells against treatment with 50 (C1), 100 (C2), 150 (C3), and 200 (C4) µg/mL of *K. fedtschenkoi* using U, untreated cells, chloroform (Cl), hexane (H), and methanol (M) extracts using the MTT assay. PS, positive control (2% Tween-20); OD, optical density. Shown is the mean ± S.D. of three independent experiments. * Represents *p*-values significantly below to <0.0001 evaluated with Tukey's test.

It is known that medicinal plants have represented an exceptional source of cytotoxic molecules with potential use against various cancers such as lung, breast, and prostate cancer. This is of broad importance in developing countries such as Mexico, since approximately 30% of Mexican patients utilize preparations from medicinal plants as a preventive, complementary, and cost-effective approach during cancer therapy [52]. Furthermore, the structural diversity of isolated molecules from medicinal plants is exploited in developed countries to develop drugs against infectious and neoplastic diseases [53,54].

According to the National Cancer Institute (NCI) of the U.S., the cytotoxicity of molecules can be considered high ($IC_{50} \leq 20$ µg/mL), moderate (IC_{50} 21–200 µg/mL), or weak (IC_{50} 201–500 µg/mL) [55,56]. We found that the hexane, chloroform, and methanol extracts were highly cytotoxic against the tested cell line, as they presented IC_{50} values of 2.090, 1.918, and 1.722 µg/mL, respectively. This result can be attributed to the extracts' polarity and their phytoconstituents.

For instance, the cytotoxicity of the hexane extract might be due to the presence of friedelin, which has been reported to inhibit the proliferation of breast cancer cells [57]. In the same regard, the cytotoxicity of the chloroform extract can be attributed to the presence of stigmasterol and tocopherol. Currently, the former is recognized as a promising cytotoxic agent for cancer therapy [58], whereas the latter can exert distinct cytotoxic effects on cell lines such as bovine endothelial cells, mouse macrophages, and human hepatocytes [59]. On the other hand, the cytotoxicity of the methanol extract might be due to synergism between the derivatives of quercetin and kaempferol, which have been reported to induce the death of different cell lines such as MDA-MB-231 and MCF-7 [60].

Non-polar solvents such as hexane and chloroform are frequently used to extract bioactive compounds such as terpenoids, fats, and oils. In contrast, polar solvents, such as methanol, are preferably used, as they are accessible, nontoxic at low concentrations, and can extract high polar compounds with strong therapeutic properties (e.g., anthocyanins, polyphenols, and flavonoids) such as antimicrobial and anticancer qualities [4,61]. Interestingly, the cytotoxicity of *Kalanchoe* species has been widely reported for polar extracts in the scientific literature. For instance, treatment with less than <40 µg/mL of *K. crenata* methanol extract was cytotoxic towards breast adenocarcinoma (MCF-7), hepatocarcinoma (HepG2), colorectal adenocarcinoma (DLD-1), and human non-small-cell lung cancer (A549) cell lines [62].

In another study, comparable effects have been documented for ethanol extracts prepared from the leaves of *K. millotii* and *K. nyikae*, which were cytotoxic against human acute lymphoblastic leukemia T (J45) and human T (H9) cell lines at IC$_{50}$ values of 503.5 and 560.5, and 846.1 and 507.6 µg/mL, respectively [63]. Against other types of cancer, such as the human ovarian cancer SKOV-3 cells, a water extract from the leaves of *K. daigremontiana* arrested the cell cycle and induced mitochondrial membrane depolarization. It was cytotoxic against the cells at 5 to 200 µg/mL [32]. Comparably to these reports, our work demonstrated that the *K. fedtschenkoi* extracts' cytotoxicity increases according to the polarity of the solvents. For instance, treatment with the methanol extract was highly cytotoxic against THP-1 cells at 1.722 µg/mL, whereas treatment with hexane extract decreased cell viability at 2.090 µg/mL. In another study, the cytotoxicity of the ethanol extract from *K. fedtschenkoi* was evaluated towards human keratinocytes cells (HaCAts); this study revealed that ethanol extract decreased cell viability at LD$_{50}$ > 250 µg/mL [25]. However, other differences between our results and other reports can be attributed to the studied species, tested concentrations, extracts' polarity, and evaluated cell lines.

2.6. Anti-Inflammatory Activity

Inflammation occurs when hazardous stimuli, such as microorganisms, toxic compounds, or damaged cells, activate immune cells [64]. Multidrug-resistant bacteria can promote inflammation processes through different mechanisms. Inflammatory responses caused by pathogenic bacteria arise from the interaction between pathogen-associated molecular patterns (PAMPs) and pattern-recognition receptors (PRRs), which are structures expressed among immune and non-immune cells [64]. PRRs recognize distinct ligands from bacterial architecture, such as flagellin, peptidoglycans, genetic material, and lipoteichoic acid [65]. The interaction between PAMPs and PRRs results in complex intracellular signaling cascades devoted to recruiting inflammatory molecules necessary to retain the progression of infection and inflammation and to initiate tissue repair [66,67].

In this study, we show that treatment with the *K. fedtschenkoi* chloroform extract significantly stimulated the secretion of the anti-inflammatory cytokine IL-10 at 50 µg/mL ($p < 0.01$). Still, it did not reduce the secretion of the pro-inflammatory cytokine IL-6 at the tested concentration (Figure 2). The anti-inflammatory activity of the chloroform extract can be attributed to the presence of stigmasterol. This widely recognized phytosterol can suppress the production of IL-6 and other pro-inflammatory cytokines such as tumor necrosis factor-α (TNF-α) [68,69].

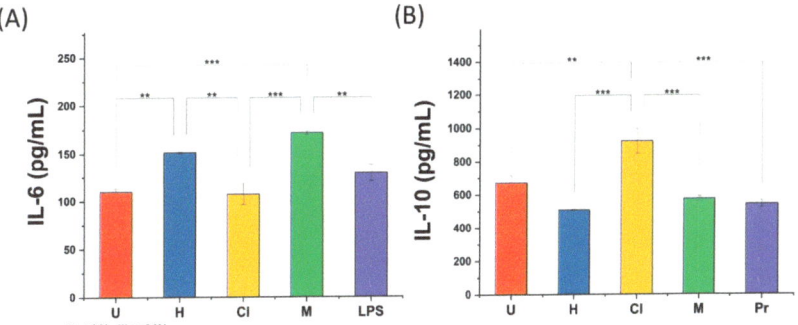

Figure 2. Immunological response of *K. fedtschenkoi* hexane (H), chloroform (Cl), and methanol (M) extracts on human-derived THP-1 cells using ELISA for (**A**) IL-6 and (**B**) IL-10. LPS, lipopolysaccharide (positive control for inflammation), untreated cells (U), and prednisone (Pr, anti-inflammatory, positive control). Shown is the mean ± S.D. of three independent experiments. ** Represents *p*-values significantly below <0.01 evaluated with Tukey's test. *** Represents *p*-values significantly below <0.001 evaluated with Tukey's test.

On the other hand, treatment with hexane and methanol extract enhanced the secretion of IL-6, which can be attributed to the cytotoxicity of these extracts. Several models have been proposed to evaluate the anti-inflammatory properties of *Kalanchoe* species. For example, it has been reported that treatment with *K. pinnata* hydroethanolic extract reduced pro-inflammatory responses among rodent models by downregulating the expression of mediators involved in inflammatory processes such as Toll-like receptors and nuclear factor kappa B (NFκB) [70]. Comparably, topical formulations prepared from *K. pinnata* and *K. brasiliensis* aqueous extracts decreased the production of TNF-α and IL-1β while enhancing the levels of IL-10 among edematogenic Swiss mice models [71].

2.7. FTIR Analysis

FTIR spectroscopy is based on the absorption of infrared light by organic and inorganic analytes. Each sample exhibits a distinctive spectrum fingerprint in this technique that can be recognized and differentiated from other molecules [72]. For plant extracts, FTIR can be used to study their chemical composition in solid and liquid samples.

Figure 3 depicts *K. fedtschenkoi* extracts exhibiting similar bands within 2916 and 2849 cm^{-1}, corresponding to the asymmetrical and symmetrical stretching of the C-H bonds from hydrocarbon chains. Comparably, hexane and methanol extracts present a broad band at 3298 cm^{-1}, related to O-H bond stretching, usually associated with phenolic compounds. This finding might suggest the presence of compounds reported among *K. fedtschenkoi* extracts, such as quercetin. Among bioactive nature products, quercetin is an abundant flavonoid in plants, fruits, and vegetables, exhibiting different peaks in the infrared region. It has been reported that flavonoids such as quercetin exhibit distinctive bands from 1610 to 1510 cm^{-1}, attributed to C=C bonds from its aromatic ring. In addition, it is documented that quercetin presents a series of bands from 1260 to 1160 cm^{-1}, which is related to the C-O bond stretch aryl ether ring of its structure [73]. Interestingly, FTIR spectra of *K. fedtschenkoi* extracts present related peaks from 1600 to 1500 cm^{-1}, which can respond to bending vibrations of C=O bonds, possibly from esters, ketones, and carboxylic acids [74]. These findings are in accord with FTIR analysis of the chemical composition of extracts from other plants, including *K. fedtschenkoi* [75,76].

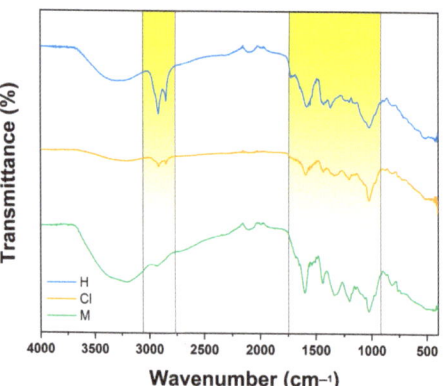

Figure 3. FTIR analysis of *K. fedtschenkoi* hexane (blue line), chloroform (yellow line), and methanol (green line).

3. Materials and Methods

3.1. Plant Material and Extract Preparation

Aerial parts of *K. fedtschenkoi* were collected in Texcoco, Mexico (19.491100418879938, −98.88525021669616). Specimens were identified by the biologist Lilián López-Chávez at the herbarium of Universidad Autónoma de Chapingo (Carr. Federal México-Texcoco, 56230, Texcoco, Estado de México) and deposited with the voucher number 36205. The collected fresh plants were dried in a ventilated and dark place at room temperature for one week. For extract preparation, 500 g of dried leaves were finely powdered utilizing a mechanical blender before being initially macerated with hexane, then with chloroform, and finally, with methanol for three days (1.5 L each). After that, the mixture was filtered, and the solvent was evaporated to dryness under reduced pressure using a Heidolph Laborota 4000 efficient rotary evaporator (Schwabach, Germany). The extracts were preserved under refrigeration until further use.

3.2. GC/MS Analysis

The phytoconstituents of *K. fedtschenkoi* extracts were identified using a Varian CP-3800 gas chromatograph coupled to a Varian 1200 quadrupole mass spectrometer [77]. Samples were injected into a Factor Four (VF-5 ms, 30 m × 0.25 mm, 0.25 µm thickness) capillary column. Helium was used as the carrier gas at a 1 mL/min flow rate. The separation was carried out by injecting 1 µL of the sample (1%) into the column at the following gradient temperature: 60 °C for 2 min, 120 °C for 16 min, 30 °C/min up to 150 °C for 15 min, 20 °C/min up to 180 °C for 15 min, 30 °C/min up to 200 °C for 10 min, 20 °C/min up to 220 °C for 15 min, 5 °C/min up to 280 °C for 20 min, and 5 °C/min up to 300 °C for 20 min. Individual components from the extracts were identified based on comparing their retention times and fragmentation patterns to the National Institute of Standards and Technology Mass Spectral (NIST-MS) database. The total area of the peaks assessed their relative percentage.

3.3. HPLC Analysis

The phytochemical content of methanol extract was studied considering reported protocols with minor modifications [26]. Briefly, HPLC analysis was performed on an Agilent Technologies 1200 series equipped with a diode array detector (DAD) utilizing reagents of HPLC grade. An RP-18 Zorbax 150 mm × 4.6 mm, 3.5 µm column was employed as a stationary phase. On the other hand, water acidified with 0.1% formic acid (A) and acetonitrile (B) were used as the mobile phase. To prepare the sample, 2 mg of extract was diluted in water/acetonitrile (5:1). To carry out analysis, 10 µL of sample was run at the following gradient: 0–20 min (0–20% B), 20–40 min (20–22% B), 40–43 min (22–30% B), 43–45 min (30–100%B), 45–50 min (100% B). Absorbance was monitored at 254 and 365 nm. The analysis was performed in triplicate.

3.4. TFC Analysis

The TFC of *K. fedtschenkoi* methanol extract was determined as published [78]. Shortly, 100 µL of methanol extract (1 mg/mL) mixed with 100 µL of 2% aluminum chloride ($AlCl_3$) was incubated for 10 min. Then, using a Cary 60 UV-Vis spectrophotometer (Agilent Technologies, Santa Clara, CA, USA), absorbance was measured at 420 nm in 1 cm quartz cuvettes. A standard curve was obtained using distinct concentrations (100 to 500 µg/mL) of a standard solution of quercetin (Qu). The TFC of methanol extract was estimated as a percentage of total quercetin equivalents per gram of extract (mg Qu/g). The experiment was performed in triplicate.

3.5. Strains and Culture Media

The antimicrobial activity of *K. fedtschenkoi* extracts was studied against a panel of pathogenic Gram-positive and -negative bacteria and yeast strains. Gram-positive bacteria included *Listeria monocytogenes* (ATCC BAA-679), methicillin-resistant *Staphylococcus aureus* (MRSA) (ATCC 700698), and *Staphylococcus aureus* (ATCC 25923). On the other hand, Gram-negative bacteria included *Acinetobacter baumannii* (ATCC BAA-747), *Escherichia coli* (ATCC 25922), and *Pseudomonas aeruginosa* (ATCC 14210) strains. Clinical isolates of *A. baumannii* and *P. aeruginosa* were also tested in this study [79]. In contrast, fungal strains included *Candida albicans* (ATCC 10231) and *Cryptococcus neoformans* (kindly provided by Dr. Karen Bartlet, University of British Columbia, Vancouver, BC, Canada). Bacteria were cultured in Mueller–Hinton broth (B&D) at 37 °C, and fungal strains were cultured in Sabouraud broth (B&D) at 28 °C. In both cases, a shaker was used.

3.6. Microdilution Assay

The antimicrobial assay was conducted as in previously published protocols [80]. Shortly, in a 96-well plate, a microdilution assay was performed to determine the minimum inhibitory concentration (MIC). In 100 µL/well of Mueller–Hinton or Sabouraud broth, the following concentrations of *K. fedtschenkoi* extracts were tested against different microbial inocula: 50, 100, 150, and 200 µg/mL. The inoculum was prepared to have a final optical density of 0.05 at 600 nm. Bacteria treated with amikacin and gentamicin were used as positive controls, whereas untreated cells and DMSO were used as negative controls. For fungi, amphotericin and terbinafine were used as positive controls. All experiments were executed in triplicate.

3.7. Cytotoxicity Assay

The cytotoxicity of 50, 100, 150, and 200 µg/mL *K. fedtschenkoi* extracts was evaluated using human-derived THP-1 monocytic cells (ATCC TIB-202), following published protocols [81]. Using RPMI 1640 (Hyclone, GE Healthcare, Logan, UT, USA) supplemented with 5% fetal bovine serum (FCS) (Hyclone) and 2 mmol L^{-1} L-glutamine (Stem Cell Technologies, Vancouver, BC, Canada), THP-1 cells were cultured. To differentiate these cells, 20 ng/mL phorbol 12-myristate 13-acetate (PMA) was used. Then, cells were dispensed in a 96-well plate with a final volume of 100 µL at a final concentration of 1×10^5 cells per well. The plate was incubated at 37 °C with 5% CO_2 for 24 h. The next day, the medium was removed and replaced with fresh medium, and *K. fedtschenkoi* extracts at concentrations mentioned above. Again, the plate was incubated at 37 °C supplemented with 5% CO_2 for 24 h. Untreated cells and DMSO served as negative controls. Cells treated with 2% Tween-20 were used as a positive control. According to the same protocol [81], the cytotoxicity of *K. fedtschenkoi* extracts was known by the 3-(4,5-dimethylthiazol-2-yl)-2,5-diphenyltetrazolium bromide (MTT) assay. The next day, 2 h prior to the end of the incubation period, 25 µL of a working solution of MTT (5 mg mL^{-1}) was added to the cells, and they were incubated for a further 4 h. To dissolve formazan, 100 µL of extraction buffer was added per well; the plate was incubated overnight at 37 °C. In a plate reader, readings were performed at 570 nm. The half-maximal inhibitory concentration (IC_{50}) was estimated by plotting the log concentrations of the extracts against the percentage of damaged cells. All experiments were performed in triplicate.

3.8. Anti-Inflammatory Assay

To study the anti-inflammatory activity of *K. fedtschenkoi* extracts, published protocols were followed [81]. In brief, THP-1 cells differentiated with PMA were used in a 96-well plate at a final concentration of 1×10^5 cells per well. Given the results obtained during the cytotoxicity assay, *K. fedtschenkoi* extracts were assayed at a final concentration of 50 µg/mL. Cells treated with 100 ng/mL of lipopolysaccharide (LPS) from *E. coli* (Sigma-Aldrich, St. Louis, MO, USA) were used as a positive control. Contrarily, cells treated with DMSO were used as a negative control. The final concentration of DMSO per well was always $\leq 1\%$. The measurement of the pro-inflammatory cytokine IL-6 and the anti-inflammatory cytokine IL-10 was executed with commercial kits (B&D) following instructions from the manufacturer. Readings were recorded using a plate reader at 450 nm. All experiments were carried out in triplicate.

3.9. FTIR Analysis

The chemical composition of *K. fedtschenkoi* extracts was studied using a Cary 630 Fourier-transform infrared (FTIR) spectrometer (Agilent Technologies, Santa Clara, CA, USA). Briefly, the detection diamond was cleaned with 10 µL of ethanol solution (100% v/v) and allowed to air dry before analyses. Background spectra were recorded without samples at room temperature. For sample analysis, 20 mg of each extract was used per reading, and the crystal was cleaned after each measurement. Measurements were recorded within the 4000 to 400 cm^{-1} wavenumber region. All readings were recorded in triplicate.

3.10. Statistical Analysis

Data from the quantitative viability analysis were subjected to two-way analysis of variance (ANOVA), followed by a Tukey's mean separation test, to determine the relationship between each treatment using OriginPro 2023 data processing software (OriginLab, Northampton, MA, USA).

4. Conclusions

This study demonstrated that *K. fedtschenkoi* chloroform and methanol extract exhibit antibacterial activity against multidrug-resistant bacteria such as *P. aeruginosa* and MRSA.

During the cytotoxicity assay, the three extracts were highly cytotoxic against THP-1 cells with $IC_{50} < 3$ μg/mL, which suggests their potential use against cancer in in vitro or in vivo models. Furthermore, statistical analysis revealed that methanol extract exhibited higher cytotoxicity ($p < 0.0001$) against the tested cell line than hexane and chloroform extract. On the other hand, treatment with *K. fedtschenkoi* chloroform extract promoted the secretion of the anti-inflammatory cytokine IL-10.

Regarding their phytochemical content, this work demonstrated that hexane and chloroform extracts are mainly comprised of fatty acids, sterols, and triterpenes. In contrast, the derivatives of flavonoids such as quercetin and kaempferol are mainly in the methanol extract. In addition, we determined that methanol extract is abundant in flavonoids, as it presented a TFC of 384.54 ± 2.25 mg Qu/g. Using FTIR spectroscopy, we also demonstrated that extracts from the leaves of *K. fedtschenkoi* contain several functional groups such as C-H, C-O, C=O, C=C, and OH groups that might be related to compounds identified during the GC/MS or HPLC analyses performed in this work.

To the best of our knowledge, this is the first report that details the antimicrobial, cytotoxic, and anti-inflammatory properties of non-polar and polar extracts from *K. fedtschenkoi*. In addition, this study expands upon the knowledge on the genus *Kalanchoe* and assesses the importance of continuing to explore its therapeutic potential.

Supplementary Materials: The following supporting information can be downloaded at: https://www.mdpi.com/article/10.3390/plants12101943/s1, Figure S1: GC/MS chromatogram of hexane extract from *K. fedtschenkoi*. Figure S2: GC/MS chromatogram of chloroform extract from *K. fedtschenkoi*. Figure S3: HPLC chromatogram of methanol extract from *K. fedtschenkoi*. Figure S4: Calibration curve of quercetin used as standard solution.

Author Contributions: Conceptualization, J.L.M.-M., H.B. and E.S.-A.; validation, A.C.L.-L., H.B., L.R.H., D.E.N.-L., E.R.L.-M. and E.S.-A.; investigation, J.L.M.-M., A.C.L.-L., D.E.N.-L and E.R.L.-M.; writing—original draft preparation, J.L.M.-M., D.E.N.-L., E.R.L.-M. and E.S.-A.; writing—review and editing, A.C.L.-L., H.B., L.R.H., D.E.N.-L., E.R.L.-M. and E.S.-A.; visualization, J.L.M.-M., H.B., D.E.N.-L. and E.R.L.-M.; supervision, H.B., L.R.H., E.R.L.-M. and E.S.-A.; project administration, H.B., E.R.L.-M. and E.S.-A. All authors have read and agreed to the published version of the manuscript.

Funding: This research received no external funding.

Data Availability Statement: The information generated in this study can be consulted with authors for correspondence from this work.

Acknowledgments: J.L.M.-M. thanks the Consejo Nacional de Ciencia y Tecnología (CONACyT) for his doctoral fellowship. J.L.M.-M. and E.S.-A. thank the Universidad de las Américas Puebla (UDLAP) for providing resources to conclude this work. The graphical abstract was created using BioRender.com; URL was accessed on 9 March 2023.

Conflicts of Interest: The authors declare no conflict of interest.

References

1. Chali, B.U.; Hasho, A.; Koricha, N.B. Preference and Practice of Traditional Medicine and Associated Factors in Jimma Town, Southwest Ethiopia. *Evid. Based Complement. Alternat. Med.* **2021**, *2021*, 9962892. [CrossRef] [PubMed]
2. Firenzuoli, F.; Gori, L. Herbal Medicine Today: Clinical and Research Issues. *Evid. Based Complement. Alternat. Med.* **2007**, *4*, 37–40. [CrossRef] [PubMed]
3. Sofowora, A.; Ogunbodede, E.; Onayade, A. The Role and Place of Medicinal Plants in the Strategies for Disease Prevention. *Afr. J. Tradit. Complement. Altern. Med.* **2013**, *10*, 210. [CrossRef] [PubMed]
4. Abubakar, A.R.; Haque, M. Preparation of Medicinal Plants: Basic Extraction and Fractionation Procedures for Experimental Purposes. *J. Pharm. Bioallied. Sci.* **2020**, *12*, 1–10. [CrossRef]
5. AMELI, J. Communicable Diseases and Outbreak Control. *Turk. J. Emerg. Med.* **2016**, *15*, 20–26.
6. Edemekong, P.F.; Huang, B. Epidemiology of Prevention of Communicable Diseases. In *StatPearls*; StatPearls Publishing: Treasure Island, FL, USA, 2023.
7. Murray, C.J.L.; Ikuta, K.S.; Sharara, F.; Swetschinski, L.; Aguilar, G.R.; Gray, A.; Han, C.; Bisignano, C.; Rao, P.; Wool, E.; et al. Global Burden of Bacterial Antimicrobial Resistance in 2019: A Systematic Analysis. *Lancet* **2022**, *399*, 629–655. [CrossRef]
8. Bloom, D.E.; Cadarette, D. Infectious Disease Threats in the Twenty-First Century: Strengthening the Global Response. *Front. Immunol.* **2019**, *10*, 549. [CrossRef]
9. Chen, Y.; Chen, X.; Liang, Z.; Fan, S.; Gao, X.; Jia, H.; Li, B.; Shi, L.; Zhai, A.; Wu, C. Epidemiology and Prediction of Multidrug-Resistant Bacteria Based on Hospital Level. *J. Glob. Antimicrob. Resist.* **2022**, *29*, 155–162. [CrossRef]
10. Budreviciute, A.; Damiati, S.; Sabir, D.K.; Onder, K.; Schuller-Goetzburg, P.; Plakys, G.; Katileviciute, A.; Khoja, S.; Kodzius, R. Management and Prevention Strategies for Non-Communicable Diseases (NCDs) and Their Risk Factors. *Front. Public Health* **2020**, *8*, 574111. [CrossRef]
11. Calcaterra, V.; Zuccotti, G. Non-Communicable Diseases and Rare Diseases: A Current and Future Public Health Challenge within Pediatrics. *Children* **2022**, *9*, 1491. [CrossRef]
12. Kabir, A.; Karim, M.N.; Islam, R.M.; Romero, L.; Billah, B. Health System Readiness for Non-Communicable Diseases at the Primary Care Level: A Systematic Review. *BMJ Open* **2022**, *12*, e060387. [CrossRef] [PubMed]
13. Xu, Z.; Deng, M. Crassulaceae. In *Identification and Control of Common Weeds: Volume 2*; Xu, Z., Deng, M., Eds.; Springer: Dordrecht, The Netherlands, 2017; pp. 475–486. ISBN 978-94-024-1157-7.
14. Thiede, J.; Eggli, U. Crassulaceae. In *Flowering Plants Eudicots: Berberidopsidales, Buxales, Crossosomatales, Fabales p.p., Geraniales, Gunnerales, Myrtales p.p., Proteales, Saxifragales, Vitales, Zygophyllales, Clusiaceae Alliance, Passifloraceae Alliance, Dilleniaceae, Huaceae, Picramniaceae, Sabiaceae*; Kubitzki, K., Ed.; The Families and Genera of Vascular Plants; Springer: Berlin/Heidelberg, Germany, 2007; pp. 83–118, ISBN 978-3-540-32219-1.
15. Hassan, M.H.A.; Elwekeel, A.; Moawad, A.; Afifi, N.; Amin, E.; Amir, D.E. Phytochemical Constituents and Biological Activity of Selected Genera of Family Crassulaceae: A Review. *S. Afr. J. Bot.* **2021**, *141*, 383–404. [CrossRef]
16. Vargas, V.; Herrera, I.; Nualart, N.; Guézou, A.; Gómez-Bellver, C.; Freire, E.; Jaramillo Díaz, P.; López-Pujol, J. The Genus Kalanchoe (Crassulaceae) in Ecuador: From Gardens to the Wild. *Plants* **2022**, *11*, 1746. [CrossRef] [PubMed]
17. Kolodziejczyk-Czepas, J.; Nowak, P.; Wachowicz, B.; Piechocka, J.; Głowacki, R.; Moniuszko-Szajwaj, B.; Stochmal, A. Antioxidant Efficacy of Kalanchoe Daigremontiana Bufadienolide-Rich Fraction in Blood Plasma In Vitro. *Pharm. Biol.* **2016**, *54*, 3182–3188. [CrossRef]
18. Phatak, R.S.; Hendre, A.S. In-Vitro Antiurolithiatic Activity of Kalanchoe Pinnata Extract. *Int. J. Pharmacogn. Phytochem. Res.* **2015**, *7*, 275–279.
19. dos Santos Nascimento, L.B.; Casanova, L.M.; Costa, S.S. Bioactive Compounds from Kalanchoe Genus Potentially Useful for the Development of New Drugs. *Life* **2023**, *13*, 646. [CrossRef]
20. Costa, S.S.; Muzitano, M.F.; Camargo, L.M.M.; Coutinho, M.A.S. Therapeutic Potential of Kalanchoe Species: Flavonoids and Other Secondary Metabolites. *Nat. Prod. Commun.* **2008**, *3*, 2151–2164. [CrossRef]
21. Verma, V.; Kumar, S.; Rani, K.; Sehgal, N.; Prakash, O. Compound Profiling in Methanol Extract of Kalanchoe Blossfeldiana (Flaming Katy) Leaves Through GC-MS Analysis and Evaluation of Its Bioactive Properties. *Glob. J. Advnced Biol. Sci.* **2015**, *1*, 38–49.
22. Schuler, V.; Suter, K.; Fürer, K.; Eberli, D.; Horst, M.; Betschart, C.; Brenneisen, R.; Hamburger, M.; Mennet, M.; Schnelle, M.; et al. Bryophyllum Pinnatum Inhibits Detrusor Contractility in Porcine Bladder Strips—A Pharmacological Study towards a New Treatment Option of Overactive Bladder. *Phytomedicine* **2012**, *19*, 947–951. [CrossRef]
23. El Abdellaoui, S.; Destandau, E.; Toribio, A.; Elfakir, C.; Lafosse, M.; Renimel, I.; André, P.; Cancellieri, P.; Landemarre, L. Bioactive Molecules in Kalanchoe Pinnata Leaves: Extraction, Purification, and Identification. *Anal. Bioanal. Chem.* **2010**, *398*, 1329–1338. [CrossRef]
24. Zakharchenko, N.S.; Belous, A.S.; Biryukova, Y.K.; Medvedeva, O.A.; Belyakova, A.V.; Masgutova, G.A.; Trubnikova, E.V.; Buryanov, Y.I.; Lebedeva, A.A. Immunomodulating and Revascularizing Activity of Kalanchoe Pinnata Synergize with Fungicide Activity of Biogenic Peptide Cecropin P1. *J. Immunol. Res.* **2017**, *2017*, 3940743. [CrossRef]
25. Richwagen, N.; Lyles, J.T.; Dale, B.L.F.; Quave, C.L. Antibacterial Activity of Kalanchoe Mortagei and K. Fedtschenkoi Against ESKAPE Pathogens. *Front. Pharmacol.* **2019**, *10*, 67. [CrossRef] [PubMed]

26. Casanova, J.M.; dos Santos Nascimento, L.B.; Casanova, L.M.; Castricini, S.D.; de Souza, J.E.E.; Yien, R.M.K.; Costa, S.S.; Tavares, E.S. Kalanchoe fedtschenkoi R. Hamet & H. Perrier, a Non-Conventional Food Plant in Brazil: HPLC-DAD-ESI-MS/MS Profile and Leaf Histochemical Location of Flavonoids. *J. Appl. Bot. Food Qual.* **2022**, *95*, 154–166. [CrossRef]
27. Poma, P.; Labbozzetta, M.; McCubrey, J.A.; Ramarosandratana, A.V.; Sajeva, M.; Zito, P.; Notarbartolo, M. Antitumor Mechanism of the Essential Oils from Two Succulent Plants in Multidrug Resistance Leukemia Cell. *Pharmaceuticals* **2019**, *12*, 124. [CrossRef] [PubMed]
28. Almeida, A.P.; Silva, S.A.G.D.; Souza, M.L.M.; Lima, L.M.T.R.R.; Rossi-Bergmann, B.; de Moraes, V.L.G.; Costa, S.S. Isolation and Chemical Analysis of a Fatty Acid Fraction of Kalanchoe Pinnata with a Potent Lymphocyte Suppressive Activity. *Planta Med.* **2000**, *66*, 134–137. [CrossRef]
29. Indriyanti, N.; Garmana, A.; Setiawan, F.; Setiawan, F. Repairing Effects of Aqueous Extract of Kalanchoe Pinnata (Lmk) Pers. on Lupus Nephritis Mice. *Pharmacogn. J.* **2018**, *10*, 548–552. [CrossRef]
30. Pereira, K.M.F.; Grecco, S.S.; Figueiredo, C.R.; Hosomi, J.K.; Nakamura, M.U.; Lago, J.H.G. Chemical Composition and Cytotoxicity of Kalanchoe Pinnata Leaves Extracts Prepared Using Accelerated System Extraction (ASE). *Nat. Prod. Commun.* **2018**, *13*, 163–166. [CrossRef]
31. Saleh, M.M.; Ghoneim, M.M.; Kottb, S.; El-Hela, A.A. Biologically Active Secondary Metabolites from Kalanchoe Tomentosa. *J. Biomed. Pharm. Res.* **2014**, *3*, 136–140.
32. Stefanowicz-Hajduk, J.; Hering, A.; Gucwa, M.; Sztormowska-Achranowicz, K.; Kowalczyk, M.; Soluch, A.; Ochocka, J.R. An In Vitro Anticancer, Antioxidant, and Phytochemical Study on Water Extract of Kalanchoe Daigremontiana Raym.-Hamet and H. Perrier. *Molecules* **2022**, *27*, 2280. [CrossRef]
33. Bhatti, M.; Kamboj, A.; Saluja, A.K. Phytochemical Screening and In-Vitro Evaluation of Antioxidant Activities of Various Extracts of Leaves and Stems of Kalanchoe Crenata. *J. Pharm. Nutr. Sci.* **2012**, *2*, 104–114. [CrossRef]
34. Fernandes, J.M.; Félix-Silva, J.; da Cunha, L.M.; dos Santos Gomes, J.A.; da Silva Siqueira, E.M.; Gimenes, L.P.; Lopes, N.P.; Soares, L.A.L.; de Freitas Fernandes-Pedrosa, M.; Zucolotto, S.M. Inhibitory Effects of Hydroethanolic Leaf Extracts of Kalanchoe Brasiliensis and Kalanchoe Pinnata (Crassulaceae) against Local Effects Induced by Bothrops Jararaca Snake Venom. *PLoS ONE* **2016**, *11*, e0168658. [CrossRef]
35. Stefanowicz-Hajduk, J.; Asztemborska, M.; Krauze-Baranowska, M.; Godlewska, S.; Gucwa, M.; Moniuszko-Szajwaj, B.; Stochmal, A.; Ochocka, J.R. Identification of Flavonoids and Bufadienolides and Cytotoxic Effects of Kalanchoe Daigremontiana Extracts on Human Cancer Cell Lines. *Planta Med.* **2020**, *86*, 239–246. [CrossRef]
36. Aisyah, L.S.; Yun, Y.F.; Julaeha, E.; Herlina, T.; Zainuddin, A.; Hermawan, W.; Supratman, U.; Hayashi, H. Flavonoids from the Fresh Leaves of Kalanchoe Tomentosa (Crassulaceae). *Open Chem. J.* **2015**, *2*, 36–39. [CrossRef]
37. Asiedu-Gyekye, I.J.; Antwi, D.A.; Bugyei, K.A.; Awortwe, C. Comparative Study of Two Kalanchoe Species: Total Flavonoid and Phenolic Contents and Antioxidant Properties. *Afr. J. Pure. Appl. Chem.* **2012**, *6*, 65–73. [CrossRef]
38. Liang, J.; Huang, X.; Ma, G. Antimicrobial Activities and Mechanisms of Extract and Components of Herbs in East Asia. *RSC Adv.* **2022**, *12*, 29197–29213. [CrossRef]
39. Álvarez-Martínez, F.J.; Barrajón-Catalán, E.; Herranz-López, M.; Micol, V. Antibacterial Plant Compounds, Extracts and Essential Oils: An Updated Review on Their Effects and Putative Mechanisms of Action. *Phytomedicine* **2021**, *90*, 153626. [CrossRef]
40. Shamsudin, N.F.; Ahmed, Q.U.; Mahmood, S.; Ali Shah, S.A.; Khatib, A.; Mukhtar, S.; Alsharif, M.A.; Parveen, H.; Zakaria, Z.A. Antibacterial Effects of Flavonoids and Their Structure-Activity Relationship Study: A Comparative Interpretation. *Molecules* **2022**, *27*, 1149. [CrossRef] [PubMed]
41. Stefanowicz-Hajduk, J.; Hering, A.; Gucwa, M.; Hałasa, R.; Soluch, A.; Kowalczyk, M.; Stochmal, A.; Ochocka, R. Biological Activities of Leaf Extracts from Selected Kalanchoe Species and Their Relationship with Bufadienolides Content. *Pharm. Biol.* **2020**, *58*, 732–740. [CrossRef]
42. Yang, D.; Wang, T.; Long, M.; Li, P. Quercetin: Its Main Pharmacological Activity and Potential Application in Clinical Medicine. *Oxidative Med. Cell. Longev.* **2020**, *2020*, e8825387. [CrossRef] [PubMed]
43. Periferakis, A.; Periferakis, K.; Badarau, I.A.; Petran, E.M.; Popa, D.C.; Caruntu, A.; Costache, R.S.; Scheau, C.; Caruntu, C.; Costache, D.O. Kaempferol: Antimicrobial Properties, Sources, Clinical, and Traditional Applications. *Int. J. Mol. Sci.* **2022**, *23*, 15054. [CrossRef]
44. Guo, Y.; Liu, Y.; Zhang, Z.; Chen, M.; Zhang, D.; Tian, C.; Liu, M.; Jiang, G. The Antibacterial Activity and Mechanism of Action of Luteolin Against Trueperella Pyogenes. *Infect. Drug Resist.* **2020**, *13*, 1697–1711. [CrossRef] [PubMed]
45. Tatsimo, S.J.N.; de Dieu Tamokou, J.; Havyarimana, L.; Csupor, D.; Forgo, P.; Hohmann, J.; Kuiate, J.-R.; Tane, P. Antimicrobial and Antioxidant Activity of Kaempferol Rhamnoside Derivatives from Bryophyllum Pinnatum. *BMC Res. Notes* **2012**, *5*, 158. [CrossRef] [PubMed]
46. El-Shamy, A.M.; Fathy, F.I.; Abdel-Rahman, E.H.; Sabry, M.M. Phytochemical, Biological and Botanical Studies of Klanchoe Blossfeldiana Poelln. *Int. J. Pharm. Photon.* **2013**, *104*, 189–205.
47. Qin, S.; Xiao, W.; Zhou, C.; Pu, Q.; Deng, X.; Lan, L.; Liang, H.; Song, X.; Wu, M. Pseudomonas Aeruginosa: Pathogenesis, Virulence Factors, Antibiotic Resistance, Interaction with Host, Technology Advances and Emerging Therapeutics. *Signal Transduct. Target.* **2022**, *7*, 199. [CrossRef]

48. Turner, N.A.; Sharma-Kuinkel, B.K.; Maskarinec, S.A.; Eichenberger, E.M.; Shah, P.P.; Carugati, M.; Holland, T.L.; Fowler, V.G. Methicillin-Resistant Staphylococcus Aureus: An Overview of Basic and Clinical Research. *Nat. Rev. Microbiol.* **2019**, *17*, 203–218. [CrossRef]
49. Mayorga, O.A.S.; da Costa, Y.F.G.; da Silva, J.B.; Scio, E.; Ferreira, A.L.P.; de Sousa, O.V.; Alves, M.S. *Kalanchoe brasiliensis* Cambess., a Promising Natural Source of Antioxidant and Antibiotic Agents against Multidrug-Resistant Pathogens for the Treatment of *Salmonella* Gastroenteritis. *Oxidative Med. Cell. Longev.* **2019**, *2019*, e9245951. [CrossRef]
50. McGaw, L.J.; Elgorashi, E.E.; Eloff, J.N. 8-Cytotoxicity of African Medicinal Plants Against Normal Animal and Human Cells. In *Toxicological Survey of African Medicinal Plants*; Kuete, V., Ed.; Elsevier: Amsterdam, The Netherlands, 2014; pp. 181–233, ISBN 978-0-12-800018-2.
51. Canga, I.; Vita, P.; Oliveira, A.I.; Castro, M.Á.; Pinho, C. In Vitro Cytotoxic Activity of African Plants: A Review. *Molecules* **2022**, *27*, 4989. [CrossRef]
52. Elizondo-Luévano, J.H.; Gomez-Flores, R.; Verde-Star, M.J.; Tamez-Guerra, P.; Romo-Sáenz, C.I.; Chávez-Montes, A.; Rodríguez-Garza, N.E.; Quintanilla-Licea, R. In Vitro Cytotoxic Activity of Methanol Extracts of Selected Medicinal Plants Traditionally Used in Mexico against Human Hepatocellular Carcinoma. *Plants* **2022**, *11*, 2862. [CrossRef]
53. Atanasov, A.G.; Zotchev, S.B.; Dirsch, V.M.; Supuran, C.T. Natural Products in Drug Discovery: Advances and Opportunities. *Nat. Rev. Drug Discov.* **2021**, *20*, 200–216. [CrossRef]
54. Rathor, L. Medicinal Plants: A Rich Source of Bioactive Molecules Used in Drug Development. In *Evidence Based Validation of Traditional Medicines: A comprehensive Approach*; Mandal, S.C., Chakraborty, R., Sen, S., Eds.; Springer: Singapore, 2021; pp. 195–209, ISBN 9789811581274.
55. Abdel-Hameed, E.-S.S.; Bazaid, S.A.; Shohayeb, M.M.; El-Sayed, M.M.; El-Wakil, E.A. Phytochemical Studies and Evaluation of Antioxidant, Anticancer and Antimicrobial Properties of *Conocarpus erectus* L. Growing in Taif, Saudi Arabia. *Eur. J. Med. Plants* **2012**, *2*, 93–112. [CrossRef]
56. Widiyastuti, Y.; Sholikhah, I.Y.M.; Haryanti, S. *Cytotoxic Activities of Ethanolic and Dichloromethane Extract of Leaves, Stems, and Flowers of Jarong [Stachytarpheta jamaicensis (L.) Vahl.] on HeLa and T47D Cancer Cell Line*; AIP Publishing LLC: Surakarta, Indonesia, 2019; p. 020101.
57. Subash-Babu, P.; Li, D.K.; Alshatwi, A.A. In Vitro Cytotoxic Potential of Friedelin in Human MCF-7 Breast Cancer Cell: Regulate Early Expression of Cdkn2a and PRb1, Neutralize Mdm2-P53 Amalgamation and Functional Stabilization of P53. *Exp. Toxicol. Pathol.* **2017**, *69*, 630–636. [CrossRef] [PubMed]
58. Zhang, X.; Wang, J.; Zhu, L.; Wang, X.; Meng, F.; Xia, L.; Zhang, H. Advances in Stigmasterol on Its Anti-Tumor Effect and Mechanism of Action. *Front. Oncol.* **2022**, *12*, 1101289. [CrossRef] [PubMed]
59. McCormick, C.C.; Parker, R.S. The Cytotoxicity of Vitamin E Is Both Vitamer- and Cell-Specific and Involves a Selectable Trait. *J. Nutr.* **2004**, *134*, 3335–3342. [CrossRef]
60. Yadegarynia, S.; Pham, A.; Ng, A.; Nguyen, D.; Lialiutska, T.; Bortolazzo, A.; Sivryuk, V.; Bremer, M.; White, J.B. Profiling Flavonoid Cytotoxicity in Human Breast Cancer Cell Lines: Determination of Structure-Function Relationships. *Nat. Prod. Commun.* **2012**, *7*, 1295–1304. [CrossRef] [PubMed]
61. Cowan, M.M. Plant Products as Antimicrobial Agents. *Clin. Microbiol. Rev.* **1999**, *12*, 564–582. [CrossRef] [PubMed]
62. Kuete, V.; Fokou, F.W.; Karaosmanoğlu, O.; Beng, V.P.; Sivas, H. Cytotoxicity of the Methanol Extracts of Elephantopus Mollis, Kalanchoe Crenata and 4 Other Cameroonian Medicinal Plants towards Human Carcinoma Cells. *BMC Complement. Altern. Med.* **2017**, *17*, 280. [CrossRef]
63. Bogucka-Kocka, A.; Zidorn, C.; Kasprzycka, M.; Szymczak, G.; Szewczyk, K. Phenolic Acid Content, Antioxidant and Cytotoxic Activities of Four Kalanchoë Species. *Saudi J. Biol. Sci.* **2018**, *25*, 622–630. [CrossRef]
64. Chen, L.; Deng, H.; Cui, H.; Fang, J.; Zuo, Z.; Deng, J.; Li, Y.; Wang, X.; Zhao, L. Inflammatory Responses and Inflammation-Associated Diseases in Organs. *Oncotarget* **2017**, *9*, 7204–7218. [CrossRef]
65. Abdullah, Z.; Knolle, P.A. Scaling of Immune Responses against Intracellular Bacterial Infection. *EMBO J.* **2014**, *33*, 2283–2294. [CrossRef]
66. Stokes, B.A.; Yadav, S.; Shokal, U.; Smith, L.C.; Eleftherianos, I. Bacterial and Fungal Pattern Recognition Receptors in Homologous Innate Signaling Pathways of Insects and Mammals. *Front. Microbiol.* **2015**, *6*, 19. [CrossRef]
67. Zhang, J.-M.; An, J. Cytokines, Inflammation and Pain. *Int. Anesth. Clin.* **2007**, *45*, 27–37. [CrossRef] [PubMed]
68. Bakrim, S.; Benkhaira, N.; Bourais, I.; Benali, T.; Lee, L.-H.; El Omari, N.; Sheikh, R.A.; Goh, K.W.; Ming, L.C.; Bouyahya, A. Health Benefits and Pharmacological Properties of Stigmasterol. *Antioxidants* **2022**, *11*, 1912. [CrossRef] [PubMed]
69. Kangsamaksin, T.; Chaithongyot, S.; Wootthichairangsan, C.; Hanchaina, R.; Tangshewinsirikul, C.; Svasti, J. Lupeol and Stigmasterol Suppress Tumor Angiogenesis and Inhibit Cholangiocarcinoma Growth in Mice via Downregulation of Tumor Necrosis Factor-α. *PLoS ONE* **2017**, *12*, e0189628. [CrossRef] [PubMed]
70. Andrade, A.W.L.; Guerra, G.C.B.; de Souza Araújo, D.F.; de Araújo Júnior, R.F.; de Araújo, A.A.; de Carvalho, T.G.; Fernandes, J.M.; Diez-Echave, P.; Hidalgo-García, L.; Rodriguez-Cabezas, M.E.; et al. Anti-Inflammatory and Chemopreventive Effects of Bryophyllum Pinnatum (Lamarck) Leaf Extract in Experimental Colitis Models in Rodents. *Front. Pharmacol.* **2020**, *11*, 998. [CrossRef] [PubMed]
71. de Araújo, E.R.D.; Félix-Silva, J.; Xavier-Santos, J.B.; Fernandes, J.M.; Guerra, G.C.B.; de Araújo, A.A.; de Souza Araújo, D.F.; de Santis Ferreira, L.; da Silva Júnior, A.A.; de Freitas Fernandes-Pedrosa, M.; et al. Local Anti-Inflammatory Activity: Topical

Formulation Containing Kalanchoe Brasiliensis and Kalanchoe Pinnata Leaf Aqueous Extract. *Biomed. Pharm.* **2019**, *113*, 108721. [CrossRef]
72. Fadlelmoula, A.; Pinho, D.; Carvalho, V.H.; Catarino, S.O.; Minas, G. Fourier Transform Infrared (FTIR) Spectroscopy to Analyse Human Blood over the Last 20 Years: A Review towards Lab-on-a-Chip Devices. *Micromachines* **2022**, *13*, 187. [CrossRef]
73. Catauro, M.; Papale, F.; Bollino, F.; Piccolella, S.; Marciano, S.; Nocera, P.; Pacifico, S. Silica/Quercetin Sol–Gel Hybrids as Antioxidant Dental Implant Materials. *Sci. Technol. Adv. Mater.* **2015**, *16*, 035001. [CrossRef]
74. Khan, N.; Jamila, N.; Ejaz, R.; Nishan, U.; Kim, K.S. Volatile Oil, Phytochemical, and Biological Activities Evaluation of Trachyspermum Ammi Seeds by Chromatographic and Spectroscopic Methods. *Anal. Lett.* **2020**, *53*, 984–1001. [CrossRef]
75. Din, S.; Hamid, S.; Yaseen, A.; Yatoo, A.M.; Ali, S.; Shamim, K.; Mahdi, W.A.; Alshehri, S.; Rehman, M.U.; Shah, W.A. Isolation and Characterization of Flavonoid Naringenin and Evaluation of Cytotoxic and Biological Efficacy of Water Lilly (*Nymphaea mexicana* Zucc.). *Plants* **2022**, *11*, 3588. [CrossRef]
76. Bhatt, N.; Mehata, M.S. A Sustainable Approach to Develop Gold Nanoparticles with Kalanchoe Fedtschenkoi and Their Interaction with Protein and Dye: Sensing and Catalytic Probe. *Plasmonics* **2023**. [CrossRef]
77. Cuevas-Cianca, S.I.; Leal, A.C.L.; Hernández, L.R.; Arreola, E.S.; Bach, H. Antimicrobial, Toxicity, and Anti-Inflammatory Activities of Buddleja Perfoliata Kunth. *Phytomed. Plus* **2022**, *2*, 100357. [CrossRef]
78. Rahim, N.A.; Roslan, M.N.F.; Muhamad, M.; Seeni, A. Antioxidant Activity, Total Phenolic and Flavonoid Content and LC–MS Profiling of Leaves Extracts of Alstonia Angustiloba. *Separations* **2022**, *9*, 234. [CrossRef]
79. Bermúdez-Jiménez, C.; Romney, M.G.; Roa-Flores, S.A.; Martínez-Castañón, G.; Bach, H. Hydrogel-Embedded Gold Nanorods Activated by Plasmonic Photothermy with Potent Antimicrobial Activity. *Nanomed. Nanotechnol. Biol. Med.* **2019**, *22*, 102093. [CrossRef] [PubMed]
80. Cruz Paredes, C.; Bolívar Balbás, P.; Gómez-Velasco, A.; Juárez, Z.N.; Sánchez Arreola, E.; Hernández, L.R.; Bach, H. Antimicrobial, Antiparasitic, Anti-Inflammatory, and Cytotoxic Activities of Lopezia Racemosa. *Sci. World J.* **2013**, *2013*, e237438. [CrossRef]
81. Juárez, Z.N.; Bach, H.; Sánchez-Arreola, E.; Bach, H.; Hernández, L.R. Protective Antifungal Activity of Essential Oils Extracted from Buddleja Perfoliata and Pelargonium Graveolens against Fungi Isolated from Stored Grains. *J. Appl. Microbiol.* **2016**, *120*, 1264–1270. [CrossRef]

Disclaimer/Publisher's Note: The statements, opinions and data contained in all publications are solely those of the individual author(s) and contributor(s) and not of MDPI and/or the editor(s). MDPI and/or the editor(s) disclaim responsibility for any injury to people or property resulting from any ideas, methods, instructions or products referred to in the content.

Article

New Garden Rose (*Rosa* × *hybrida*) Genotypes with Intensely Colored Flowers as Rich Sources of Bioactive Compounds

Nataša Simin [1], Nemanja Živanović [1], Biljana Božanić Tanjga [2], Marija Lesjak [1], Tijana Narandžić [3] and Mirjana Ljubojević [3,*]

[1] Faculty of Sciences, University of Novi Sad, Trg Dositeja Obradovića 3, 21000 Novi Sad, Serbia
[2] Breeding Company 'Pheno Geno Roses', Maršala Tita 75, 23326 Ostojićevo, Serbia
[3] Faculty of Agriculture, University of Novi Sad, Trg Dositeja Obradovića 8, 21000 Novi Sad, Serbia
* Correspondence: mirjana.ljubojevic@polj.uns.ac.rs; Tel.: +381-21-4853-251

Abstract: Garden roses, known as *Rosa* × *hybrida*, hold a prominent position as one of the most important and economically valuable plants in horticulture. Additionally, their products—essential oil, rose water, concrete, and concentrate—find extensive use in the cosmetic, pharmaceutical, and food industries, due to their specific fragrances and potential health benefits. Rose flowers are rich in biologically active compounds, such as phenolics, flavonoids, anthocyanins, and carotenoids. This study aims to investigate the potential of five new garden rose genotypes with intensely colored flowers to serve as sources of biologically active compounds. Phenolic profile was evaluated by determination of total phenolic (TPC), flavonoid (TFC), and monomeric anthocyanins (TAC) contents and LC-MS/MS analysis of selected compounds. Antioxidant activity was evaluated via DPPH and FRAP assays, neuroprotective potential via acetylcholinesterase inhibition assay, and antidiabetic activity via α-amylase and α-glucosidase inhibition assays. The flowers of investigated genotypes were rich in phenolics (TPC varied from 148 to 260 mg galic acid eq/g de, TFC from 19.9 to 59.7 mg quercetin eq/g de, and TAC from 2.21 to 13.1 mg cyanidin 3-*O*-glucoside eq/g de). Four out of five genotypes had higher TPC than extract of *R. damascene*, the most famous rose cultivar. The dominant flavonoids in all investigated genotypes were glycosides of quercetin and kaempferol. The extracts showed high antioxidant activity comparable to synthetic antioxidant BHT, very high α-glucosidase inhibitory potential, moderate neuroprotective activity, and low potential to inhibit α-amylase.

Keywords: garden roses; *Rosa* × *hybrida*; polyphenols; antioxidant; FRAP; DPPH; acetylcholinesterase inhibition

Citation: Simin, N.; Živanović, N.; Božanić Tanjga, B.; Lesjak, M.; Narandžić, T.; Ljubojević, M. New Garden Rose (*Rosa* × *hybrida*) Genotypes with Intensely Colored Flowers as Rich Sources of Bioactive Compounds. *Plants* **2024**, *13*, 424. https://doi.org/10.3390/plants13030424

Academic Editors: Luis Ricardo Hernández, Eugenio Sánchez-Arreola and Edgar López-Mena

Received: 11 January 2024
Revised: 28 January 2024
Accepted: 30 January 2024
Published: 31 January 2024

Copyright: © 2024 by the authors. Licensee MDPI, Basel, Switzerland. This article is an open access article distributed under the terms and conditions of the Creative Commons Attribution (CC BY) license (https://creativecommons.org/licenses/by/4.0/).

1. Introduction

In the realm of horticultural research, special attention is dedicated to the creation of new genotypes of flowering plants with intense flower colors, including red, violet, and purple, as their beauty adds a burst of life and vibrancy to any garden. Garden roses (*Rosa* × *hybrida*) stand out as specimens with captivating beauty and a rich palette of intense colors.

The pigments responsible for the red, violet, and purple hues of rose petals are primarily anthocyanins, a subgroup of flavonoid glycosides characterized by the presence of flavylium cation. Anthocyanins are water-soluble pigments known for their role as natural pigments in plants, contributing to the red, purple, blue, and violet colors of the flowers, depending on the pH environment. Beyond contributing to coloration, plant phenolics, particularly anthocyanins, contribute to the potential health benefits provided by plants. Numerous studies have suggested that anthocyanins derived from rose petals showcase robust anti-inflammatory, antioxidant, anticancer, antimicrobial, and antiallergic properties [1]. In addition to anthocyanins, rose hips and petals are rich in other polyphenols, essential oil, essential fatty acids, minerals (Ca, Mg, K, S, Si, Se, Mn and Fe), and vitamins

A, C, and E [2]. Our recent findings supported the hypothesis that rose petals of new cultivars of *Rosa × hybrida* are a rich source of biochemically active compounds and have high antioxidant and neuroprotective activity [3]. Thus, besides their role in ornamental horticulture, roses could potentially find applications in various industries, including the cosmetic, pharmaceutical and food industry.

Over the past decade, the 'Pheno Geno Roses' company in the Netherlands has successfully developed an extensive assortment of new garden rose varieties [4]. This achievement involved strategic hybridization, employing controlled pollination of parent plants with desirable traits and generating offspring with enhanced characteristics. A primary breeding objective was the identification of genotypes possessing favorable sensory attributes, particularly focusing on fragrance, with potential applications in the cosmetic and perfumery sectors. Standardized sensory analysis of the fragrance profiles was examined, extinguishing five out of the numerous genotypes evaluated (Figure 1) due to their notable fragrance and the presence of large, intensely colored flowers.

Figure 1. Investigated rose genotypes. PA—'Pure Aroma'; MIF—'Mina Frayla'; AA—'Adore Aroma'; AR—'Andre Rieu'; MIL—'Mileva Frayla'.

The objective of this research was to further enhance the value of these rose genotypes by analyzing the composition of biologically active polyphenols and exploring potential health benefits. Hence, the specific aims included the evaluation of the following: (a) morpholocagil attributes concerning the abundance of flowering shoots and flowers, as well as the size and quantity of flower petals; (b) intensity of fragrance and the scoring using a sensory panel for major top, heart, and base fragrance notes, following the olfactory pyramid; (c) phytochemical composition related to polyphenols; (d) antioxidant capacity using The Ferric Reducing Antioxidant Potential (FRAP) and 2,2-diphenyl-1-picrylhydrazyl (DPPH)assays; (e) neuroprotective activity thought determination of the potential to inhibit acetylcholinesterase (AChE-IP); and (f) antidiabetic effect thought α-amylase and α-glucosidase inhibitory potential determination. Phytochemical profile determination included analysis of total phenolics (TPC), flavonoids (TFC) and monomeric anthocyanins (TAC) contents, and quantification of individual phenolic compounds.

Determining the chemical composition of new rose genotypes and assessing their biological activities will enable the identification of cultivars with the greatest potential for health benefits, alongside their valuable sensory properties. This will potentially enhance the market value of these genotypes, making them more attractive to potential customers from the cosmetic and pharmaceutical industries.

2. Results

2.1. Morphological Traits

Tetraploid (4n) garden rose cultivars included 'Pure Aroma' (PA), 'Adore Aroma' (AA), 'Andre Rieu' (AR), 'Mina Frayla' (MIF), and 'Mileva Frayla' (MIL).

The growth type labeled as shrub did not differ in the investigated cultivars, while the semi-upright growth habit was noted in 'PA' and 'MIL', and upright in the remaining rose cultivars (Table 1). Plant height reached from 69.6cm in 'AA' to 90.4 cm in 'MIL', showing a significant difference according to the statistical test. Leaf color varied from a medium in 'AR' and 'MIL' to dark intensity of green color in 'PA', 'AA', and 'MIF'. Leaf anthocyanin coloration was present in all investigated genotypes while glossiness of the upper side was weak or very weak in three out of five genotypes. Leaf length and width also showed significant variation according to Tukey's HSD test, taking values from 3.58 cm to 5.68 cm and 2.66 cm to 3.88 cm, respectively, with both properties being minimal in 'MIF' and maximal in 'MIL'.

Table 1. Investigated vegetative characteristics of five rose genotypes.

Cultivar/Trait	PA	AA	AR	MIF	MIL
			Plant		
GT	Shrub	Shrub	Shrub	Shrub	Shrub
GH	Semi-upright	Upright	Upright	Upright	Semi-upright
Height (cm)	80.8 [b]	69.6 [d]	75.0 [c]	79.8 [b]	90.4 [a]
			Leaf		
IGC	Dark	Dark	Medium	Dark	Medium
LAC	Present	Present	Present	Present	Present
GUS	Weak	Medium	Weak	Medium	Very weak
Length (cm)	4.76 [b]	4.24 [c]	4.56 [b]	3.58 [d]	5.68 [a]
Width (cm)	2.98 [b]	2.86 [b,c]	3.72 [a]	2.66 [c]	3.88 [a]

Glossiness of upper side—GUS; growth habit (excluding climbers)—GH; growth type—GT; intensity of green color (upper side)—IGC; leaf anthocyanin coloration—LAC; PA—'Pure Aroma'; MIF—'Mina Frayla'; AA—'Adore Aroma'; AR—'Andre Rieu'; MIL—'Mileva Frayla'. Mean values designated with the same letter were not significantly different according to Tukey's HSD test ($p \leq 0.05$).

As to the flowering shoot qualitative characteristics (Table 2), shoots were absent in all rose cultivars except in 'AA' with 1.82 shoots on average. Correspondingly, flowering laterals were present in all other cultivars, achieving values from 1.62 and 1.64 in 'MIF' and 'AR' to 3.44 in 'MIL', and the number of flowers per lateral achieving values from 2.94 in 'PA' to 8.72in 'MIL'. Flower type was uniform in all investigated cultivars, varying from medium purple and medium violet to medium purple red and dark purple. The color of the flower center was red–purple in all investigated cultivars except for 'MIL'. The flower shape was rounded in 'AA', 'MIL', and 'AR', while irregularly rounded in others. Upper part profiles were described as flat in all five cultivars, while the lower part profile differed only for 'MIL' (flat), while concave in the rest of the cultivars. Fragrance was marked as strong (label 5) for all cultivars. Regarding the quantitative characteristics, the number of petals per flower significantly differed, taking values from 65.8 in 'MIF' to almost double in 'AA' (114.4 petals), with the petals achieving the highest length in 'AR' and the highest value for width in 'PA'.

Table 2. Flowering characteristics of the analyzed rose genotypes.

Cultivar/Trait	PA	AA	AR	MIF	MIL
		Flowering Shoot			
NFS	0 [b]	1.82 [a]	0 [b]	0 [b]	0 [b]
FL	Present	Absent	Present	Present	Present
NFL	1.83 [b]	0 [d]	1.64 [c]	1.62 [c]	3.44 [a]
NF/L	2.94 [c]	0 [d]	3.86 [b]	3.68 [b]	8.72 [a]
		Flower			
TP	Double	Double	Double	Double	Double
CG	Medium purple	Medium violet	Dark purple	Medium purple red	Medium purple red
CC	Red purple	Red purple	Red purple	Red purple	Pink
SH	Irregularly rounded	Rounded	Rounded	Irregularly rounded	Rounded
PUP	Flat	Flat	Flat	Flat	Flat
PLP	Concave	Concave	Concave	Concave	Flat
FG	5	5	5	5	5
NOP	96.2 [c]	114.4 [a]	107.4 [b]	65.8 [d]	94.4 [c]
DM (cm)	5.86 [b]	6.54 [a]	6.66 [a]	6.50 [a]	6.56 [a]
PL (cm)	4.40 [b]	4.48 [b]	4.84 [a]	3.72 [c]	3.32 [d]
PW (cm)	4.24 [a]	3.58 [b]	3.36 [b]	3.39 [b]	3.24 [b]

Color group—CG; Color of center (only varieties with flower type double)—CC; Flower diameter—DM; Flowering laterals—FL; Fragrance—FG on the 1–5 scale; Number of flowering laterals—NFL; Number of flowers per lateral (Only varieties with flowering laterals)—NF/L; Number of flowering shoots (Only varieties with no flowering laterals)—NFS; Number of petals—NOP; Profile of lower part—PLP; Shape—SH; Petals length—PL; Petals width—PW; Profile of upper part—PUP; type—TP; Mean values designated with the same letter were not significantly different according to Tukey's Honest Significant Difference test ($p \leq 0.05$).

2.2. Sensory Evaluation

The fragrance was described as 5—strong in all five cultivars and further scored by the panelists for the presence of fragrance components (Figures 2–5). Apart from receiving high scores for their overall fragrances, cultivars were characterized as a blend of various top, heart, and base notes. Specifically, the top notes showcased a prevalence of fruity aromas, with panelists linking them to various common fruit species such as orange, lemon, apple, mandarin, and grapefruit, supplemented with mint fragrance in 'AA' and anise and sweet fennel in 'MIF'. Bergamot and eucalyptus scored lowest (mainly around 5%) while scores for sweet fennel varied from 0.9 to 10.7%. All cultivars seem to be very complex since panelists recorded notes from all 11 groups (Figure 2).

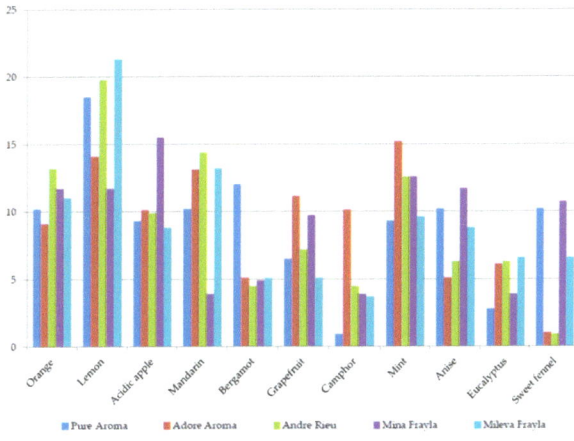

Figure 2. Panel scoring for top notes belonging to citrus or minty (aromatic) fragrance components.

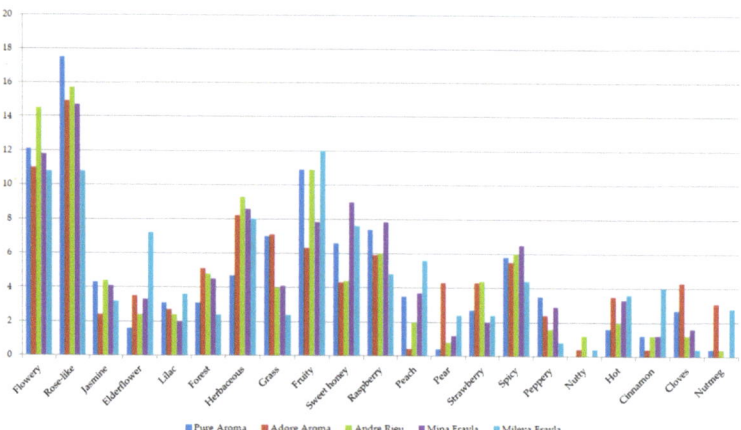

Figure 3. Panel scoring for middle notes belonging to floral, green, fruity, or spicy fragrance components.

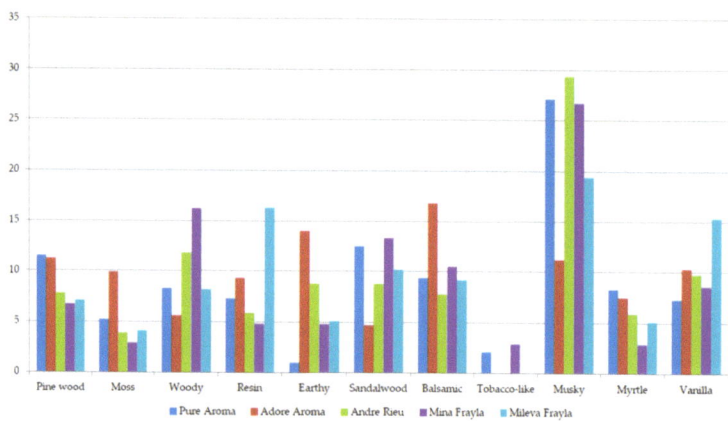

Figure 4. Panel scoring for base notes belonging to woody, earthy, or balsamic fragrance components.

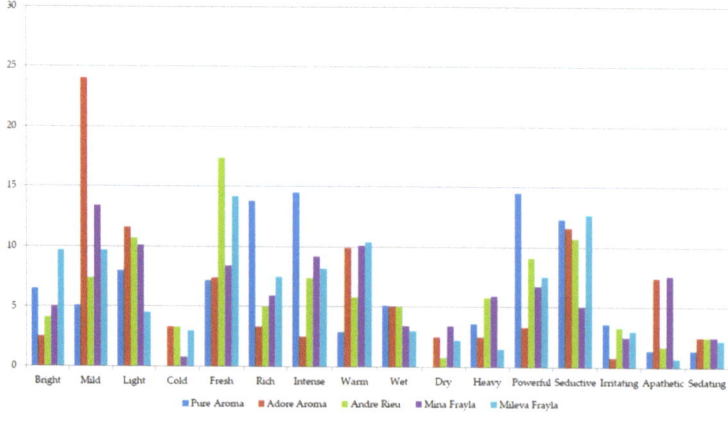

Figure 5. Panel scoring for general impression by the present fragrance components.

Middle notes belonging to 21 different categories showed a higher variation with only three groups reaching over the threshold of 10%—flowery(with elderflower dominant in 'MIL'), fruity (among which raspberry and strawberry dominate) and typical rose-like notes. Spicy fragrances scored lowest, with the majority of components reaching less than 5%, while nutty and nutmeg (in a range 0–3%) can be considered as a mixed fragrance and not as a certain association with any of the investigated cultivars (Figure 3).

Regarding the base notes belonging to 11 different categories, eight exceeded the 10% threshold, with musky fragrance notably scored in all five cultivars (from 11.2% in 'AA' to 29.4% in 'AR'). Interestingly, different peaks can be noted according to Figure 4—woody fragrance in 'MIF', resin and vanilla in 'MIL', balsamic in 'AA', and musky in 'PA', 'AR', and 'MIF'.

Panelists were asked to express the first emotional association upon the flower smelling, which yielded the results presented in the Figure 5. Most of the times panelists expressed association on mild in 'AA', fresh in 'AR', rich, intense, seductive, and powerful in 'PA' as well as fresh and seductive in 'MIL'. Regarding 'MIF', associations were divided among mild, light, fresh, intense, and warm, with around 10% of the scores. Associations on cold, dry, irritating, or sedating effects were the least noted (below 5%).

2.3. Chemical Profile of Methanol Extracts of Rose Petals

The chemical analysis of methanol extracts from rose petals involved the determination of the phenolic profile through the measurement of total phenolics content (TPC), total flavonoids content (TFC), and total monomeric anthocyanins content (TAC) using spectrophotometric techniques. Furthermore, the quantification of quinic acid and 44 selected phenolic compounds was conducted using LC-MS/MS.

The results obtained indicated that the studied rose genotypes are rich in phenolic compounds (as depicted in Table 3 and Figure 6). Nevertheless, there was significant variability observed in TPC, TFC, and TAC among different genotypes. The samples AR and PA were the richest in TPC (260 mg GAE/g de and 249 mg GAE/g de, respectively), while the sample AA has the lowest TPC (148 mg GAE/g de). When expressed in terms of 1 g of fresh petal weight, the TPC ranged from 14.0 mg GAE/g fw in the AA sample to 26.8 mg GAE/g fw in the PA. The TFC of examined extracts ranged from 23.7 mg QE/g de to 59.8 mg QE/g de and decreased in the following order: AA > MIF > MIL > PA > AR. If expressed per gram of fresh petals, the TFC was in the range from 2.36 mg QE/g fw to 5.70 mg QE/g fw. The results obtained for TAC indicated that all samples contain a significant number of anthocyanins, ranging from 2.21 mg CE/g de in the MIL sample to 13.1 mg CE/g de in the AR genotype. If expressed per gram of fresh petals, the TAC ranged from 0.188 mg CE/g fw to 1.31 mg CE/g fw.

Table 3. Contents of total phenolics, total flavonoids, and total anthocyanins in rose petal methanol extracts.

Genotype	TPC		TFC		TAC	
	mg GAE/g de	mg GAE/g fw	mg QE/g de	mg QE/g fw	mg CE/g de	µg CE/g fw
PA	249 ± 13.0 a [a]	26.8 ± 1.40 a	43.7 ± 2.52 b	4.70 ± 0.271 c	9.19 ± 0.193 b	0.990 ± 0.021 b
AA	148 ± 14.6 c	14.0 ± 1.31 c	59.8 ± 3.73 a	5.65 ± 0.335 b	7.55 ± 0.317 c	0.714 ± 0.028 c
AR	260 ± 5.01 a	25.9 ± 0.50 a	23.7 ± 0.215 c	2.36 ± 0.021 e	13.1 ± 0.615 a	1.31 ± 0.061 a
MIF	182 ± 15.8 bc	18.2 ± 1.58 b	56.8 ± 4.90 a	5.70 ± 0.492 a	2.24 ± 0.019 d	0.225 ± 0.002 d
MIL	193 ± 18.3 b	16.4 ± 1.56 bc	44.0 ± 2.81 b	3.75 ± 0.239 d	2.21 ± 0.008 d	0.188 ± 0.001 e

[a] values are means ± standard error. Means within each column with different letters (a–e) differ significantly according to Tukey's HSD test ($p \leq 0.05$); AA—'Adore Aroma'; AR—'Andre Rieu'; CE—cyanidin-3-O-glucoside equivalents; de—dry extract; fw—fresh weight; GAE—gallic acid equivalents; MIF—'Mina Frayla'; MIL—'Mileva Frayla'; PA—'Pure Aroma'; QE—quercetin equivalents; TAC—total anthocyanin content; TFC total flavonoid content; TPC total phenolic content.

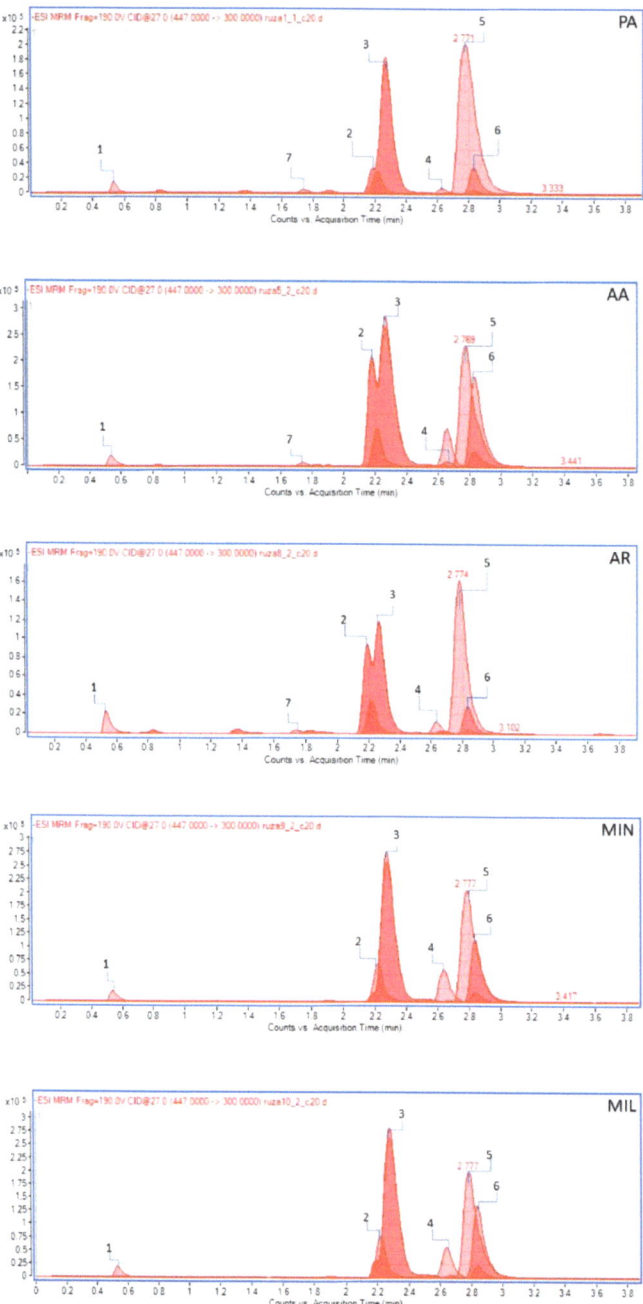

Figure 6. LC-MS/MS chromatograms of selected phenolic compounds in methanol extracts of rose petals. PA—Pure Aroma, AA—Adore Aroma, AR—Andre Rieu, MIN—Mina Frayla, MIL—Mileva Frayla. Only the most pronounces peaks are marked: 1—Quinic acid, 2—Quercetin-3-O-galactoside, 3—Quercetin-3-O-Glucoside, 4—Rutin, 5—Quercitrin, 6—Kaempferol-3-O-Glucoside, 7—p-Coumaric acid.

The results of LC-MS/MS quantitative analysis indicated that out of the 45 compounds targeted for quantification, only 20 were detected (Table 4). Corresponding chromatograms are shown on Figure 6. Quercetin 3-O-glycosides, specifically quercitrin (quercetin 3-O-rhamnoside), quercetin 3-O-glucoside, quercetin 3-O-galactoside and rutin (quercetin 3-O-rutinoside), as well as kaempferol-3-O-glucoside, were identified as the predominant flavonoids in all investigated samples. It was not possible to quantify individual amounts of quercetin 3-O-glucoside and quercetin 3-O-galactosid, but only the total amount (quercetin 3-O-glucoside + quercetin 3-O-galactosid), since their peaks in the chromatogram were overlapped. Flavonoid aglycones—catechin, quercetin and kaempferol—were also present in all samples, but in significantly lower amounts. The quinic acid content was found to be very high in all investigated samples, ranging from 13.4 mg/g de in AA genotype to 23.9 mg/g de in AR. Among benzoic acid derivatives, low amounts of protocatechuic (12.8–60.7 µg/g de) and gallic acids (21.7–36.4 µg/g de) were found. Chlorogenic and p-coumaric acids were the only hydroxycinnamic acids found in analyzed samples, but their quantities were very low (up to 5.98 µg/g de). The following compounds were analyzed but not detected in any of the examined extracts or were below the limit of quantitation in all samples: 2,5-dihydroxybenzoic acid, epigallocatechin gallate, aesculetin, caffeic acid, vanillic acid, syringic acid, umbelliferone, scopoletin, ferulic acid, sinapic acid, hyperoside, apiin, o-coumaric acid, myricetin, secoisolariciresinol, 3,4-dimethoxycinnamic acid, baicalin, daidzein, matairesinol, cinnamic acid, luteolin, genistein, apigenin, baicalein, and amentoflavone.

Table 4. Contents of quinic acid and selected phenolic compounds in methanol extracts of rose petals.

Genotype	Content [µg/g de] [a]				
	PA	AA	AR	MIF	MIL
Quinic acid	16,837 [b] ± 1684 bc [c]	13,380 ± 1338 c	23,929 ± 2393 a	14,960 ± 1496 c	21,541 ± 2154 ab
Protocatechuic acid	43.9 ± 3.51 b	60.7 ± 4.86 a	51.6 ± 4.13 b	15.3 ± 1.22 c	12.8 ± 1.02 c
p-Coumaric acid	0.209 ± 0.019 b	2.92 ± 0.262 a	0.111 ± 0.010 b	0.137 ± 0.012 b	0.173 ± 0.016 b
Gallic acid	22.6 ± 2.03 cd	29.3 ± 2.64 bc	21.7 ± 1.95 d	30.4 ± 2.74 ab	36.4 ± 3.27 a
Naringenin	0.335 ± 0.023 b	0.717 ± 0.050 a	0.651 ± 0.046 a	<0.3 [d]	0.341 ± 0.024 b
Luteolin	0.520 ± 0.026 a	0.523 ± 0.026 a	0.546 ± 0.027 a	0.520 ± 0.026 a	0.440 ± 0.022 b
Kaempferol	2.48 ± 0.174 c	9.52 ± 0.666 a	1.61 ± 0.113 c	5.79 ± 0.405 b	6.41 ± 0.448 b
Catechin	19.2 ± 1.92 b	144 ± 14.4 a	133 ± 13.3 a	30.9 ± 3.09 b	26.5 ± 2.65 b
Epicatechin	0.969 ± 0.097 c	1.92 ± 0.192 b	5.17 ± 0.517 a	/ [e]	/
Chrysoeriol	0.171 ± 0.005 c	<0.075	0.170 ± 0.005 c	0.960 ± 0.029 b	2.10 ± 0.063 a
Quercetin	31.4 ± 9.43 ab	59.2 ± 17.8 a	53.3 ± 16.0 a	14.9 ± 4.48 b	17.1 ± 5.14 b
Chlorogenic acid	0.603 ± 0.030 d	1.37 ± 0.069 c	0.502 ± 0.025 d	3.02 ± 0.151 b	5.98 ± 0.299 a
Apigenin-7-O-glucoside	<0.075	0.131 ± 0.007a	0.116 ± 0.006 b	<0.075	<0.075
Vitexin	<0.075	<0.075	<0.075	0.102 ± 0.005 b	0.132 ± 0.007 a
Kaempferol-3-O-glucoside	2348 ± 93.9 d	23,965 ± 959 a	1147 ± 45.9 d	11,454 ± 458 c	14,275 ± 571 b
Luteolin-7-O-glucoside	1.10 ± 0.033 a	/	0.824 ± 0.025 b	/	/
Quercitrin	13,945 ± 837 a	10,654 ± 639 b	2978 ± 179 d	9338 ± 560 bc	8564 ± 514 c
Quercetin-3-O-glucoside + Quercetin-3-O-galactoside	8267 ± 496 d	19,949 ± 1197 a	3981 ± 239 e	11,969 ± 718 c	14,709 ± 883 b
Rutin	2235 ± 67.0 b	2193 ± 65.8 b	703 ± 21.1 c	2188 ± 65.6 b	2663 ± 79.9 a
Total phenolics (mg/g de) [f]	43.76	70.45	33.01	50.01	61.86

[a] Results are given as content (µg/g of dry extract) ± standard error of repeatability (as determined via method validation); [b] The values higher than 10 are marked with bold letters; [c] means within each row with different letters (a–e) differ significantly according to Tukey's HSD test ($p \leq 0.05$); [d] below limit of quantitation (LoQ); [e] not detected; [f] Sum of the contents of all detected phenolic compounds using LC-MS/MS; AA—'Adore Aroma'; AR—'Andre Rieu'; de—dry extract; MIF—'Mina Frayla'; MIL—'Mileva Frayla'; PA—'Pure Aroma'.

Principal Component Analysis (PCA) was conducted on a dataset comprising 10 dominant compounds present in the extracts, each in amounts exceeding 10 µg/g (Figure 7). The first and second principal components (PC1 and PC2) accounted for 83.2%

and 13.9% of the total variance, respectively, which indicates significant metabolic differences among the investigated rose genotypes. The analysis revealed a certain level of grouping of the samples AA and MIL in the upper right quadrants of the biplot. This could mainly be attributed to the high loadings of kaempferol-3-O-glucoside and quercetin-3-O-glucoside + quercetin-3-O-galactoside. A common feature for samples AR and PA is significantly lower amounts of these glycosides compared to the other samples. AR is separated in the left upper quadrant of the biplot due to extremely high content of quinic acid compared to other samples. The sample PA stands out from the others, primarily due to the significantly higher amount of quercitrin present in this sample. The sample MIF is characterized by moderate levels of all analyzed compounds and this is the main reason for separation of this sample in the lower right quadrant of the biplot. The similar conclusions can be made from the the dendogram obtainedusing hierarchical clustering analysis of the same data (using the Ward's method, where closeness was measured by Euclidean distance, Figure 8).

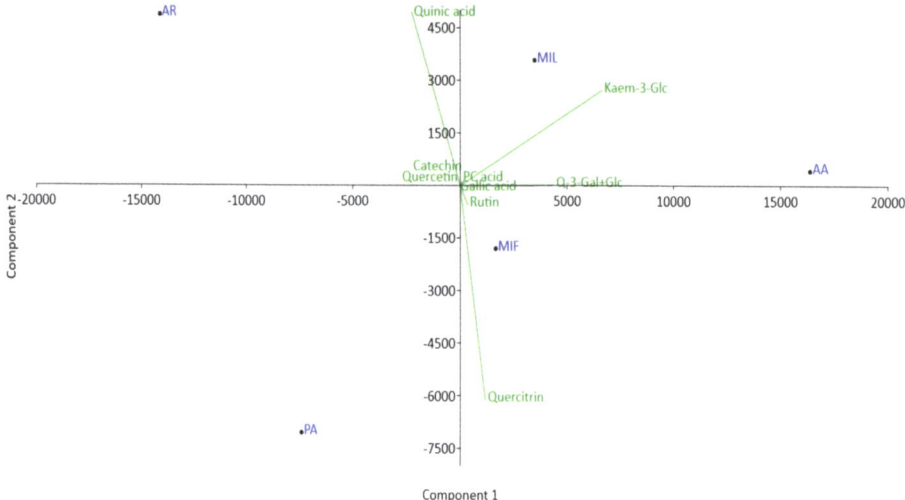

Figure 7. Biplot of a principal component analysis (PC1 vs PC2) analysis of 10 dominant compounds in rose petal methanol extracts. AA—'Adore Aroma'; AR—'Andre Rieu'; MIF—'Mina Frayla'; MIL—'Mileva Frayla'; PA—'Pure Aroma'.

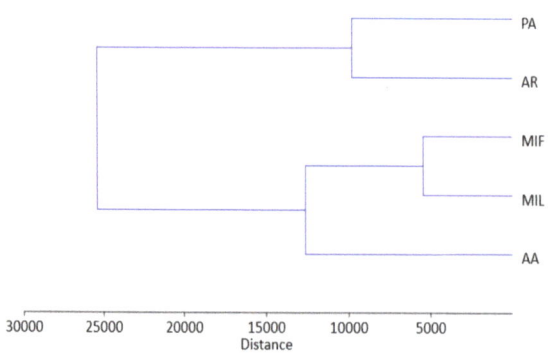

Figure 8. Dendrogram obtained using hierarchical clustering analysis of 10 dominant compounds in rose petal methanol extracts. AA—'Adore Aroma'; AR—'Andre Rieu'; MIF—'Mina Frayla'; MIL—'Mileva Frayla'; PA—'Pure Aroma'.

2.4. Methanol Extracts' Antioxidant Activity of Rose Petals

The antioxidant capacity of methanol extracts from rose petals was evaluated using two assays: FRAP and DPPH assay.

All examined rose petal extracts in the current investigation demonstrated noteworthy antioxidant activity (refer to Table 5). A robust correlation was observed between the results of the two antioxidant assays carried out ($R^2 = 0.983$). In the FRAP assay, the activity of the extracts was in the range from 132 mg AAE/g de to 209 mg AAE/g de. The strongest ferric ion reducing ability was expressed by PA and AR extracts. The IC_{50} values in the DPPH assay fell within a very narrow range (17.7–27.8 µg/mL) for all investigated genotypes, with the highest activity observed in the extracts of PA and MIL.

Table 5. Antioxidant, neuroprotective, and antidiabetic activity of rose petal methanol extracts.

Genotype	FRAP mg AAE /g de	DPPH IC_{50} µg/mL	AChE-IP ng EE/g de	α-Amylase-IP I (%)	α-Amylase-IP mg ACAE /g de	α-Glucosidase-IP I (%)	α-Glucosidase-IP g ACAE /g de
PA	209 ± 10.1 a [a]	18.6 ± 1.25 c	18.3 ± 0.561 a	11.5 ± 1.62 a	95.6 ± 13.6 a	87.1 ± 4.11 a	292 ± 7.20 b
AA	132 ± 7.40 d	27.8 ± 1.01 a	10.2 ± 0.117 e	2.40 ± 0.435 b	19.5 ± 3.66 b	47.2 ± 4.35 b	38.6 ± 6.98 d
AR	198 ± 12.8 ab	22.7 ± 0.16 b	16.4 ± 0.405 b	13.6 ± 1.99 a	113 ± 16.7 a	86.3 ± 1.19 a	339 ± 39.4 a
MIF	175 ± 6.24 c	24.0 ± 1.19 b	12.4 ± 0.132 d	0.94 ± 0.156 c	7.3 ± 1.32 c	57.1 ± 5.70 b	60.7 ± 16.4 c
MIL	177 ± 1.96 bc	17.7 ± 1.27 c	14.7 ± 1.16 c	0.00 ± 0.00 d	0.00 ± 0.00 d	54.5 ± 1.01 b	52.7 ± 2.39 c

[a] means within each row with different letters (a–e) differ significantly according to Tukey's HSD test ($p \leq 0.05$); AA—'Adore Aroma'; AAE—ascorbic acid equivalents; ACAE—acarbose equivalents; AChE-IP—the potential to inhibit acetylcholinesterase; AR—'Andre Rieu'; de—dry extract; EE—eserine equivalents; FRAP—Ferric reducing antioxidant potential; I—inhibition; IC_{50} (DPPH)—the concentration of the extract that neutralizes 50% of DPPH radicals; MIF—'Mina Frayla'; MIL—'Mileva Frayla'; PA—'Pure Aroma'; α-amylase-IP—the potential to inhibit α-amylase; α-glucosidase-IP—the potential to inhibit α-glucosidase.

2.5. Methanol Extracts' Neuroprotective Activity of Rose Petals

The neuroprotective activity of the rose petal extracts was estimated by measuring their ability to inhibit acetylcholinesterase (AChE). Since alkaloid physostigmine (eserine), a highly potent AChE inhibitor, was used as a positive control, the results were expressed as nanograms of eserine equivalents per gram of dry extract (EE/g de; Table 5). The AChE inhibitory activity of the extracts at a concentration of 50 µg/mL was considerable and ranged from 29% to 40%. When expressed in eserine equivalents, anti-AChE activity ranged from 10.2–18.3 ng EE/g de. The highest activity was exhibited by the extract of PA genotype (18.3 ng EE/g de).

2.6. Methanol Extracts' Antidiabetic Activity of Rose Petals

The methanol extracts of rose petals were assessed for their potential antidiabetic properties through the evaluation of their inhibitory effects on α-amylase and α-glucosidase enzymes. In the α-amylase assay, the results were expressed as the percentage of inhibition for the extracts at a concentration of 278 µg/mL and as mg of acarbose equivalents (ACAE)/g de (Table 5). The results in α-glucosidase assay were expressed as percentage of inhibition for the extracts at a concentration of 2.5 µg/mL and as g of ACAE/g de (Table 5). The extracts expressed much higher inhibitory activity towards α-glucosidase than against α-amylase. Even at a very low concentration (2.5 µg/mL), samples PA and AR achieved nearly 90% inhibition of α-glucosidase. Samples AA, MIL and MIF exhibited slightly lower inhibitory activity against α-glucosidase (47.2–57.1%). The extracts exhibited significantly higher activity than the well-known α-glucosidase inhibitor, acarbose, as evident from the results expressed as g ACAE/g de. Inhibitory potential of the extracts against α-amylase was negligible (0–13.6% of inhibition at a concentration of 167 µg/mL).

3. Discussion

The creation of new rose genotypes can be driven by various factors. One of the most important drivers is to fulfill market demands by developing roses with enhanced characteristics, such as color variations, fragrance, size, and petal structure, to meet specific aesthetic preferences [5]. Other driving forces include the creation of new genotypes with extended blooming periods and resistance to common diseases, and which are well-adapted to specific climatic conditions or regions [5]. Beyond ornamental purposes, roses are essential raw materials in the cosmetic and pharmaceutical sectors, which underscore the economic and cultural significance of these plants [6]. Among all plant parts, rose flowers and fruits (rose hips) are the most utilized in the production of a wide range of products in these industries.

Rose petals are rich in essential oil which contributes to the pleasant fragrance associated with roses [7]. Rose petal extracts or distillates (rose oils) are often incorporated into various cosmetic products, including perfumes, lotions, creams, and facial cleansers for their rich floral and appealing scent and potential skincare benefits [5,8]. Rose water, a by-product obtained during the steam distillation of rose petals, is commonly used in the cosmetic industry as a toner, facial mist, or ingredient in skincare products due to its hydrating and anti-inflammatory properties [9,10]. Rosehip seed oil, derived from rose hips is rich in sterols, tocopherols, and essential fatty acids and it is commonly used in skincare products for its nourishing and regenerative properties [11].

Rose petal extracts are also known for their potential health benefits, including antioxidant, anti-inflammatory, anti-elastase, analgesic, anticonvulsant, antimicrobial, and antidiabetic properties [12–15]. Thus, rose-derived compounds are used in the pharmaceutical industry in the formulation of diverse medicinal products and dietary supplements. Additionally, the fragrance of roses is often used in aromatherapy, contributing to stress relief and relaxation [16]. Roses are also edible plants that have been used in culinary applications for centuries. Recently, the concept of edible flowers has gained increased recognition, and roses are now utilized in the preparation of a variety of food products including jams, salads, ice creams, juices, and wines [17,18].

Considering all the information mentioned, the breeding and cultivation of roses with targeted chemical profiles, aiming to enhance their fragrance or provide significant amounts of health-beneficial compounds, is of utmost interest to the cosmetic and pharmaceutical industries. This is especially true for those sectors that prioritize the use of natural ingredients.

In this study, petals of five novel garden rose genotypes, obtained through planned hybridization and distinguishing themselves by exhibiting exceptionally pleasant fragrances and large, intensely pink, violet, and red flowers, have been chosen for detailed examination of their chemical composition and biological activities to recommend them as potentially valuable candidates for use in cosmetic and pharmaceutical industries. Although the cultivar 'Mileva Frayla' possesses superior morphological characteristics, including the largest number of flowering laterals, the highest number of flowers per lateral, and relatively large petals, the selection must also take into account the biochemical profile, since it is responsible for the significant biological activities expressed by all cultivars explored in this study. Furthermore, the sensory analysis revealed different panelists' and future consumers' attitudes towards the fragrance components and their pleasantness. Fruity top notes, followed by flowery, fruity and rose-like middle components, supplemented with heavier base notes belonging to woody, musky, and balsamic, comprised unique petals' aromas in all five investigated roses. Woody notes previously detected in peaches [19] tend to be positive attributes in consumers' preferences, which was confirmed in our study with 'Mina Frayla'. Complex compositions yielded interesting emotional associations with 'Mina Frayla' being scored as mild, light, fresh, intense, and warm. According to Wendin et al. [20] sensory analysis might be incorporated in consumer-oriented breeding and marketing of horticultural plants possessing unique fragrance compositions that might improve consumers' health and well-being. Sensory analysis of edible *Begonia* × *tuberhybrida*, *Tropaeolum majus*, *Calendula officinalis*, *Rosa*, *Hemerocallis*, and *Tagetes patula* species, designed as ones with appropriately large flowers showed significant differences in appearance, fragrance, consistency, overall taste, and similar [21], marking the importance of all listed properties toward acceptance by consumers.

The petals of all investigated rose genotypes were very rich in phenolic compounds (TPC fell within the range from 148 mg GAE/g de to 260 mg GAE/g de). Plant phenolic compounds, also referred to as polyphenols, constitute a diverse group of naturally occurring chemicals characterized by their phenolic structure, featuring one or more phenol rings with hydroxyl groups. These compounds are extensively present throughout the plant kingdom and can be broadly categorized into four classes: phenolic acids, flavonoids, stilbenes, and lignans. Phenolic acids are further classified into hydroxybenzoic and hydroxycinnamic acids [22]. Polyphenols display a range of biological activities, encompassing antioxidant, anti-inflammatory, anti-cancer, and cardioprotective effects, thereby exhibiting potential health benefits [22]. The composition of polyphenols in plants can vary depending on factors such as plant type, variety, growing conditions, and maturity state [23]. The findings from previous examinations of methanol extracts from *R. brunonii*, *R. baurboniana*, and *R. damascena* (254 mg GAE/g de, 178 mg GAE/g de, and 145 mg GAE/g de, respectively) [24], methanol extracts from six edible cultivars of *R.* × *hybrida* (91.4–217 mg GAE/g de) [3], as well as ethanol extracts from nine cultivars of *R.* × *hybrida* (7.99–29.79 mg/g FW) [25], closely resembled the results obtained in this study. Notably, the 'Andre Rieu' cultivar exhibited even higher TPC than all other listed samples. It's worth mentioning that the TPC in rose petals, particularly in the AR and PA samples, is comparable to that found in phenolic-rich fruits like blackberry and blueberry, while being much higher than levels observed in vegetables such as carrot and tomato [26].

Flavonoids, a diverse subgroup within the plant polyphenols, are characterized by the presence of two aromatic rings (A and B) connected by a chain of three carbon atoms. These compounds play a crucial role as pigments in flowers and fruits. Among them, anthocyanins stand out as a subgroup responsible for red, purple, blue, and violet tones. Beyond their function in providing vibrant colors, flavonoids, including anthocyanins, serve as effective absorbers of UV radiation and potent antioxidants, thereby shielding plants from the adverse effects of environmental stressors [27]. Numerous studies have underscored potent anti-inflammatory, anticancer, antimicrobial, and antiallergic properties of anthocyanins extracted from rose petals, making them valuable for applications in pharmaceuticals, functional foods, and cosmetics [1]. The TFC in the examined rose cultivars in this study was notably high, ranging from 23.7 mg QE/g de to 59.8 mg QE/g de. Comparable levels of TFC were previously identified in the ethanol extract of white rose (23.7 mg catechin equivalents per gram of dry extract) [14], methanol extracts of six edible cultivars of *R.* × *hybrida* (17.3–56.3 mg QE/g de) [3], and ethanol extracts of nine cultivars of *R.* × *hybrida* (0.786–5.31 mg catechin equivalents per gram of fresh weight) [25]. When examining the results of TFC and TPC, it is evident that the TFC for each sample is significantly less than its TPC. This indicates that only a minor fraction of the TPC is identified as flavonoids, while the extracts also contain substantial quantities of phenolic compounds from other categories.

Previous studies in the literature suggest that varieties with red and pink colors primarily consist of cyanidin glycosides, which are responsible for these specific hues [1]. In our study, where all investigated cultivars had purple, red, and violet-colored flowers, the TAC was high as expected, ranging from 2.21 to 13.1 mg CE/g. Samples AR and PA, exhibiting the most intense colors, were found to be the most abundant in anthocyanins. The TAC values of these samples significantly surpassed those values determined for pink/red/purple flowers of edible rose cultivars of *R.* × *hybrida* ('Lavander Vaza', 'Marija Frayla' and 'Theo Clevers') from our previous research (3.23–6.66 mg CE/g de) [3]. However, some rose varieties belonging to *R.* × *hybrida* exhibited anthocyanin levels as much as five times higher (5.03 mg CE/g fw), as demonstrated in a study by Yang et al. [25], compared to the richest cultivar in our study (AR), which recorded 1.31 mg CE/g fw.

LC-MS/MS quantitative analysis revealed that, among 45 investigated compounds, glycosides of quercetin (quercetin 3-*O*-glucoside, quercetin 3-*O*-galactoside, quercitrin and rutin), along with kaempferol-3-O-glucoside and quinic acid, stand out as dominant compounds in all examined rose cultivars. However, significant variations in the quantitative composition of these compounds were found among the samples, rather than differences in their qualitative aspects. The 'Adore Aroma' genotype exhibited the highest contents of the investigated flavonoid glycosides, while 'Andre Rieu' is characterized by the lowest content of these compounds along with the highest level of quinic acid.

In a study by Mikanagi et al. [28], that investigated 120 taxa within subgenus *Rosa*, the prevalent flavonoids were recognized as kaempferol 3-O-glycosides and quercetin 3-O-glycosides. Listed findings align with the results of our study. However, the content of kaempferol 3-O-glucoside in six edible cultivars of *R.* × *hybrida* (60.9–193 µg/g de) in our previous research [3] was notably lower than in the cultivars investigated in this study (1147–23965 µg/g de).

Oxidative stress, triggered by an excessive generation of free radicals within the body, can exert harmful impacts on cells. This is due to the fact that these reactive free radicals have the potential to harm primary biomolecules such as lipids, proteins, carbohydrates, and nucleic acids, resulting in cellular dysfunction and the possibility of tissue and organ damage. Oxidative stress is implicated in a diverse array of health challenges, encompassing aging, as well as the initiation and progression of inflammatory conditions, atherosclerosis, and diverse chronic diseases. These conditions include cardiovascular diseases, neurodegenerative disorders such as Alzheimer's and Parkinson's, and cancer [29]. The consumption of antioxidants or their topical application stands as a preventive measure, capable of preventing or reducing oxidative stress and its associated adverse effects [30]. Antioxidants are commonly added to cosmetic formulations to help prevent unwanted chemical changes to the product that may occur due to oxidation. This protection against oxidation is crucial in maintaining the stability and quality of cosmetic formulations [31]. On the other hand, antioxidants in topical cosmetics help protect the skin against oxidative damage caused by free radicals and UV radiation [31]. This protective function is associated with slowing down the aging process. Nowadays, the use of plant-derived natural antioxidants in cosmetics is preferred over synthetic antioxidants [32]. Plant extracts generally contain a mixture of natural compounds, which could have synergetic effects; therefore, they can have better effects and less toxicity than synthetic antioxidants [32].

Both assays (DPPH and FRAP) employed in the present study to assess the antioxidant potential of rose petal extracts are based on a single electron transfer mechanism. In the DPPH assay, antioxidants convert the DPPH radical into a stable molecule, while in the FRAP assay, they reduce ferric ion (Fe^{3+}) to ferrous ion (Fe^{2+}). Simultaneously, the antioxidants are transformed into a relatively stable and unreactive radical form. The rose petal extracts from all examined genotypes exhibited significant antioxidant activity, surpassing the efficacy of vitamin C, a recognized potent antioxidant with an IC_{50} of 30.0 µg/mL in the DPPH assay [24], and slightly lower than the synthetic antioxidant BHA (IC_{50} of 11.08 µg/mL), extensively used in the food and cosmetic industries [33]. The extracts' activity was comparable to that of previously tested methanol extracts from *R. brunonii*, *R. baurboniana*, and *R. damascene* (with IC_{50} values in the DPPH assay being 35.2 µg/mL, 25.0 µg/mL, and 21.4 µg/mL, respectively) [24]. Since new rose genotypes investigated in this study exhibited high antioxidant activity, they are promising candidates for inclusion in cosmetic products and pharmaceutical formulations. AChE catalyzes the hydrolysis of acetylcholine (ACh), an endogenous neurotransmitter, converting it into acetic acid and choline. Thus, AChE serves to terminate synaptic transmission in chemical synapses of the cholinergic type within the central nervous system, autonomic ganglia, and neuromuscular junctions. Reduced concentrations of AChE are frequently noted in neurodegenerative disorders like Alzheimer's disease and various forms of dementia. Inhibition of AChE leads to enhancement of Ach level within synapse, consequently increasing cholinergic signaling and enhancing cognitive functions, encompassing learning, memory, behavior, and emotional responses. Therefore, reversible inhibitors of AChE, such as galantamine, rivastigmine, and donepezil are often used in treatments of dementia. However, these drugs could have detrimental adverse effects including gastrointestinal issues, fatigue, cramps, and sinus node dysfunction [34]. Hence, scientists are persistently searching for novel AChE inhibitors that not only effectively inhibit the enzyme, but also come with fewer unwanted side effects. Various plant phenolics were proven to successfully inhibit AChE, thus having neuroprotective activity [35]. The findings of this investigation validate these results, as the substantial correlation coefficient ($R^2 = 0.9948$) observed between TPC and anti-AChE activity in rose extracts suggested that a higher polyphenolic content significantly contributes to enhanced anti-AChE activity. Since these compounds, beside AChE inhibitory activity, often exhibit additional pharmacological properties, especially antioxidants, they could be applied in multi-target strategies for combating the onset and progression of Alzheimer's disease and other neurological conditions. The AChE inhibitory activity of new rose genotypes investigated in this study was moderate (29–40%) at a concentration of 50 µg/mL, with the 'Pure Aroma' genotype exhibiting the highest activity. The PA extract stands out from others due to significantly higher content of quercitrin, indicating that this compound could contribute to the activity. However, the activity of PA is still lower than the activity of 'Eveline Wide' genotype of *R* × *hybrida* (69.4%) investigated in our previous study [3]. In comparison to the activity of different varieties of *R. damascena* from Turkey (IC_{50} values in the range from 3.9 µg/mL to 32.0 µg/mL), new cultivars investigated in the present study showed comparable activity.

Diabetes mellitus (DM) is a chronic metabolic disease characterized by elevated blood glucose levels, which has become a widespread health concern worldwide. It is considered as a top 10 cause of death globally, contributing to approximately 1.6 million deaths [36]. Prolonged hyperglycemia results in increased generation of reactive oxygen species (ROS) and consequently, the onset of oxidative stress and inflammation. These conditions are implicated in the impairment of the pancreatic β-cells and DM complications [37]. One of the strategies for lowering blood glucose levels, which is used in the treatment of DM, is slowing down the absorption of carbohydrates following food intake. This can be achieved by inhibiting the activity of α-amylase and α-glucosidase, the enzymes in the gastrointestinal tract involved in the breakdown of complex carbohydrates into simpler sugars (such as glucose) during the digestive process [37]. Inhibitors of α-glucosidase and α-amylase, such as acarbose, miglitol, and voglibose, are often used in the management of DM. However, these drugs have unwanted side effects, such as flatulence, diarrhea, and abdominal distention and pain, due to bacterial action on undigested carbohydrates [38]. Extracts of new rose genotypes investigated in this study expressed extremely high inhibitory potential to α-glucosidase, significantly higher than that of acarbose, and negligible activity against α-amylase. Even at a very low concentration (2.5 µg/mL), samples PA and AR achieved nearly 90% inhibition of α-glucosidase, while samples AA, MIL, and MIF exhibited slightly lower inhibitory activity (47.2–57.1%). Moreover, the results of this study suggest that among the investigated polyphenol types, anthocyanins exhibit the highest potential for inhibiting α-glucosidase, as opposed to phenols and flavonoids. Specifically, the correlation coefficient observed between TAC and anti-α-glucosidase activity ($R^2 = 0.8748$) was considerably higher than those observed for TPC and TFC and anti-α-glucosidase activity ($R^2 = 0.7634$ and $R^2 = 0.3568$, respectively). In the study of Gholamhoseinian et al. [39], methanol extract of the flowers of *R. damascene* expressed anti-α-glucosidase activity in vitro (at a concentration of 2 µg/mL inhibited AchE for 98%), comparable to the activity of AR and PA genotypes in the present study. In the same study [39], the activity of *R. damascene* extract was confirmed in vivo (in normal and diabetic rats), where the extract significantly suppressed the elevation of blood glucose after the administration of the high maltose diet. This indicates the great potential of *R. damascene*, as well as the extracts of new genotypes investigated in the present study, to be used to suppress postprandial hyperglycemia in diabetic patients, making them good candidates for the development of new formulations for alternative and/or complementary management of Diabetes mellitus type 2. They could also be applied to control obesity by reducing the food efficiency ratio, particularly the intake of carbohydrates.

4. Materials and Methods

4.1. Plant Material

The plant material utilized in the experiments (depicted in Figure 1) comprised five tetraploid (4n) garden rose cultivars named 'Pure Aroma' (PA), 'Adore Aroma' (AA), 'Andre Rieu' (AR), 'Mina Frayla' (MIF), and 'Mileva Frayla' (MIL), predominantly available on market in Serbia, Hungary, Poland, France, Italy, Germany, Netherlands, and the UK. All these genotypes belong to the $R \times hybrida$ species and were cultivated using conventional breeding practices, devoid of any chemical protection.

The garden rose specimens were of a biennial age and thrived in the outdoor environmental settings at the 'Pheno Geno Roses' private company in Temerin, Northern Serbia (coordinates: 45°24′19″ N 19°53′13″ E/45.105166° N 19.886833° E). The area is distinguished by a conventional continental climate, featuring notably warm summers and frigid winters.

The research field, measuring 30 m in length and 20 m in width, was created during the autumn of 2017 through on-site bud grafting. There were 150 grafted plants for each cultivar, with a 10 cm spacing between plants and a 1 m gap between rows.

In late August 2021, forty flowers from each cultivar were gathered. The freshly picked blooms were preserved by freezing at −80 °C until subjected to analysis.

4.2. Morphological Traits

The morphological characterization during the full blossom included both descriptive and metrical properties following the UPOV protocol [40] for roses (Rosa L.). The methodology was previously described in detail by Simin et al. [3].

4.3. Sensory Analysis

Fragrance evaluation as a qualitative trait was performed by 45 panel specialists (of different genders, seniority, specialties, and interests in roses). Detailed methodology regarding fragrance components divided into top, heart, and base notes, smelling time, and replications, as well as data approximation was published by Simin et al. [3].

4.4. Preparation of Methanol Extracts

Methanol extracts of rose petals were prepared via maceration of frozen plant material with 80% MeOH (in a 1:10 ratio) over 48 h at room temperature. The resulting macerate underwent filtration, and the maceration process was repeated once more. Subsequently, the macerates were evaporated to dryness under vacuum at 35 °C. The resulting dry extracts were dissolved in DMSO to reach a final concentration of 200 mg/mL and were stored at −20 °C until the time of analysis. The obtained extracts were used for chemical composition analysis and assessing biological activities, including antioxidant, neuroprotective, and antidiabetic.

4.5. Chemical Characterization of Rose Petal Methanol Extracts

4.5.1. Determination of Total Phenolic Content

TPC was determined by Folin–Ciocalteu (FC) assay, described previously by Lesjak et al. [41].

4.5.2. Determination of Total Flavonoid Content

Determination of TFC was conducted using the aluminum chloride assay method as outlined by Lesjak et al. [41].

4.5.3. Determination of Total Monomeric Anthocyanin Content

The determination of TAC was carried out utilizing the pH differential method, following the procedure previously published [42], with adaptations suitable for 96-well microplates.

4.5.4. Quantitative Analysis of Selected Compounds

The content of quinic acid and 44 selected phenolic compounds (14 phenolic acids, 25 flavonoids, three coumarins, and two lignans) was investigated using LC-MS/MS according to the previously described method [33].

4.6. Antioxidant Potential

DPPH and FRAP Assays

The capability of the extracts to counteract the DPPH radical was assessed following the method outlined by Lesjak et al. [41]. Furthermore, FRAP assay was conducted following the procedure also described by Lesjak et al. [41].

4.7. Neuroprotective Activity

The neuroprotective potential of the extracts was evaluated by determining AChE-IP through Ellman's method, with specific modifications detailed in a previous study by Pintać et al. [43].

4.8. Antidiabetic Activity

The antidiabetic activity was examined by measuring the potential to inhibit α-amylase and α-glucosidase.

Inhibition of α-amylase was examined with a method used by Yang et al. [44], adapted for a 96-well plate. Briefly, 90μL of alpha-amylase (0.1 μg/mL in 20 mM phosphate buffer pH 6.9) were added to 80μL 0.05% starch dissolved in 20 mM phosphate buffer pH 6.9 along with 10 μL of extract (5.0 mg/mL) or standard (acarbose, 0.125–3.0 mg/mL). In the blank probe, α-amylase was replaced with 20 mM phosphate buffer pH 6.9, and in the control 20 mM phosphate buffer pH 6.9 was added instead of the sample. The 96-well plate was then incubated for 10min at 37 °C with constant shaking. After incubation, the reaction was stopped by the addition of 100 μL of cold 1 M HCl, after which, 20 μL of Lugol solution was added and absorbance was measured at 620 nm. All tests were done in triplicate and results were expressed as % of inhibition and mg of ACAE/g de.

Inhibition of α-glucosidase was examined using a modified procedure of Palanisamy et al. [45]. Briefly, 100 μL of 0.1 M phosphate buffer pH 6.8 was mixed with 10 μL of α-glucosidase (0.1 U/mL in 0.1 M phosphate buffer pH 6.8), 20 μL of extract (20 μg/mL) or standard (acarbose, 0.310–10.0 mg/mL), and 20 μL of p-nitrophenyl α-D-glucoside. In the blank probe, α-glucosidase was substituted with 0.1 M phosphate buffer pH 6.8, and in the control, instead of the sample, 20 μL of 0.1 M phosphate buffer was added. The 96-well plate was then incubated for 15min at 37 °C with constant shaking. After incubation for 15 min, 80 μL of 0.2 M Na_2CO_3 and absorbance was measured at 400 nm. All tests were done in triplicate and the results were expressed as % of inhibition and g ACAE/g de.

4.9. Statistical Analysis

The data of all spectrophotometric measurements were analyzed using one-way ANOVA followed by the post hoc Tukey's Honest Significant Difference (HSD) test for multiple comparisons of means in order to determine whether the data obtained for different rose genotypes differed significantly between each other (Real Statistics Resource Pack add in for Excel 2013). Statistical significance was set at $p < 0.05$. Principal component analysis (PCA) and hierarchical clustering analysis of results of LC-MS/MS analysis was done using Past version 4.11.

5. Conclusions

This intersection of horticulture and industry showcases the potential for innovation in creating roses that not only serve ornamental purposes but also contribute to the formulation of diverse and beneficial products in the cosmetic and pharmaceutical fields. The objective of this research was to further enhance the value of these rose genotypes by analyzing the composition of biologically active polyphenols and exploring potential health benefits. Morphologically, the cultivar MIL is distinguished by having the largest number of flowering laterals, the highest number of flowers per lateral, and relatively large petals, but the significant biological activities were expressed by all cultivars explored in this study.

The petals of all investigated rose genotypes were very rich in phenolic compounds (TPC ranging between 148 mg GAE/g de and 260 mg GAE/g de). Furthermore, LC-MS/MS quantitative analysis revealed that, among 45 investigated compounds, glycosides of quercetin (quercetin 3-O-glucoside, quercetin 3-O-galactoside, quercitrin and rutin), along with kaempferol-3-O-glucoside and quinic acid, were dominant compounds in all examined rose cultivars. Additionally, the extracts of all investigated genotypes expressed very high antioxidant activity, much higher than the activity of vitamin C, which is known as a potent antioxidant (IC_{50} in DPPH assay is 30.0 μg/mL). Due to the very high antioxidant activity observed for methanol extracts of petals from the new rose genotypes investigated in this study, they could be recommended as promising candidates for application as an ingredient of cosmetic products, as well as of pharmaceutical formulations.

One particularly important result was the AChE inhibitory activity (29–40%) at a concentration of 50 μg/mL, with the 'Pure Aroma' genotype exhibiting the highest activity. The PA extract stands out from others due to the significantly higher content of quercitrin, indicating that this compound could contribute to neuroprotective activity. Extracts of new rose genotypes investigated in this study expressed extremely high inhibitory potential to α-glucosidase, significantly higher than that of acarbose, and negligible activity against α-amylase. Even at a very low concentration (2.5 μg/mL), samples PA and AR achieved nearly 90% inhibition of α-glucosidase, while other samples exhibited slightly lower inhibitory activity (47.2–57.1%). This indicates great potential of the investigated genotypes to be used as anti-diabetic agents.

Due to their valuable health-promoting properties, the studied new rose genotypes could potentially act as drugs in the prophylaxis and treatment of many diseases. A great challenge for future research in pharmacology and medical fields could be the study of these roses in preclinical and clinical trials.

Author Contributions: Conceptualization, N.S. and M.L. (Marija Lesjak); methodology, M.L. (Marija Lesjak), M.L. (Mirjana Ljubojević) and N.Ž.; software, N.S. and N.Ž; validation, N.Ž.; investigation, B.B.T., T.N. and N.Ž.; resources, B.B.T.; data curation, N.S. and N.Ž.; writing—Original draft preparation, N.S.; writing—Review and editing, M.L. (Marija Lesjak), B.B.T. and M.L. (Mirjana Ljubojević); project administration, M.L. (Mirjana Ljubojević); funding acquisition, M.L. (Mirjana Ljubojević) and B.B.T. All authors have read and agreed to the published version of the manuscript.

Funding: This research was funded and conducted within the frame of a four-year project entitled 'Biochemically assisted garden roses' selection aiming towards the increased quality and marketability of producers in Vojvodina', grant number 142-451-3481/2023-01/01, financed by the Provincial Secretariat for Higher Education and Scientific Research, Autonomous Province of Vojvodina, Republic of Serbia.

Data Availability Statement: The datasets used and/or analyzed during the current study are available from the corresponding author on a reasonable request.

Acknowledgments: Authors express sincere gratitude to the 'Pheno Geno Roses' company for the generous provision of rose samples of good quality, which significantly contributed to the success of this study.

Conflicts of Interest: Because of the perception of a conflict of interest and in the interest of full transparency, author Biljana Božanić Tanjga is disclosing the relationship with Breeding Company 'Pheno Geno Roses'. Funder Provincial Secretariat for Higher Education and Scientific Research, Autonomous Province of Vojvodina, Republic of Serbia, and employees of the Breeding company 'Pheno Geno Roses' had no influence on the trial design, results acquisition, data processing, or interpretation and delivery of the conclusions.

References

1. Kumari, P.; Raju, D.V.S.; Prasad, K.V.; Saha, S.; Panwar, S.; Paul, S.; Banyal, N.; Bains, A.; Chawla, P.; Fogarasi, M.; et al. Characterization of anthocyanins and their antioxidant activities in Indian rose varieties (*Rosa* × *hybrida*) using HPLC. *Antioxidants* **2022**, *11*, 2032. [CrossRef]
2. Patel, S. Rose hip as an underutilized functional food: Evidence-based review. *Trends Food Sci. Technol.* **2017**, *63*, 29–38. [CrossRef]
3. Simin, N.; Lesjak, M.; Živanović, N.; Božanić Tanjga, B.; Orčić, D.; Ljubojević, M. Morphological Characters, Phytochemical Profile and Biological Activities of Novel Garden Roses Edible Cultivars. *Horticulturae* **2023**, *9*, 1082. [CrossRef]
4. Vukosavljev, M.; Stranjanac, I.; van Dongen, B.W.P.; Voorrips, R.E.; Miric, M.; Bozanic Tanjga, B.; Arens, P.; Smulders, M.J.M. A novel source of food—Garden rose petals. *Acta Hortic.* **2023**, *1362*, 165–171. [CrossRef]
5. Datta, S.K. Breeding of new ornamental varieties: Rose. *Curr. Sci.* **2018**, *114*, 1194–1206. [CrossRef]
6. Mileva, M.; Ilieva, Y.; Jovtchev, G.; Gateva, S.; Zaharieva, M.M.; Georgieva, A.; Dimitrova, L.; Dobreva, A.; Angelova, T.; Vilhelmova-Ilieva, N.; et al. Rose flowers—A delicate perfume or a natural healer? *Biomolecules* **2021**, *11*, 127. [CrossRef] [PubMed]
7. Yousefi, B.; Jaimand, K. Chemical variation in the essential oil of Iranian *Rosa damascena* landraces under semi-arid and cool conditions. *Int. J. Hortic. Sci. Technol.* **2018**, *5*, 81–92.
8. Baser, K.H.C.; Kurkcuoglu, M.; Ozek, T. Turkish rose oil: Recent results. *Perfum. Flavorist* **2003**, *28*, 34–42.
9. Tanjga, B.B.; Lončar, B.; Aćimović, M.; Kiprovski, B.; Šovljanski, O.; Tomić, A.; Traviċić, V.; Cvetković, M.; Raičević, V.; Zeremski, T. Volatile profile of garden rose (*Rosa hybrida*) hydrosol and evaluation of its biological activity in vitro. *Horticulturae* **2022**, *8*, 895. [CrossRef]
10. Bayhan, G.I.; Gumus, T.; Alan, B.; Savas, I.K.; Cam, S.A.; Sahin, E.A.; Arslan, S.O. Influence of *Rosa damascena* hydrosol on skin flora (contact culture) after hand-rubbing. *GMS Hyg. Infect. Control* **2020**, *15*, Doc21.
11. Popović-Djordjević, J.; Špirović-Trifunović, B.; Pećinar, I.; de Oliveira, L.F.C.; Krstić, Đ.; Mihajlović, D.; Akšić, M.F.; Simal-Gandara, J. Fatty acids in seed oil of wild and cultivated rosehip (*Rosa canina* L.) from different locations in Serbia. *Ind. Crops Prod.* **2023**, *191*, 115797. [CrossRef]
12. Mawarni, E.; Ginting, C.N.; Chiuman, L.; Girsang, E.; Handayani, R.A.S.; Widowati, W. Antioxidant and elastase inhibitor potential of petals and receptacle of rose flower (*Rosa damascena*). *Pharm. Sci. Res. (PSR)* **2020**, *7*, 105–113.
13. Boskabady, M.H.; Shafei, M.N.; Saberi, Z.; Amini, S. Pharmacological effects of *Rosa damascena*. *Iran. J. Basic. Med. Sci.* **2011**, *14*, 295–307.
14. Choi, J.K.; Lee, Y.B.; Lee, K.H.; Im, H.C.; Kim, Y.B.; Choi, E.K.; Joo, S.S.; Jang, S.K.; Han, N.S.; Kim, C.H. Extraction conditions for phenolic compounds with antioxidant activities from white rose petals. *J. Appl. Biol. Chem.* **2015**, *58*, 117–124. [CrossRef]
15. Ghavam, M.; Afzali, A.; Manconi, M.; Bacchetta, G.; Manca, M.L. Variability in chemical composition and antimicrobial activity of essential oil of *Rosa* × *damascena* Herrm. from mountainous regions of Iran. *Chem. Biol. Technol. Agric.* **2021**, *8*, 22. [CrossRef]
16. Dagli, R.; Avcu, M.; Metin, M.; Kiymaz, S.; Ciftci, H. The effects of aromatherapy using rose oil (*Rosa damascena* Mill.) on preoperative anxiety: A prospective randomized clinical trial. *Eur. J. Integr. Med.* **2019**, *26*, 37–42. [CrossRef]
17. Mlcek, J.; Rop, O. Fresh edible flowers of ornamental plants—A new source of nutraceutical foods. *Trends Food Sci. Technol.* **2011**, *22*, 561–569. [CrossRef]
18. Hegde, A.S.; Gupta, S.; Sharma, S.; Srivatsan, V.; Kumari, P. Edible rose flowers: A doorway to gastronomic and nutraceutical research. *Food Res. Int.* **2022**, *162*, 111977. [CrossRef]
19. Delgado, C.; Crisosto, G.M.; Heymann, H.; Crisosto, C.H. Determining the primary drivers of liking to predict consumers' acceptance of fresh nectarines and peaches. *J. Food Sci.* **2013**, *78*, S605–S614. [CrossRef] [PubMed]
20. Wendin, K.; Pálsdóttir, A.M.; Spendrup, S.; Mårtensson, L. Odor Perception and Descriptions of Rose-Scented Geranium Pelargonium graveolens 'Dr. Westerlund'—Sensory and Chemical Analyses. *Molecules* **2023**, *28*, 4511. [CrossRef] [PubMed]
21. Mlcek, J.; Plaskova, A.; Jurikova, T.; Sochor, J.; Baron, M.; Ercisli, S. Chemical, Nutritional and Sensory Characteristics of Six Ornamental Edible Flowers Species. *Foods* **2021**, *10*, 2053. [CrossRef]
22. Pandey, K.B.; Rizvi, S.I. Plant polyphenols as dietary antioxidants in human health and disease. *Oxid. Med. Cell Longev.* **2009**, *2*, 270–278. [CrossRef]
23. Faller, A.L.K.; Fialho, E. Polyphenol content and antioxidant capacity in organic and conventional plant foods. *J. Food Comp. Anal.* **2010**, *23*, 561–568. [CrossRef]
24. Kumar, N.; Bhandari, P.; Singh, B.; Shamsher, S.B. Antioxidant activity and ultra-performance LC-electrospray ionization-quadrupole time-of-flight mass spectrometry for phenolics-based fingerprinting of Rose species: *Rosa damascena*, *Rosa bourboniana* and *Rosa brunonii*. *Food Chem. Toxicol.* **2009**, *47*, 361–367. [CrossRef]
25. Yang, H.; Shin, Y. Antioxidant compounds and activities of edible roses (*Rosa hybrida* spp.) from different cultivars grown in Korea. *Appl. Biol. Chem.* **2017**, *60*, 129–136. [CrossRef]
26. Lutz, M.; Hernández, J.; Henríquez, C. Phenolic content and antioxidant capacity in fresh and dry fruits and vegetables grown in Chile. *CyTA—J. Food* **2015**, *13*, 541–547.
27. Panche, A.N.; Diwan, A.D.; Chandra, S.R. Flavonoids: An overview. *J. Nutr. Sci.* **2016**, *5*, e47. [CrossRef]
28. Mikanagi, Y.; Yokoi, M.; Ueda, Y.; Saito, N. Flower flavonol and anthocyanin distribution in subgenus *Rosa*. *Biochem. Syst. Ecol.* **1995**, *23*, 183–200. [CrossRef]

29. Beara, I.; Torović, L.J.; Pintać, D.; Majkić, T.; Orčić, D.; Mimica-Dukić, N.; Lesjak, M. Polyphenolic profile, antioxidant and neuroprotective potency of grape juices and wines from Fruška Gora region (Serbia). *Int. J. Food Prop.* **2017**, *20*, S2552–S2568. [CrossRef]
30. Sharifi-Rad, M.; Anil Kumar, N.V.; Zucca, P.; Varoni, E.M.; Dini, L.; Panzarini, E.; Rajkovic, J.; Tsouh Fokou, P.V.; Azzini, E.; Peluso, I.; et al. Lifestyle, oxidative stress, and antioxidants: Back and forth in the pathophysiology of chronic diseases. *Front. Physiol.* **2020**, *11*, 694. [CrossRef] [PubMed]
31. Kusumawati, I.; Indrayanto, G. Natural Antioxidants in Cosmetics. In *Studies in Natural Products Chemistry*; Atta-ur-Rahman, Ed.; Elsevier: Amsterdam, The Netherlands, 2013; Volume 40, pp. 485–505.
32. Chermahini, S.H.; Adibah, F.; Majid, A.; Sarmidi, M.R. Cosmeceutical value of herbal extracts as natural ingredients and novel technologies in anti-aging. *J. Med. Plants Res.* **2011**, *5*, 3074–3077.
33. Simin, N.; Orcic, D.; Cetojevic-Simin, D.; Mimica-Dukic, N.; Anackov, G.; Beara, I.; Mitic-Culafic, D.; Bozin, B. Phenolic profile, antioxidant, anti-inflammatory and cytotoxic activities of small yellow onion (*Allium flavum* L. subsp. *flavum*, Alliaceae). *LWT—Food Sci. Technol.* **2013**, *54*, 139–146. [CrossRef]
34. Conti Filho, C.E.; Loss, L.B.; Marcolongo-Pereira, C.; Rossoni Junior, J.V.; Barcelos, R.M.; Chiarelli-Neto, O.; da Silva, B.S.; Passamani Ambrosio, R.; Castro, F.C.A.Q.; Teixeira, S.F.; et al. Advances in Alzheimer's disease's pharmacological treatment. *Front. Pharmacol.* **2023**, *14*, 1101452. [CrossRef]
35. Murray, A.P.; Faraoni, M.B.; Castro, M.J.; Alza, N.P.; Cavallaro, V. Natural AChE inhibitors from plants and their contribution to Alzheimer's disease therapy. *Curr. Neuropharmacol.* **2013**, *11*, 388–413. [CrossRef]
36. Oguntibeju, O.O. Type 2 diabetes mellitus, oxidative stress and inflammation: Examining the links. *Int. J. Physiol. Pathophysiol. Pharmacol.* **2019**, *11*, 45–63.
37. Oboh, G.; Ogunsuyi, O.B.; Ogunbadejo, M.D.; Adefegha, S.A. Influence of gallic acid on α-amylase and α-glucosidase inhibitory properties of acarbose. *J. Food Drug Anal.* **2016**, *24*, 627–634. [CrossRef]
38. Ghani, U. Re-exploring promising α-glucosidase inhibitors for potential development into oral anti-diabetic drugs: Finding needle in the haystack. *Eur. J. Med. Chem.* **2015**, *103*, 133–162. [CrossRef] [PubMed]
39. Gholamhoseinian, A.; Fallah, H.; Sharififar, F. Inhibitory effect of methanol extract of *Rosa damascena* Mill. flowers on α-glucosidase activity and postprandial hyperglycemia in normal and diabetic rats. *Phytomedicine* **2009**, *16*, 935–941. [CrossRef]
40. UPOV. *Guidelines for the Conduct of Tests Distinctness, Uniformity and Stability—Rosa L.*; International Union for The Protection of New Varieties of Plants: Geneva, Switzerland, 2010.
41. Lesjak, M.M.; Beara, I.N.; Orčić, D.Z.; Anačkov, G.T.; Balog, K.J.; Francišković, M.M.; Mimica-Dukić, N.M. *Juniperus sibirica* Burgsdorf. as a novel source of antioxidant and anti-inflammatory agents. *Food Chem.* **2011**, *124*, 850–856. [CrossRef]
42. Lee, J.; Durst, R.W.; Wrolstad, R.E. Determination of total monomeric anthocyanin pigment content of fruit juices, beverages, natural colorants, and wines by the pH differential method: Collaborative study. *J. AOAC Int.* **2005**, *88*, 1269–1278. [CrossRef]
43. Pintać, D.; Četojević-Simin, D.; Berežni, B.; Orčić, D.; Mimica-Dukić, N.; Lesjak, M. Investigation of the chemical composition and biological activity of edible grapevine (*Vitis vinifera* L.) leaf varieties. *Food Chem.* **2019**, *286*, 686–695. [CrossRef] [PubMed]
44. Yang, X.W.; Huang, M.Z.; Jin, Y.S.; Sun, L.N.; Song, Y.; Chen, H.S. Phenolics from *Bidens bipinnata* and their amylase inhibitory properties. *Fitoterapia* **2012**, *83*, 1169–1175. [CrossRef] [PubMed]
45. Palanisamy, U.D.; Ling, L.T.; Manaharan, T.; Appleton, D. Rapid isolation of geraniin from *Nephelium lappaceum* rind waste and its anti-hyperglycemic activity. *Food Chem.* **2011**, *127*, 21–27. [CrossRef]

Disclaimer/Publisher's Note: The statements, opinions and data contained in all publications are solely those of the individual author(s) and contributor(s) and not of MDPI and/or the editor(s). MDPI and/or the editor(s) disclaim responsibility for any injury to people or property resulting from any ideas, methods, instructions or products referred to in the content.

Article

Structure–Activity Relationship of Natural Dihydrochalcones and Chalcones, and Their Respective Oxyalkylated Derivatives as Anti-*Saprolegnia* Agents

Alejandro Madrid [1,*], Evelyn Muñoz [1], Valentina Silva [1], Manuel Martínez [1], Susana Flores [1], Francisca Valdés [1], David Cabezas-González [2] and Iván Montenegro [3]

1. Laboratorio de Productos Naturales y Síntesis Orgánica (LPNSO), Departamento de Ciencias y Geografía, Facultad de Ciencias Naturales y Exactas, Universidad de Playa Ancha, Avda. Leopoldo Carvallo 270, Playa Ancha, Valparaíso 2340000, Chile; evdmunoz@gmail.com (E.M.); silvapedrerosv@gmail.com (V.S.); manuel.m.lobos@gmail.com (M.M.); s.flores.gonzalez@gmail.com (S.F.); fvaldesnavarro@gmail.com (F.V.)
2. Instituto de Química y Bioquímica, Facultad de Ciencias, Universidad de Valparaíso, Av. Gran Bretaña 1111, Valparaíso 2360102, Chile; david.cabezas@postgrado.uv.cl
3. Escuela de Obstetricia y Puericultura, Facultad de medicina, Campus de la Salud, Universidad de Valparaíso, Angamos 655, Reñaca, Viña del Mar 2520000, Chile; ivan.montenegro@uv.cl
* Correspondence: alejandro.madrid@upla.cl; Tel.: +56-032-250-0526

Citation: Madrid, A.; Muñoz, E.; Silva, V.; Martínez, M.; Flores, S.; Valdés, F.; Cabezas-González, D.; Montenegro, I. Structure–Activity Relationship of Natural Dihydrochalcones and Chalcones, and Their Respective Oxyalkylated Derivatives as Anti-*Saprolegnia* Agents. *Plants* **2024**, *13*, 1976. https://doi.org/10.3390/plants13141976

Academic Editors: Luis Ricardo Hernández, Eugenio Sánchez-Arreola and Edgar R. López-Mena

Received: 26 June 2024
Revised: 17 July 2024
Accepted: 18 July 2024
Published: 19 July 2024

Copyright: © 2024 by the authors. Licensee MDPI, Basel, Switzerland. This article is an open access article distributed under the terms and conditions of the Creative Commons Attribution (CC BY) license (https://creativecommons.org/licenses/by/4.0/).

Abstract: *Saprolegnia* sp. is a pathogenic oomycete responsible for severe economic losses in aquaculture. To date, there is no treatment for its control that is effective and does not pose a threat to the environment and human health. In this research, two dihydrochalcones **1** and **2**, and three chalcones **3–5**, isolated from the resinous plant *Adesmia balsamica*, as well as their synthesized oxyalkylated derivatives **6–29** already reported and a new synthesized series of oxyalkylchalcones **30–35**, were evaluated for their anti-*saprolegnia* activity and structure–activity relationship as potential control and treatment agents for strains of *Saprolegnia parasitica* and *S. australis*. Among the molecules tested, natural 2′,4′-dihydroxychalcone (**3**) and new oxyalkylchalcone **34** were the most potent anti*saprolegnia* agents against both strains, even with better results than the commercial control bronopol. On the other hand, the structure–activity relationship study indicates that the contributions of steric and electrostatic fields are important to enhance the activity of the compounds, thus the presence of bulky substituents favors the activity.

Keywords: oxyalkylchalcones; *Adesmia balsamica*; *Saprolegnia parasitica*; *Saprolegnia australis*; SAR studies

1. Introduction

Species belonging to the genus *Saprolegnia* were long considered aquatic fungi; currently, it is known that they belong to the kingdom Chromista, which includes brown algae and diatoms [1]. *Saprolegnia* belongs to the division of Oomicota, which houses mostly saprophytic or parasitoid microorganisms. The species of this division have zoospores with barbed and smooth flagella [2]. The Saprolegniaceae Family has, as general characteristics, a highly branched coenocytic mycelium and a cell wall made up mainly of glucans and cellulose, in which septa appear to separate the reproductive organs. Its habitat is mainly aquatic, although it can also live in humid soils. It is tolerant to large temperature ranges from 3 to 33 °C, and its relative salinity tolerance range is approximately 1.75% NaCl [3].

Saprolegnia generally travels in colonies consisting of one or more species. They first form a mass of hyphae that, when it grows sufficiently, can be seen with the naked eye, forming a mycelium. Colonies are generally white, turning gray depending on the presence of bacteria or other microorganisms [4]. They release their zoospores into the aquatic environment, which remain dormant until they find a suitable substrate where they begin the infectious process, causing saprolegniosis in fish, amphibians, and crustaceans, both in wild and cultivated environments, threatening biodiversity and food security around the world [5]. One of the most destructive oomycete pathogens in freshwater ecosystems is *Saprolegnia parasitica*, which mainly affects salmonids, from eggs to adult fish. It is characterized by visible patches of white or gray filamentous mycelium on the skin and fins. In severe cases, hyphae invade epidermal tissues, muscles, and blood vessels, ultimately causing the death of infected fish due to osmoregulatory failure and respiratory failure [6]. On the other hand, *S. australis* is responsible for significant mortality of salmonid eggs and fry in fish farms in distant locations, such as Chile and Spain [7], also affecting native North American crustaceans, such as crayfish (*Orconectes propinquus*) [8]; however, this is not the most frequently isolated species from freshwater samples and fish farms [7], but it is no less aggressive.

Saprolegnia infection in fish, species-independent, can cause enormous losses in terms of millions of dollars to the aquaculture industry annually [5]. Currently, there are no effective treatments to control *Saprolegnia*. Malachite green was formerly used, but its use has been discontinued worldwide due to its persistence in fish tissues and the environment [9]; later, its successor, formalin, was also banned [10]. The use of veterinary drugs has also declined for the same reason, as they can cause adverse effects to the environment and human health, as well as bacterial resistance due to excessive use of antibiotics [4]. Due to the harmful effects of chemotherapeutic agents, it is necessary to investigate new treatment alternatives that are effective and safe for the environment and the health of the human population. An alternative to current treatments is the use of medicinal plants, such as *Adesmia balsamica* Bertero ex Colla (Fabaceae), known in the vernacular as "paramela de Puangue" and "jarilla" [11]; it is a native and endemic resinous shrub that grows between Aconcagua and Valparaíso provinces in the Valparaíso Region, Chile [12].

Historically, the relevance of "paramelas" has been due to its various uses in traditional medicine to treat rheumatic pains, hair loss [13], colds, digestive disorders [14], and amenorrhea [15], and as an aphrodisiac [16]. *A. balsamica* has the ability to generate large numbers of resinous exudates. It is known that resinous exudates consist of a variety of non-polar compounds, mostly terpenoids and flavonoids [17]. Flavonoids found in exudates are usually chalcone aglycones, flavanones and flavones, which exhibit increased hydrophobicity due to the presence of methoxylated or dehydroxylated substituents on the A- and B-rings compared to the internal tissue forms [18]. Given the large number of bioactive compounds that resinous exudates possess, such as flavonoids and particularly chalcones, they are presented as an alternative for the control of oomycetes, especially considering their lower production cost, biodegradability and non-toxic behavior for the environment.

Chalcones are secondary metabolites belonging to the flavonoid family. Found naturally in fruits, spices, teas, and soy-based foods, they have received much attention due to their intriguing and potentially beneficial properties [19]. Both natural and synthetic chalcones show high pharmacological potential since they have various biological activities, such as antibacterial [20], anti-inflammatory [21], analgesic [22], antiviral [23], and immunomodulatory [24], among others. Oxy-prenylated chalcones are an important subclass of natural chalcones [25]. In recent years, it has been shown that they have relevant biological activities, such as antimicrobial activity, which would be provided by the O-alkyl chain that increases the lipophilicity of the compounds, allowing greater interaction with cell membranes [26]. On the other hand, for O-substituted chalcones as xanthohumol derivatives, anticancer and antioxidant properties have been described in in vitro models [27].

Previously, our study group reported the anti-oomycete activity of natural dihydrochalcones: $2',4'$-dihydroxydihydrochalcone (**1**) and dihydroisorcordoin (**2**), as well as their respective chalcones: 2,4-dihydroxychalcone (**3**) and isocordoin (**4**) (see Figure 1). In turn, our group has described for these four natural molecules their respective oxyalkylated derivatives, which have shown strong inhibition of various strains of *Saprolegnia* sp. [25–31]. In this context, the aim of this study is to report the presence of $2',4'$-dihydroxy-$5'$ prenylchalcone (**5**) in the resinous exudate of *A. balsamica*, as well as the synthesis of a new series of oxyalkylated derivatives from it, to subsequently perform new anti-oomycete assays against two strains of saprolegniales, *S. parasitica* and *S. australis*, for both the previously described natural and synthetic compounds and chalcone **5** and its synthetic derivatives. In addition, a structure–activity relationship study of all the molecules tested was carried out in order to elucidate the behavior of these types of natural molecule and their oxyalkylated derivatives against *Saprolegnia* strains.

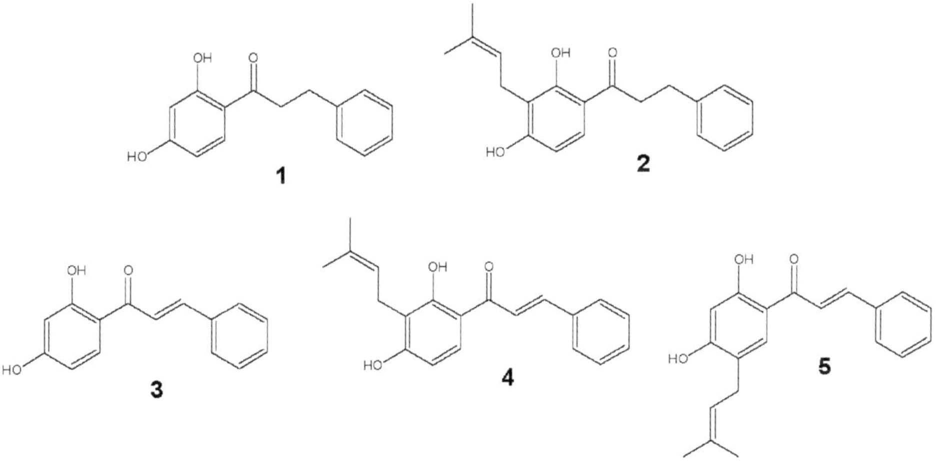

Figure 1. Natural dihydrochalcones **1**, **2** and chalcones **3**–**5** obtained from *Adesmia balsamica*.

2. Results and Discussion

2.1. Natural Dihydrochalcones and Chalcones from the Resinous Exudate of Adesmia balsamica

Compounds **1** to **4** were available in the laboratory due to the previous research mentioned above. Compound **5** was isolated in a 3.25% yield from the resinous exudate of *A. balsamica*. Its structure was determined by NMR and the spectroscopic data were compared with the report of this chalcone previously isolated from the aerial parts of *Lonchocarpus cultratus* [32]. On the other hand, these natural dihydrochalcones and chalcones constitute an interesting biosynthetic series, as shown in Figure 1, since they all present a dihydroxylation at the 2′ and 4′ position of the A-ring and the unsubstituted B-ring in their structure.

2.2. Synthesis of 4-Oxyalkylchalcones Derivates from Compound 5

As described above, all the oxyalkylated derivatives (**6**–**29**, Scheme 1) of the natural molecules **1**–**4**, which are available in our laboratory, were synthesized by a Williamson reaction [25–31].

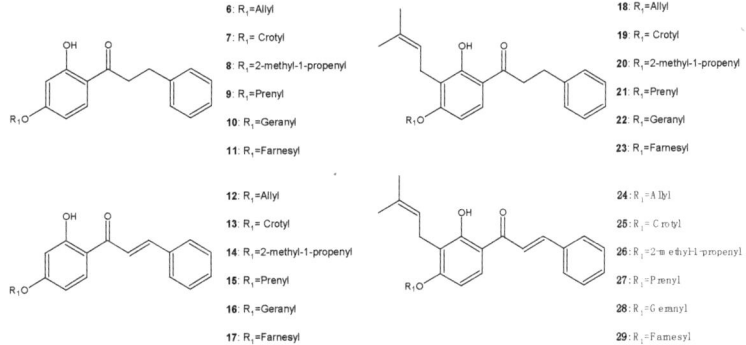

Scheme 1. Structure of known dihydrochalcone and chalcone derivatives **6**–**29**.

The new series of 4-oxyalkylchalcones **30–35** (Scheme 2), were synthesized through a nucleophilic substitution of the natural chalcone **5** with a series of alkyl bromides by selective alkylation on the hydroxyl located at C-4′. To optimize the preparation of the 4′-mono-O-alkylated molecules, different reaction conditions were evaluated, considering the type of solvent, reaction time and temperature, as well as the stoichiometric ratio of the reactants. The conditions that yielded the best products were those with a reaction stoichiometry of 1:1.2:0.5 (natural chalcone/alkyl bromide/potassium carbonate), anhydrous acetone as solvent, and a temperature of 75 °C at reflux for 6 h. The yields obtained for these products ranged from 57.3% for compound **35** to 80% for compound **30**.

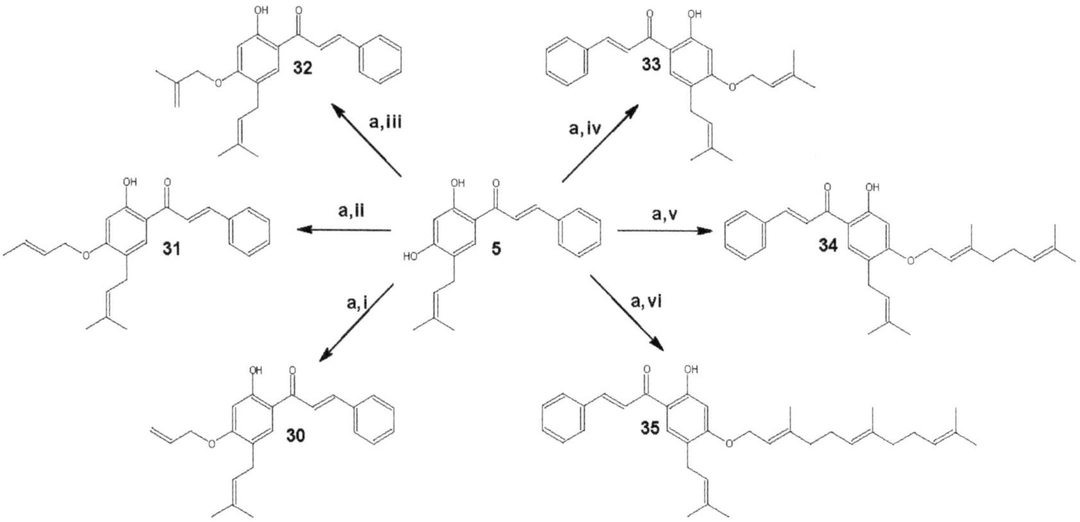

Scheme 2. Synthesis of **30–35**. Reagents and conditions: (**a**) K_2CO_3, dry acetone, reflux at 75 °C for 6 h; (**i**) allyl bromide; (**ii**) crotyl bromide; (**iii**) 2-methyl-1-propenyl bromide; (**iv**) prenyl bromide; (**v**) geranyl bromide; (**vi**) farnesyl bromide.

The identification of the new 4-oxyalkylchalcones **30–35** depends on the determination of a doublet type signal in the proton spectrum corresponding to the C-prenylation on the 5′-carbon of the A-ring of the chalcone, which has a typical chemical shift (δ) for a prenylic chain attached to an aromatic system between 3.0 and 3.05 ppm, which integrates for two protons and presents a coupling constant value (J) from 7.0 to 7.2 Hz signal correlated with a signal with δ between 28 and 29 ppm corresponding to the allylic carbon atom of the prenylic chain. The pattern of aromatic substitution of ring A should also be considered, where two singlet signals are observed at 7.85 and 6.40 ppm, corresponding to H-6′ and H-3′, respectively, for chalcones with an unsaturated alpha-beta system, indicating a *"para"* substitution of the aromatic system A. Finally, it is worth mentioning that the O-prenylation signal corresponds to a doublet having a chemical shift (δ) typical of an allylic ether-type molecule between 4.50 and 4.60 ppm, which integrates for two protons and has a constant value of coupling (J) of 6.2–6.6 Hz signal correlated with a signal with δ between 65 and 72 ppm corresponding to the allylic carbon atom of each alkyl halide.

2.3. Anti-Oomycete Activity against S. parasitica and S. australis

To evaluate the anti-oomycete activity of the compounds, the microdilution method described above was used. The commercially used antifungals bronopol, fluconazole and a naturally occurring compound safrole were used as controls.

The results shown in Table 1 demonstrate that compounds **3** and **34** are the most active compounds at 72 h, compared to the controls used. The minimum inhibitory concentration

(MIC) value of compound **3** was 52.0 µmol/L against both strains, while the MIC value of compound **34** was 112.5 µmol/L against *S. parasitica* and 56.2 µmol/L against *S. australis*. The minimum oomyceticide concentration (MOC) value for compound **3** was 208.1 µmol/L against *S. parasitica* and 166.4 µmol/L against *S. australis*, while for compound **34** it was 112.5 µmol/L against *S. parasitica* and 56.2 µmol/L against *S. australis*, These results show that compounds **3** and **34** have better activity even than the controls used, such as bronopol, a broad spectrum biocide, which has been used as an effective and economically acceptable alternative in the treatment and control of *Saprolegnia* sp. [31].

Table 1. Minimum inhibitory concentration (MIC) and minimum oomyceticide concentration (MOC) values of evaluated compounds at 72 h.

Compounds	MIC [a] (µmol/L)		MOC [b] (µmol/L)		Compounds	MIC [a] (µmol/L)		MOC [b] (µmol/L)	
	S. parasitica	*S. australis*	*S. parasitica*	*S. australis*		*S. parasitica*	*S. australis*	*S. parasitica*	*S. australis*
1	515.9	412.8	619.2	515.9	20	679.4	594.5	>679.4	679.4
2	322.2	322.2	322.2	322.2	21	648.6	175	>648.6	648.6
3	52.0	52.0	208.1	166.4	22	531.2	464.8	531.2	464.8
4	>648.6	648.6	>648.6	>648.6	23	281.1	224.9	149.9	281.1
5	>648.5	648.5	>648.5	>648.5	24	286.9	215.2	358.7	430.5
6	531.3	442.8	708.4	619.9	25	275.9	206.9	344.9	275.9
7	>674.8	674.8	>674.8	674.8	26	344.9	344.9	344.9	344.9
8	421.8	421.8	506.1	421.8	27	398.4	332.0	332.0	332.0
9	>644.3	644.3	>644.3	644.3	28	>449.8	>449.8	>449.8	>449.8
10	>528.4	528.4	>528.4	528.4	29	292.6	146.3	292.6	195.1
11	>447.8	>447.8	>447.8	>447.8	30	>573.9	573.9	>573.9	>573.9
12	356.7	285.3	427.9	356.7	31	>550.3	550.3	>550.3	>550.3
13	342.9	342.9	411.5	342.9	32	551.8	551.8	>551.8	>551.8
14	411.5	411.5	411.5	411.5	33	265.6	265.6	125	265.6
15	>528.4	264.2	>528.4	330.3	34	112.5	56.2	112.5	56.2
16	>447.8	>447.8	>447.8	>447.8	35	292.6	243.8	292.6	243.8
17	242.8	194.3	242.8	194.3	Bronopol	>1000	875.0	>1000	1000
18	713.5	1624.3	713.5	624.3	Fluconazole	>653.0	>653.0	>653.0	>653.0
19	>679.5	679.5	>679.5	679.5	Safrole	1078.9	924.8	1078.9	924.8

[a,b] Each value represents the mean of three experiments ($p < 0.05$), performed in quadruplicate.

In the case of the other compounds evaluated, we could classify them as having moderate activity in relation to compounds **3** and **34**, but higher than the activity of the controls. In this classification are compounds **2, 24, 25** and **33**, as can be seen in Table 1. The rest of the 4-oxyalkylated derivatives showed lower anti-*Saprolegnia* activity than the controls used.

Inhibition of mycelial growth was observed by fluorescence microscopy and inverted microscopy. Dehydration and necrosis of the mycelium, as well as the formation of crystals and precipitates, which caused a decrease in the formation of zoosporangium, were considered important damages to the mycelium. The results showed that compound **3** is that which produces a higher percentage of mycelial damage to the *S. parasitica* strains, while compound **1** produces mycelial damage to both *Saprolegnia* strains and reaches 100% damage to *S. australis*.

A similar effect, although on a smaller scale, was observed for compounds **2, 8, 11** and **34**, with percentages of mycelial damage close to or above 50% for both *Saprolegnia* strains. These values are significantly higher than the control used, as shown in Table 2.

Compounds **4, 5, 7, 16, 19, 20, 21, 26, 28, 30, 31** and **32** showed no activity against any of the strains.

Table 3 shows the cell membrane damage caused by the compounds. This type of test is based on the direct effect of the compounds on sterol formation in fungal or mycobacterial cells and is positive if there is an increase in absorbance at 260 nm in cells treated with the test compounds. Sodium dodecyl sulfate (SDS) is used to compare the effect observed with this chaotropic agent, which causes 100% cell lysis. This assay is performed in triplicate. In this assay, again, the most active compounds were **3** and **34**, showing significantly better activity than the control. Compounds **1, 2** and **33** also showed moderate activity on both strains, as shown.

Table 2. Percentage of mycelial growth inhibition (MGI) by evaluated compounds at 48 h.

Compounds	MGI (%)		Compounds	MGI (%)	
	S. parasitica	S. australis		S. parasitica	S. australis
1	55	100	19	0	0
2	43	45	20	0	0
3	100	0	21	0	0
4	0	0	22	36	38
5	0	0	23	40	13
6	13	36	24	31	33
7	0	0	25	25	30
8	50	43	26	0	0
9	0	38	27	28	30
10	10	13	28	0	0
11	50	55	29	37	39
12	15	17	30	0	0
13	10	12	31	0	0
14	8	10	32	0	0
15	12	14	33	39	41
16	0	0	34	55	58
17	19	23	D	20	23
18	12	37	Bronopol	0	33

Each value represents the mean of three experiments ($p < 0.05$), performed in quadruplicate.

Table 3. Percentage of damage produced by evaluated compounds.

Compounds	Damage (%) [a]		Compounds	Damage (%) [a]	
	S. parasitica	S. australis		S. parasitica	S. australis
1	40	50	19	0	0
2	43	45	20	0	0
3	100	25	21	10	0
4	0	0	22	23	27
5	0	0	23	30	35
6	0	0	24	31	33
7	0	0	25	25	30
8	30	40	26	35	38
9	0	0	27	28	30
10	0	0	28	0	0
11	0	0	29	37	39
12	15	17	30	0	0
13	10	12	31	0	0
14	8	10	32	0	0
15	12	14	33	40	44
16	0	0	34	58	60
17	19	23	35	27	29
18	16	0	Bronopol	0	33

[a] Damage produced by evaluated compounds compared to the damage produced by SDS. This was utilized at a final concentration of 2%, which produced 100% of cell lysis. The assay was performed in triplicate. Results presented are the means of values from at least two independent assays.

2.4. 3D-QSAR

2.4.1. Statistical Results of the Models

The results of the internal validation and r^2_{pred} are presented in Table 4. The q^2 values must be greater than 0.5, and in both cases, this threshold is surpassed. The r^2_{pred} value must exceed 0.6. In the first model, an r^2_{pred} of 0.763 was achieved, while the other model resulted in an r^2_{pred} of 0.831, indicating a high predictive capacity of the model. The predictive values of each model, along with their corresponding residual values, are presented in Table 5. The residual values, obtained by comparing experimental values with predicted values, do not exceed one logarithmic unit. This supports the concept of the model's predictive capacity. Regarding the contributions of the steric and electrostatic fields, in the first model, the percentage of steric contribution is 58.7%, and the electrostatic contribution is 41.3%. As shown in Table 4, the second model follows the same trend with a larger difference (79.5% and 20.5%, respectively). These results offer insights into the criteria to be employed for the design of new structures.

Table 4. Statistical results of the 3D-QSAR CoMFA–steric-electrostatic (CoMFA–SE) analysis.

CoMFA–SE	N	q^2	r^2_{NCV}	SEE	F	SEP	r^2_{pred}	% Contribution	
								Steric	Electrostatic
S. parasitica	1	0.547	0.729	0.08	45718	0.103	0.763	0.587	0.413
S. australis	2	0.623	0.917	0.05	88692	0.099	0.831	0.795	0.205

Table 5. Predicted pIC$_{50}$ activity values and residuals of the 3D-QSAR CoMFA Models.

	pIC$_{50}$ S. parasitica				pIC$_{50}$ S. australis		
Compound	Exp	Pred	Residual	Compound	Exp	Pred	Residual
*5	3.1881	3.270	−0.082	*5	3.188	3.191	−0.003
6	3.2747	3.159	0.116	6	3.354	3.147	0.207
7	3.1708	3.187	−0.016	7	3.171	3.200	−0.029
*8	3.3749	3.197	0.178	*8	3.375	3.174	0.201
9	3.1909	3.185	0.006	9	3.191	3.188	0.003
*10	3.2770	3.289	−0.012	*10	3.277	3.312	−0.035
11	3.3489	3.353	−0.004	11	3.349	3.300	0.049
18	3.1466	3.157	−0.010	18	3.205	3.172	0.033
19	3.1678	3.184	−0.016	19	3.168	3.224	−0.056
20	3.1678	3.196	−0.028	20	3.226	3.208	0.018
21	3.1881	3.186	0.002	21	3.246	3.223	0.023
22	3.2747	3.293	−0.018	22	3.333	3.342	−0.009
*23	3.5511	3.344	0.207	*23	3.648	3.321	0.327
32	3.2582	3.310	−0.052	32	3.258	3.320	−0.062
33	3.5758	3.393	0.183	33	3.576	3.498	0.078
*34	3.9490	3.491	0.458	*34	4.250	3.637	0.613
35	3.5338	3.559	−0.025	35	3.613	3.635	−0.022

* Test set.

2.4.2. Contour Maps of CoMFA–SE Models and SAR

In Figures 2 and 3, we present the resulting contour maps of the 3D-QSAR CoMFA steric and electrostatic field models (CoMFA–SE) for S. parasitica and S. australis, respectively. The steric contour maps for S. parasitica are displayed in Figure 2A,B. In panel A, the most active molecule in the model (34, pIC$_{50}$ = 3.949; 112.5 µmol/L) is depicted, while the least active is represented in panel B (18, pIC$_{50}$ = 3.147; 713.5 µmol/L). In the steric maps, the green poly-hedra indicate that the presence of relatively bulky substituents is favorable for activity. Compound 34 has two bulky groups at the 4′ and 5′ position of ring A, compared to compound 18, which has one less bulky group than compound 34 at the 4′ position. This difference in activity between these structures may be explained by this observation. As indicated in Table 4, steric contribution holds great relevance for activity. It is suggested to further explore bulky groups in the regions demarcated by the green poly-hedra (4′ and 5′). The electrostatic contour map is presented in Figure 2C,D, with 3C depicting the most active molecule in the model and D representing the least active (compounds 34 and 18, respectively). In this case, the red poly-hedra suggest that the presence of electronegative groups is favorable while the blue poly-hedra indicate that electropositive groups are favorable. In both cases, the presence of the oxygen atom located in the 4′ position of the ring A is favorable. In the isopentyl group region (5′ of the ring A) of compound 34, groups with electronegative atoms can be exposed.

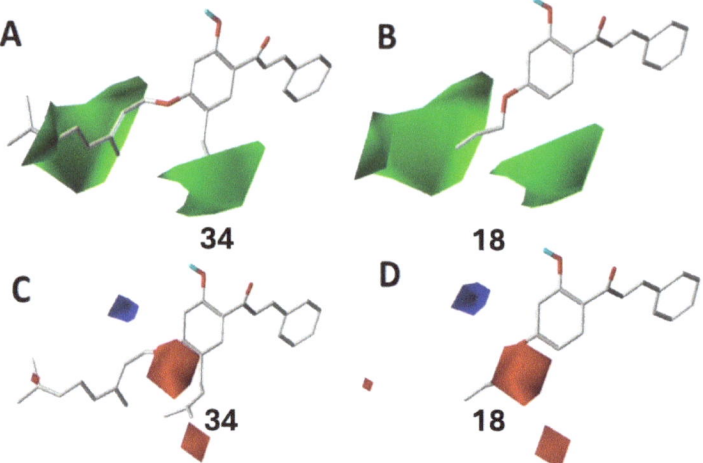

Figure 2. 3D–QSAR CoMFA of *Saprolegnia parasitica*. In (**A,B**), the steric contour maps are presented, while, in panels (**C,D**), the electrostatic maps are displayed.

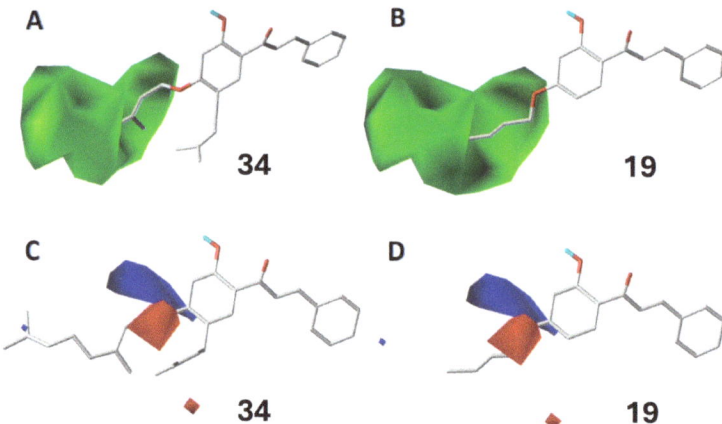

Figure 3. 3D–QSAR CoMFA of *Saprolegnia australis*. In (**A,B**), the steric contour maps are presented, while, in panels (**C,D**), the electrostatic maps are displayed.

In Figure 3A,B, we have the steric contour map for the *S. australis* model. In panel A, the most active compound in the model (**34**, pIC$_{50}$ = 4.250; 112.5 µmol/L) is depicted, while the least active is represented in panel B (**19**, pIC$_{50}$ = 3.168; 679.5 µmol/L). We observe that the trend seen in the model of *S. parasitica*, as depicted in Figure 3, is repeated. The bulkier group is of greater relevance to the activity. On the other hand, in the electrostatic contour map shown in Figure 3C,D, the same trend is visualized. Figure 3C depicts the most active molecule in the model, while Figure 3D represents the least active compounds (**34** and **19**, respectively). At position **3′**, other groups with electropositive characteristics or low electronegativity can be explored. One option is to add iso-groups or other small hydrocarbon branches. The results discussed in Figure 2 are also applicable to Figure 3.

3. Materials and Methods

3.1. General

All chemicals and positive controls were obtained from Aldrich (St. Louis, MO, USA) and were used without further purification. Both natural **1–4** and synthetic molecules **6–29** were available in our laboratory. All reactions were monitored by thin layer chromatography (TLC) on TLC precoated silica gel 60 F254 glass-backed plates (Merck KGaA, Darmstadt, Germany). Flash column chromatography was performed on silica gel (200–300 mesh) (Merck KGaA, Darmstadt, Germany). The ^{1}H and ^{13}C spectra were recorded in CDCl$_3$ solutions and are referenced to the residual peaks of CHCl$_3$ at δ = 7.26 ppm and δ = 77.0 ppm for ^{1}H and ^{13}C, respectively, on an Avance 400 Digital NMR spectrometer (Bruker, Rheinstetten, Germany) operating at 400.1 MHz for ^{1}H and 100.6 MHz for ^{13}C. HRMS were recorded in a MAT 95 XL mass spectrometer (Thermo Finnigan, Bremen, Germany).

3.2. Isolation of Natural Chalcone from the Resinous Exudate of Adesmia balsamica

The resinous exudate of *A. balsamica* was obtained according to the method previously described by Flores et al. [25], which consists of immersing the fresh branches and leaves of *A. balsamica* (200 g) in dichloromethane for 45 s. The solvent is then removed under reduced pressure, leaving only the extraction products. The resulting resinous exudate (10 g) was fractionated through silica gel chromatographic columns using hexane-ethyl acetate in increasing order of polarity. Compounds **5** were obtained with yields of 3.25%. This procedure was repeated until the critical mass necessary for the development of the subsequent syntheses was obtained.

3.3. Synthesis of 4-Oxyalkylchalcones

A solution of natural chalcone **5** (1 mmol), alkyl halides (1.2 mmol), and K$_2$CO$_3$ (1.5 mmol) in anhydrous acetone (10 mL) was refluxed for 6 h at 75 °C. The end of the reaction is verified by TLC, and then the mixture is poured into ice water (20 mL) and extracted with ethyl acetate (3 × 25 mL). The resulting organic phase was dried over anhydrous potassium sulfate and filtered. The solvent is then evaporated under reduced pressure. All synthesized products were separated and purified by column chromatography (CC) eluting with mixtures of petroleum ether/ethyl acetate of increasing polarity (9.0:1.0 →5.8:4.2). The progress in the separation of the synthetic compounds was analyzed by TLC [33]. The structural determination of the natural chalcone **5** and their derivatives **30–35** was confirmed from their spectroscopic properties by NMR. Structural determinations of the previously reported compounds **1–29** can be found in the Supplementary Material.

(2*E*)-1-[4-(allyloxy)-2-hydroxy-5-(prenyl)phenyl]-3-phenylprop-2-en-1-one (**30**): The compound was isolated as a yellow solid in a yield of 80.0% and a melting point of 94–95 °C. ^{1}H NMR (400 MHz, CDCl$_3$): δ 13.38 (*s*, 1H, **2′**-OH), 7.86 (*s*, 1H, H-**6′**); 7.78 (*d*, *J* = 9.1 Hz, 1H, H-8); 7.62 (*m*, 2H, H-7 and H-4); 7.42 (*m*, 4H, H-2, H-3, H-5 and H-6); 6.47 (*s*, 1H, H-**3′**); 6.07 (*m*, 1H, H-**2″**); 5.44 (*m*, 1H, H-**3″**b); 5.30 (*m*, 1H, H-**3″**a); 5.26 (*m*, 1H, H-**8′**); 4.65 (*d*, *J* = 5.1 Hz, 2H, H-**1″**); 3.03 (*d*, *J* = 7.1 Hz, 1H, H-**7′**); 1.68 (*s*, 3H, H-10′ and H-11′). ^{13}C NMR (100 MHz, CDCl$_3$): δ 191.3 (C-9); 163.2 (C-**2′**); 162.5 (C-**4′**); 144.1 (C-7); 135.1 (C-1); 132.7 (C-**2″**); 131.8 (C-**9′**); 130.5 (C-4); 129.0 (C-**6′**); 129.0 (C-2 and C-6); 128.5 (C-3 and C-5); 121.9 (C-**8′**); 120.6 (C-8); 118.0 (C-**5′**) 117.6 (C-**3″**); 114.6 (C-1′); 118.4 (C-**3″**); 103.2 (C-**3′**); 70.0 (C-**1″**); 25.8 (C-**7′**); 21.8 (C-10′); 17.9 (C-11′). HRMS: [M + H]$^+$ ion m/z 349.1809 (calcd for C$_{23}$H$_{24}$O$_3$, 349.1804).

(2*E*)-1-[4-(crotyloxy)-2-hydroxy-5-(prenyl)phenyl]-3-phenylprop-2-en-1-one (**31**): The compound was isolated as a yellow solid in a yield of 77.1% and a melting point of 93–95 °C. ^{1}H NMR (400 MHz, CDCl$_3$): δ 13.38 (*s*, 1H, **2′**-OH), 7.85 (*s*, 1H, H-**6′**); 7.77 (*d*, *J* = 9.1 Hz, 1H, H-8); 7.63 (*m*, 2H, H-7 and H-4); 7.42 (*m*, 4H, H-2, H-3, H-5 and H-6); 6.40 (*s*, 1H, H-**3′**); 5.85 (*m*, 1H, H-**2″**); 5.72 (*m*, 1H, H-**3″**); 5.25 (*m*, 1H, H-**8′**); 4.57 (*d*, *J* = 5.9 Hz, 2H, H-**1″**); 3.05 (*d*, *J* = 7.0 Hz, 1H, H-**7′**); 1.68 (*s*, 3H, H-10′ and H-11′); 1.58 (*s*, 3H, H-**4″**). ^{13}C NMR (100 MHz, CDCl$_3$): δ 191.6 (C-9); 163.2 (C-**2′**); 162.6 (C-**4′**); 144.1 (C-7); 134.9 (C-1); 131.7 (C-**9′**); 130.5 (C-4); 130.3 (C-**6′**); 129.1 (C-**2″**); 129.0 (C-2 and C-6); 128.5 (C-3 and C-5); 125.6 (C-**3″**); 122.1

(C-**8**′); 120.7 (C-**5**′); 117.8 (C-8); 115.2 (C-**1**′); 103.7 (C-**3**′); 69.5 (C-**1**″); 25.8 (C-**7**′); 21.8 (C-**10**′); 17.9 (C-**11**′); 17.8 (C-**4**″). HRMS: [M + H]$^+$ ion m/z 363.1962 (calcd for $C_{24}H_{26}O_3$, 363.1960).

(2E)-1-[4-(2-methylpropenyloxy)-2-hydroxy-5-(prenyl)phenyl]-3-phenylprop-2-en-1-one (**32**): The compound was isolated as a yellow solid in a yield of 76.2% and a melting point of 97–98 °C. ^{1}H NMR (400 MHz, CDCl$_3$): δ 13.34 (s, 1H, **2**′-OH), 7.86 (s, 1H, H-**6**′); 7.77 (d, J = 9.0 Hz, 1H, H-8); 7.63 (m, 2H, H-7 and H-4); 7.42 (m, 4H, H-2, H-3, H-5 and H-6); 6.37 (s, 1H, H-**3**′); 5.26 (m, 1H, H-**2**″); 5.11 (s, 1H, H-3b″); 5.01 (s, 1H, H-3a″); 4.55 (s, 2H, H-**1**″); 3.03 (d, J = 7.1 Hz, 1H, H-**7**′); 1.85 (s, 3H, H-**4**″); 1.80 (s, 3H, H-10′); 1.68 (s, 3H, H-11′). ^{13}C NMR (100 MHz, CDCl$_3$): δ 191.6 (C-9); 163.2 (C-**2**′); 162.8 (C-**4**′); 143.9 (C-7); 141.3 (C-**2**″); 134.9 (C-1); 131.3 (C-**9**′); 130.5 (C-4); 129.1 (C-**6**′); 129.0 (C-2 and C-6); 128.5 (C-3 and C-5); 122.0 (C-**8**′); 120.6 (C-**5**′); 117.8 (C-8) 117.8 (C-**3**′); 114.6 (C-1); 113.0 (C-**3**″); 103.2 (C-**3**′); 71.3 (C-**1**″); 25.8 (C-**7**′); 21.8 (C-**10**′); 19.3 (C-**4**″); 17.8 (C-**11**′). HRMS: [M + H]$^+$ ion m/z 363.1969 (calcd for $C_{24}H_{26}O_3$, 363.1960).

(2E)-1-[4-(prenyloxy)-2-hydroxy-5-(prenyl)phenyl]-3-phenylprop-2-en-1-one (**33**): The compound was isolated as a yellow solid in a yield of 79.0 % and a melting point of 102–104 °C. ^{1}H NMR (400 MHz, CDCl$_3$): δ 13.33 (s, 1H, **2**′-OH), 7.85 (s, 1H, H-**6**′); 7.77 (d, J = 9.1 Hz, 1H, H-8); 7.63 (m, 2H, H-7 and H-4); 7.42 (m, 4H, H-2, H-3, H-5 and H-6); 6.41 (s, 1H, H-**3**′); 5.48 (m, 1H, H-**2**″); 5.25 (m, 1H, H-**8**′); 4.62 (d, J = 6.6 Hz, 2H, H-**1**″); 3.04 (d, J = 7.1 Hz, 1H, H-**7**′); 1.80 (s, 3H, H-**5**″); 1.78 (s, 3H, H-**4**″ and H-10′); 1.76 (s, 3H, H-11′). ^{13}C NMR (100 MHz, CDCl$_3$): δ 191.6 (C-9); 163.2 (C-**2**′); 162.8 (C-**4**′); 144.0 (C-7); 138.1 (C-**3**″); 134.9 (C-1); 131.7 (C-**6**′); 130.5 (C-**9**′); 129.1 (C-4); 129.0 (C-2 and C-6); 128.5 (C-3 and C-5); 122.1 (C-**8**′); 120.7 (C-8); 119.5 (C-**2**″) 117.9 (C-**5**′); 114.5 (C-**1**′); 103.3 (C-**3**′); 65.2 (C-**1**″); 25.8 (C-**7**′); 25.7 (C-**5**″); 21.8 (C-**10**′); 18.3 (C-**4**″); 17.8 (C-**11**′). HRMS: [M + H]$^+$ ion m/z 377.2124 (calcd for $C_{25}H_{28}O_3$, 377.2117).

(2E)-1-[4-(geranyloxy)-2-hydroxy-5-(prenyl)phenyl]-3-phenylprop-2-en-1-one (**34**): The compound was isolated as a yellow solid in a yield of 59.0% and a melting point of 81–83 °C. ^{1}H NMR (400 MHz, CDCl$_3$): δ 13.34 (s, 1H, **2**′-OH), 7.85 (s, 1H, H-**6**′); 7.77 (d, J = 9.1 Hz, 1H, H-8); 7.63 (m, 2H, H-7 and H-4); 7.42 (m, 4H, H-2, H-3, H-5 and H-6); 6.41 (s, 1H, H-**3**′); 5.48 (m, 1H, H-**2**″); 5.25 (m, 1H, H-**8**′); 4.62 (d, J = 6.6 Hz, 2H, H-**1**″); 3.05 (d, J = 7.1 Hz, 1H, H-**7**′); 2.10 (m, 4H, H-**4**″ and H-**5**″); 1.79 (s, 3H, H-**7**″); 1.75 (s, 3H, H-10′); 1.68 (s, 3H, H-11′); 1.61 (s, 3H, H-10″); 1.56 (s, 3H, H-**9**″). ^{13}C NMR (100 MHz, CDCl$_3$): δ 191.6 (C-9); 163.1 (C-**2**′); 162.8 (C-**4**′); 143.9 (C-7); 141.2 (C-**3**″); 134.9 (C-1); 131.9 (C-**6**′); 131.7 (C-**7**″); 130.5 (C-**9**′); 129.1 (C-4); 129.0 (C-2 and C-6); 128.5 (C-3 and C-5); 123.7 (C-**6**″); 122.1 (C-**8**′); 120.7 (C-**5**′); 119.3 (C-**2**″); 117.8 (C-8); 114.4 (C-**1**′); 103.1 (C-**3**′); 65.1 (C-**1**″); 39.3 (C-**4**″); 26.3 (C-**7**′); 25.8 (C-**5**″); 25.6 (C-**10**′); 21.8 (C-**10**″); 17.8 (C-**11**′); 17.7 (C-**9**″); 16.7 (C-**8**″). HRMS: [M + H]$^+$ ion m/z 445.2753 (calcd for $C_{30}H_{36}O_3$, 445.2743).

(2E)-1-[4-(farnesyloxy)-2-hydroxy-5-(prenyl)phenyl]-3-phenylprop-2-en-1-one (**35**): The compound was isolated as a yellow solid in a yield of 57.3% and a melting point of 74–76 °C. ^{1}H NMR (400 MHz, CDCl$_3$): δ 13.36 (s, 1H, **2**′-OH), 7.85 (s, 1H, H-**6**′); 7.77 (d, J = 9.1 Hz, 1H, H-8); 7.63 (m, 2H, H-7 and H-4); 7.42 (m, 4H, H-2, H-3, H-5 and H-6); 6.41 (s, 1H, H-**3**′); 5.48 (m, 1H, H-**2**″); 5.25 (m, 1H, H-**8**′); 5.08 (m, 2H, H-**6**″ and H-**10**″); 4.65 (d, J = 6.6 Hz, 2H, H-**1**″); 3.03 (d, J = 7.0 Hz, 1H, H-**7**′); 2.09 (m, 4H, H-**4**″, H-**5**″, H-**8**″ and H-**9**″); 1.79 (s, 3H, H-**12**″); 1.75 (s, 3H, H-10′); 1.67 (s, 6H, H-11′ and H-**15**″); 1.62 (s, 6H, H-**13**″ and H-**14**″). ^{13}C NMR (100 MHz, CDCl$_3$): δ 192.1 (C-9); 163.2 (C-**2**′); 162.9 (C-**4**′); 144.0 (C-7); 141.3 (C-**3**″);135.3 (C-**7**″); 134.9 (C-1); 131.5 (C-9′ and C-**11**″); 131.3 (C-**6**′); 130.5 (C-4); 129.0 (C-2 and C-6); 128.5 (C-3 and C-5); 124.3 (C-**6**″); 123.6 (C-**10**″); 122.1 (C-**8**′); 120.7 (C-**5**′); 119.3 (C-**2**″) 116.0 (C-8); 114.4 (C-**1**′); 103.3 (C-**3**′); 65.1 (C-**1**″); 39.6 (C-**4**″); 39.4 (C-**8**″); 26.7 (C-**7**′); 26.2 (C-**9**″); 25.8 (C-**5**″); 25.7 (C-**12**″); 21.9 (C-**10**′); 17.9 (C-**15**″); 17.7 (C-**11**′); 16.7 (C-**13**″); 16.0 (C-**14**″). HRMS: [M + H]$^+$ ion m/z 513.3376 (calcd for $C_{35}H_{44}O_3$, 513.3369).

3.4. Determination of Anti-Oomycete Activity against S. parasitica and S. australis

3.4.1. Oomycete Isolate and Culture Conditions

The strains used in the study were obtained from infected Atlantic salmon captured in Puerto Montt, Chile. They were isolated using previously reported methods [34]. The strains used were ATCC 42030 *S. parasitica* and 38487 *S. australis*. Hyphal growth cultures were performed in petri dishes in yeast dextrose (DY) agar medium and a mixture of antibiotics (oxytetracycline 80%, flumequine 100% and florfenicol 80%) to eliminate the accompanying bacterial biota. The cultures were purified by successive transfers (at least 5 times) using pieces of agar (10 × 10 mm) obtained from the periphery of the growing colony. Plates were then incubated at 18 °C and sub-cultured every 3 days. A reference population was established on a DY gradient and maintained at 4 °C.

3.4.2. Determination of Minimum Inhibitory Concentration (MIC) and Spore's Germination Inhibition Test

The anti-oomycete activities of all tested compounds were evaluated by dilution test at final concentrations of 3.125, 6.25, 12.5, 25.0, 75.0, 100.0, 125.0, 150.0 and 200.0 μg/L in Griffin sporulation medium [35]. Bronopol, fluconazole and safrole were used as a positive control, while a 1% EtOH/Tween 20 solution was considered as a negative control.

To perform the spore germination inhibition test, the isolated *Saprolegnia* sp. strains are cultured on potato dextrose agar (PDA) plates for 7 to 14 days. After this period, spores were collected from the sporulated colonies and suspended in sterile distilled water (SDW). The concentration of spores in the suspension is determined with a hemocytometer and adjusted to approximately 1×10^4 CFU/mL. Then, 10 μL of the spore suspension were placed on Petri dishes containing the required concentration of compounds to be evaluated in 10 mL PDA and incubated at 25 °C for 72 h. The minimum oomycete concentration (MOC) was determined as the lowest concentration at which the tested chemical compounds prevented visible growth or germination of spores.

3.4.3. Mycelial Growth Inhibition Test

The in vitro anti-oomycete activities of the evaluated compounds were performed based on the inhibition rate of mycelia growth [31]. The diameter of mycelial growth was measured after inoculation at 25 °C for 48 h. The growth inhibition rate was calculated as shown in Equation (1):

$$\%IR = 100 \times \frac{(x-y)}{(x-z)} \tag{1}$$

where *IR* is the growth inhibition rate; *x*, mycelial growth in control; and the growth of the mycelium in the sample; and *z*, the average diameter of rape seeds.

3.4.4. Measurement of Cell Membrane Lysis

Cell membrane lysis was evaluated according to Lunde [36], measuring the 260 nm absorbing materials released into the medium once the membrane was broken. Cells were cultured with shaking at 30 °C until early stationary phase, then washed twice and diluted to approximately 5×10^7 CFU/mL with cold MOPS buffer, pH 6.0. Cells were aliquoted into tubes and compounds were added from stock solutions at a 100-fold concentration. Cells will be incubated stationary at 30 °C and centrifuged at $8000 \times g$ for 5 min in microcentrifuge tubes. Lego supernatants were collected for analysis. The results presented are the means of the values of at least three independent tests.

3.5. 3D-QSAR
3.5.1. CoMFA Method

CoMFA studies were performed with Sybyl X software version 1.2 [37] installed in a Windows 10 environment on a PC with an AMD Ryzen 7 7700X 8-Core. To acquire the best conformers for each molecule, every compound was drawn in ChemDraw and subjected to a preliminary geometry optimization using MM2 molecular mechanics as is implemented in ChemBio3D software version 12.0. The mol2 structures were imported to Sybyl and MMFF94 charges were assigned to each atom. The minimized structures were superimposed by the atom-by-atom. In addition, a minimization was carried out based on the Powell method [38].

To derive the CoMFA descriptor fields, the aligned training set molecules were placed in a three-dimensional cubic lattice with a grid spacing of 2 Å in the x, y, and z directions such that the entire set was included in it. The CoMFA steric and electrostatic field energies were calculated using a sp^3 carbon probe atom with a van der Waals radius of 1.52 Å and a charge of +1.0. Cut-off values for both steric and electrostatic fields were set to 30.0 kcal/mol.

3.5.2. Internal Validation and Partial Least Squares (PLS) Analysis

PLS analysis was used to construct a linear correlation between the CoMFA descriptors (independent variables) and the activity values (dependent variables) [39]. To select the best model, the cross-validation analysis was performed using the leave-one-out (LOO) method (and sample distance PLS (SAMPLS)), which generates the square of the cross-validation coefficient (q^2) and the optimum number of components (N). The non-cross validation was performed with a column filter value of 2.0 to speed up the analysis and reduce the noise. The q^2 which is a measure of the internal quality of the models, was obtained according to the following Equation (2):

$$q^2 = 1 - \frac{\sum(y_i - y_{pred})^2}{\sum(y_i - \bar{y})^2} \qquad (2)$$

where y_i, $\bar{y}$, and y_{pred} are observed, mean, and predicted activity in the training set, respectively.

3.6. Statistical Analysis

The data was reported as mean of at least two independent assays. In the case of a parametric distribution, a One-way ANOVA test was used. If the distribution was not parametric, a Kruskal-Wallis ANOVA test was used. In both cases, the trust level was 95% using the program STATISTICA 7.0.

4. Conclusions

The structure–activity relationship of a series of natural dihydrochalcones and chalcones and their respective oxyalkylated derivatives was studied. In this study, the structural identification of compound **5**, obtained from the resinous exudate of *A. balsamica*, was reported for the first time. Six new derivatives **30–35** were synthesized from the natural compound. Compounds **1** to **35** showed good anti-oomycete activity against two strains of *Saprolegnia*, highlighting the natural compound **3** and the new synthetic compound **34**. Among the most active compounds, the SAR study indicates that the activity of compound **34** would be given by the contributions of steric and electrostatic fields, given by the two bulky groups in the 4′ (O-geranyl) and 5′ (C-prenyl) positions of ring A, unlike compound **3**, which only presents two hydroxyl groups in ring A. In conclusion, this family of compounds stands out as potential anti-*Saprolegnia* agents, which can be used as a real alternative to commercial products.

Supplementary Materials: The following supporting information can be downloaded at https://www.mdpi.com/article/10.3390/plants13141976/s1, S1. Spectroscopic data of natural dihydrochalcones and chalcones **1–5**. S2. Spectroscopic data and structures of known dihydrochalcones and chalcones derivatives **6–29**. S3. ^1H, ^{13}C NMR of new compounds **30–35**.

Author Contributions: A.M. supervised the whole study. V.S., M.M., F.V. and S.F. performed the isolation and synthesis of all compounds. I.M. conceived and designed the biologic experiments; I.M. performed the biologic experiments. E.M. and D.C.-G. conceived and designed the computational methodologies. A.M., E.M. and I.M. collaborated in the discussion and interpretation of the results. E.M. and A.M. wrote the manuscript. All authors have read and agreed to the published version of the manuscript.

Funding: This research was funded by FONDECYT Grant number 1230311.

Institutional Review Board Statement: Not applicable.

Informed Consent Statement: Not applicable.

Data Availability Statement: All data are available on the manuscript or in the Supplementary Material for the scientific community.

Acknowledgments: The authors thank the project Fondecyt Post Doctoral N°3230296 awarded by the National Research and Development Agency (ANID) Chile and the APC was funded by Universidad de Playa Ancha, Plan de Fortalecimiento Universidades Estatales-Ministerio de Educación, Convenio UPA 1999.

Conflicts of Interest: The authors declare no conflicts of interest.

References

1. Van West, P. *Saprolegnia parasitica*, an oomycete pathogen with a fishy appetite: New challenges for an old problem. *Mycologist* **2006**, *20*, 99–104. [CrossRef]
2. Barde, R.D. Clinical and pathological investigations in ulcer disease of *Cyrinus carpio* caused by Aeromonas hydrophila. *Int. J. Health Sci.* **2022**, *6*, 519–3526. [CrossRef]
3. Zaror, L.; Collado, L.; Bohle, H.; Landskron, E.; Montaña, J.; Avendaño, F. *Saprolegnia parasitica* in salmon and trout from southern Chile. *Arch. Med. Vet.* **2004**, *36*, 71–78. [CrossRef]
4. Pavić, D.; Grbin, D.; Hudina, S.; Zmrzljak, U.; Miljanović, A.; Košir, R.; Varga, F.; Ćurko, J.; Marčić, Z.; Bielen, A. Tracing the oomycete pathogen *Saprolegnia parasitica* in aquaculture and the environment. *Sci. Rep.* **2022**, *12*, 16646. [CrossRef] [PubMed]
5. Aly, S.M.; Elatta, M.A.; Nasr, A.; Fathi, M. Efficacy of garlic and cinnamon as an alternative to chemotherapeutic agents in controlling *Saprolegnia* infection in *Nile tilapia*. *Aquaculture and Fisheries. Sci. Rep.* **2023**, *9*, 18013. [CrossRef]
6. Korkea-Aho, T.; Wiklund, T.; Engblom, C.; Vainikka, A.; Viljamaa-Dirks, S. Detection and Quantification of the Oomycete *Saprolegnia parasitica* in Aquaculture Environments. *Microorganisms* **2022**, *10*, 2186. [CrossRef] [PubMed]
7. Rezinciuc, S.; Sandoval-Sierra, J.W.; Diéguez-Uribeondo, J. Molecular identification of a bronopol tolerant strain of *Saprolegnia australis* causing egg and fry mortality in farmed brown trout, *Salmo trutta*. *Fungal Biol.* **2014**, *118*, 591–600. [CrossRef] [PubMed]
8. Krugner-Higby, L.; Haak, D.; Johnson, P.T.J.; Shields, J.D.; Jones, W., III; Reece, K.S.; Meinke, T.; Gendron, A.; Rusak, J. Ulcerative disease outbreak in crayfish *Orconectes propinquus* linked to *Saprolegnia australis* in Big Muskellunge Lake, Wisconsin. *Dis. Aquat. Org.* **2010**, *91*, 57–66. [CrossRef] [PubMed]
9. Tedesco, P.; Fioravanti, M.L.; Galuppi, R. In vitro activity of chemicals and commercial products against *Saprolegnia parasitica* and *Saprolegnia delica* strains. *J. Fish. Dis.* **2019**, *42*, 237–248. [CrossRef]
10. Chanu, K.V.; Thakuria, D.; Pant, V.; Bisht, S.; Tandel, R.S. Development of multiplex PCR assay for species-specific detection and identification of *Saprolegnia parasitica*. *Biotechnol. Rep.* **2022**, *9*, 758. [CrossRef]
11. Hechenleitner, V.; Gardner, M.F.; Thomas, P.I.; Echeverria, C.; Escobar, B.; Brownless, P.; Martinez, C. *Plantas amenazadas del Centro-Sur de Chile. Distribución, Conservación y Propagación*, Primera Edición; Universidad Austral de Chile y Real Jardín Botánico de Edimburgo: Valdivia, Chile, 2005; pp. 42–43.
12. Burkart, A. Sinopsis del género sudamericano de Leguminosas Adesmia D.C. (Contribución al estudio del género Adesmia, VII). *Darwiniana* **1967**, *14*, 463–568. Available online: https://www.jstor.org/stable/23213858 (accessed on 2 May 2024).
13. Martínez, R. *Apuntes Sobre la Vegetación del Lago Cholila*. Publ. Técnica no 1; Universidad Nacional Del Nordeste—UNNE: Corrientes, Argentina, 1980; pp. 1–22.
14. Montes, A.L.; Peltz, L. Esencias de plantas aromáticas del Parque Nacional Nahuel Huapi y sus aledaños. 2. *Adesmia boronioides* Hooker, o paramela. *An. Soc. Cient. Argent.* **1963**, *175*, 91–101.
15. Muñoz, M.; Barrera, E.; Meza, I. *El uso medicinal y alimenticio de plantas nativas y naturalizadas en Chile*; Publicación Ocasional n°33; Museo Nacional de Historia Natural: Santiago, Chile, 1981; p. 54.
16. Montes, M.; Wilkomirsky, T. *Medicina Tradicional Chilena*; Editorial de la Universidad de Concepción: Concepción, Chile, 1985; pp. 104–105.
17. Campos, A.M.; Lissi, E.; Chavez, M.; Modak, B. Antioxidant activity in heterogeneous and homogeneous system of the resinous exudates from *Heliotropium stenophylum* and *H. sinuatum* and of 3-O-methylgalangin their main component. *Bol. Latinoam. Caribe Plant. Med. Aromat.* **2012**, *11*, 549–555. Available online: https://www.redalyc.org/articulo.oa?id=85624607007 (accessed on 2 May 2024).
18. Midiwo, J.; Omoto, F.; Yenesew, A.; Akala, H.; Wangui, J.; Liyala, P.; Wasunna, C.; Waters, N. The first 9-hydroxyhomoisoflavanone, and antiplasmodial chalcones, from the aerial exudates of *Polygonum senegalense*. *Arkivoc* **2007**, *9*, 21–27. Available online: https://www.arkat-usa.org/get-file/23074/ (accessed on 3 May 2024). [CrossRef]
19. Elkanzi, N.A.A.; Hrichi, H.; Alolayan, R.A.; Derafa, W.; Zahou, F.M.; Bakr, R.B. Synthesis of Chalcones Derivatives and Their Biological Activities: A Review. *ACS Omega* **2022**, *7*, 27769–27786. [CrossRef] [PubMed]
20. Qin, H.L.; Zhang, Z.W.; Lekkala, R.; Alsulami, H.; Rakesh, K. Chalcone hybrids as privileged scaffolds in antimalarial drug discovery: A key review. *Eur. J. Med. Chem.* **2020**, *193*, 112215. [CrossRef]

21. Rashid, H.; Xu, Y.; Ahmad, N.; Muhammad, Y.; Wang, L. Promising anti-inflammatory effects of chalcones via inhibition of cyclooxygenase, prostaglandin E2, inducible NO synthase and nuclear factor κb activities. *Bioorganic Chem.* **2019**, *87*, 335–365. [CrossRef]
22. Lakshminarayanan, B.; Kannappan, N.; Subburaju, T. Synthesis and biological evaluation of novel chalcones with methanesulfonyl end as potent analgesic and anti-inflammatory agents. *Int. J. Pharm. Res. Biosci.* **2020**, *11*, 4974–4981. [CrossRef]
23. Duran, N.; Polat, M.F.; Aktas, D.A.; Alagoz, M.A.; Ay, E.; Cimen, F.; Tek, E.; Anil, B.; Burmaoglu, S.; Algul, O. New chalcone derivatives as effective against SARS-CoV-2 agent. *Int. J. Clin. Pract.* **2021**, *75*, 14846. [CrossRef]
24. Bhoj, P.; Togre, N.; Bahekar, S.; Goswami, K.; Chandak, H.; Patil, M. Immunomodulatory Activity of Sulfonamide Chalcone Compounds in Mice Infected with Filarial Parasite *Brugia malayi*. *Indian. J. Clin. Biochem.* **2019**, *34*, 225–229. [CrossRef]
25. Flores, S.; Montenegro, I.; Villena, J.; Cuellar, M.; Werner, E.; Godoy, P.; Madrid, A. Synthesis and Evaluation of Novel Oxyalkylated Derivatives of 2′,4′-Dihydroxychalcone as Anti-Oomycete Agents against Bronopol Resistant Strains of *Saprolegnia* sp. *Int. J. Mol. Sci.* **2016**, *17*, 1366. [CrossRef] [PubMed]
26. Montenegro, I.; Muñoz, O.; Villena, J.; Werner, E.; Mellado, M.; Ramírez, I.; Caro, N.; Flores, S.; Madrid, A. Structure-Activity Relationship of Dialkoxychalcones to Combat Fish Pathogen *Saprolegnia australis*. *Molecules* **2018**, *23*, 1377. [CrossRef] [PubMed]
27. Żołnierczyk, A.K.; Baczyńska, D.; Potaniec, B.; Kozłowska, J.; Grabarczyk, M.; Woźniak, E.; Anioł, M. Antiproliferative and antioxidant activity of xanthohumol acyl derivatives. *Med. Chem. Res.* **2017**, *26*, 1764–1771. [CrossRef]
28. Villena, J.; Montenegro, I.; Said, B.; Werner, E.; Flores, S.; Madrid, A. Ultrasound assisted synthesis and cytotoxicity evaluation of known 2′,4′-dihydroxychalcone derivatives against cancer cell lines. *Food Chem. Toxicol.* **2021**, *148*, 111969. [CrossRef] [PubMed]
29. Montenegro, I.; Madrid, A. Synthesis of dihydroisorcordoin derivatives and their in vitro anti-oomycete activities. *Nat. Prod. Res.* **2019**, *33*, 1214–1217. [CrossRef] [PubMed]
30. Werner, E.; Montenegro, I.; Said, B.; Godoy, P.; Besoain, X.; Caro, N.; Madrid, A. Synthesis and Anti-*Saprolegnia* Activity of New 2′,4′-Dihydroxydihydrochalcone Derivatives. *Antibiotics* **2020**, *9*, 317. [CrossRef]
31. Escobar, B.; Montenegro, I.; Villena, J.; Werner, E.; Godoy, P.; Olguín, Y.; Madrid, A. Hemi-synthesis, and anti-oomycete activity of analogues of isocordoin. *Molecules* **2017**, *22*, 968. [CrossRef] [PubMed]
32. Madrid Villegas, A.; Espinoza Catalán, L.; Montenegro Venegas, I.; Villena García, J.; Carrasco Altamirano, H. New Catechol Derivatives of Safrole and Their Antiproliferative Activity towards Breast Cancer Cells. *Molecules* **2011**, *16*, 4632–4641. [CrossRef] [PubMed]
33. da Silva, E.M.B.M.; Ruiz, A.L.T.G.; de Carvalho, J.E.; Pomini, A.M.; Pastorini, L.H.; Oliveira Santin, S.M. Antiproliferative activity and chemical constituents of *Lonchocarpus cultratus* (Fabaceae). *Nat. Prod. Res.* **2019**, *35*, 2056–2059. [CrossRef]
34. Willoughby, L.G.; Roberts, R.J. Improved methodology for isolation of Aphanomyces fungal pathogen of epizootic ulcerative syndrome (EUS) in Asian fish. *J. Fish. Dis.* **1994**, *17*, 541–543. [CrossRef]
35. Griffin, D.H. Achlya bisexualis. In *Lower Fungi in the Laboratory*; Fuller, M.S., Ed.; Southeastern Publishing Corporation: Athens, GA, USA, 1978; pp. 67–68.
36. Lunde, C.; Kubo, I. Effect of Polygodial on the Mitochondrial ATPase of Saccharomyces cerevisiae. *Antimicrob. Agents Chemother.* **2000**, *44*, 1943–1953. [CrossRef] [PubMed]
37. Córdova-Sintjago, T.; Villa, N.; Fang, L.; Booth, R.G. Aromatic interactions impact ligand binding and function at serotonin 5-HT2C G protein-coupled receptors: Receptor homology modeling, ligand docking, and molecular dynamics results validated by experimental studies. *Mol. Phys.* **2014**, *112*, 398–407. [CrossRef] [PubMed]
38. Powell, M.J.D. An efficient method for finding the minimum of a function of several variables without calculating derivatives. *Comput. J.* **1964**, *7*, 155–162. [CrossRef]
39. Clark, M.; Cramer, R.D.; van Opdenbosch, N. Validation of the general purpose Tripos 5.2 force field. *J. Comput. Chem.* **1989**, *10*, 982–1012. [CrossRef]

Disclaimer/Publisher's Note: The statements, opinions and data contained in all publications are solely those of the individual author(s) and contributor(s) and not of MDPI and/or the editor(s). MDPI and/or the editor(s) disclaim responsibility for any injury to people or property resulting from any ideas, methods, instructions or products referred to in the content.

Article

Cosmetic Preservative Potential and Chemical Composition of *Lafoensia replicata* Pohl. Leaves

Débora Machado de Lima [1], Anna Lívia Oliveira Santos [2], Matheus Reis Santos de Melo [3], Denise Crispim Tavares [3], Carlos Henrique Gomes Martins [2] and Raquel Maria Ferreira Sousa [1,*]

[1] Chemistry Instituto, Federal University of Uberlândia, Av. João Naves de Ávila 2121, Uberlândia 38400-902, MG, Brazil
[2] Institute of Biomedical Sciences, Federal University of Uberlândia, Av. João Naves de Ávila 2121, Uberlândia 38400-902, MG, Brazil
[3] University of Franca, Av. Dr. Armando Salles Oliveira, 201, Franca 14404-600, SP, Brazil
* Correspondence: rsousa@ufu.br

Abstract: The study evaluated the preservative potential of *Lafoensia replicata* Pohl. leaf extracts in cosmetics, highlighting their antioxidant, antimicrobial, and in vitro cytotoxic activities for ethanolic extract prepared by the maceration and tincture method. Total phenol content showed a higher phenol concentration in ethanolic extract and tinctures, and by LC-MS/MS-ESI-QTOF analysis, flavonoids, hydrolyzed tannins, and phenolic acids were identified. The ethanolic extract and tincture showed high antimicrobial activity against *Staphylococcus aureus*, *Pseudomonas aeruginosa*, and *Candida albicans* (MIC < 50 µg mL^{-1}), high antioxidant activity (EC$_{50}$ < 50 µg mL^{-1} in the DPPH method, and results > 450 µmol trolox equivalent in the ABTS and FRAP method), and low cytotoxicity in human keratinocytes (IC$_{50}$ > 350 µg mL^{-1}). The results suggest these extracts could be an alternative to synthetic preservatives in the cosmetic industry.

Keywords: antioxidant; antimicrobial; cytotoxic; cosmetics; Lytraceae

1. Introduction

The use of cosmetics (e.g., soaps, shampoos, skin creams, etc.) in people's daily hygiene routine is widespread. The market for these products is constantly growing and is particularly important for the economy [1,2]. With the rise of social media, an increasing number of consumers search the internet for health and well-being information. Easy access to receipts has prompted some individuals to create cosmetics using natural ingredients. This process, commonly called DIY (do it yourself), may lead the general population to believe that these homemade cosmetics are safer than commercially produced ones, mainly when natural products are used [3].

Generally, cosmetics are susceptible to degradation, such as changes in the organoleptic characteristics (color, texture, and smell), fungal proliferation, pH variation, and viscosity. To prevent some of these unpleasant effects, chemical preservatives are added to inhibit the growth of microorganisms during the cosmetic's shelf life. Esters of 4-hydroxybenzoic acid—known as methyl, ethyl, propyl, and butylparaben—are the most widely used preservatives in the cosmetics industry. They are chosen due to their ability to act against a broad spectrum of microorganisms (i.e., Gram-positive and Gram-negative bacteria and fungi). Moreover, they are easily soluble and have good sensitivity to pH variations. However, adverse effects, such as sensitization through contact and their ability to interfere with the endocrine system [4,5], might occur when parabens are employed. In addition, due to the rise of pro-ecological trends and the search for a sustainable lifestyle, cosmetics producers are increasingly looking for alternatives to replace synthetic preservatives with compounds of natural origin.

Natural products have been used for medicinal purposes and skin care since ancient civilizations. Plant extracts are used due to their antioxidant capacity (natural preservative), pigmentation, and inhibition of microbial activity, which can also be beneficial in preventing various diseases [6]. Furthermore, several companies use plant extracts such as *Aloe vera*, *Persea americana*, *Bambusoideae*, and *Matricaria chamomilla* to formulate moisturizers, shampoos, soaps, conditioners, and hair care products. The compounds in these extracts need to present significant antimicrobial and antioxidant activities to act as preservatives.

Cerrado is one of South America's largest biomes, occupying around 23% of the Brazilian territory. It holds 12,829 native species and plays an important social role, as many populations survive on its natural resources and have traditional knowledge of its biodiversity [7]. The genus Lafoensia is among the species found in Cerrado, and some metabolites have already been isolated. However, despite its wide variety of species, few studies are related to the genus. Most studies are about *L. pacari*, used in folk medicine to treat gastric ulcers, scarring, tonics, back pain, and cancer, and with some compounds [8–10].

Another species that belongs to this genus is *L. replicata* Pohl., which is often confused with *L. pacari*, as they are visually very similar, differing in the crests less pronounced in *L. replicata*. There are no studies about the chemical composition or biological activity of *L. replicata* Pohl. Ethnopharmacological studies with the local population from eastern Maranhão (northeast region of Brazil) show the use of *L. replicata* in liver disease, inflammation, healing, curing, or relieving kidney problems, gastritis, high blood pressure, headaches, and stomach aches [11].

Previous studies have demonstrated that *L. replicata* contains tannins and a high concentration of total phenols, which may suggest potent antioxidant, anti-inflammatory, and healing properties [12,13]. As no previous study has been found for this plant, and with the growing demand for natural products to replace parabens, *Lafoensia replicata* emerges as a promising alternative for use in cosmetic formulations, where these activities are fundamental for effective preservative action.

This study evaluated the preservative potential of *L. replicata* leaf extract prepared by maceration and tincture through cytotoxicity, antioxidant, and antimicrobial activity tests. In addition, the chemical constituents of the extracts were obtained using mass spectrometry.

2. Material and Methods

2.1. Plant Material

Leaves from *L. replicata* were collected on the Brazilian Federal Road BR-497 (18°59'20.3" S 48°25'14.5" W) in June 2021. Dr. Taciana B. Cavalcanti, from the Brazilian Agricultural Research Corporation (EMBRAPA), confirmed the plant's identification. The voucher specimen was deposited in the Herbarium of the Federal University of Uberlândia (HUFU 82057) and the CEN Herbarium of the Embrapa Genetic Resources and Biotechnology (CEN 121433). This study was registered in the National System for the Management of Genetic Heritage and Associated Traditional Knowledge (SisGen) to access the plant material (AFB587D).

2.2. Extract Preparation

2.2.1. Ethanolic Extracts

The leaves were dried in an incubator (BOD Nova Ética, São Paulo, Brazil, model 411/FPD 155 L) with air circulation at 35 °C. The dried powdered leaves were initially extracted with *n*-hexane by maceration at room temperature for 48 h. The *n*-hexane extract (HE) was filtered, and a rotary evaporator removed the remaining solvent at 40 °C under reduced pressure. The same procedure was performed sequentially with 98% ethanol, obtaining the ethanolic extract (EE). The dry extract was stored in a glass flask at −5 °C.

2.2.2. Tincture

The tincture of *L. replicata* leaves was obtained through the maceration process [14] with 98% ethanol. The dried and crushed plant material was placed in contact with a solvent for 30 days at room temperature and protected from light. Two types of extracts were prepared. One was filtered and removed using a rotary evaporator at reduced pressure at 40 °C to obtain a dried tincture (DT). The other one was filtered and stored at room temperature in an amber glass bottle to get the commercial tincture (CT).

2.3. Total Phenolic Content

The total phenolic content was determined according to a methodology described by Quaresma et al. [15]. In a conical glass test tube was added 0.5 mL of a solution of the extracts/tinctures (250 µg mL^{-1}, methanol), 2.5 mL of the Folin–Ciocalteu solution (10% m v^{-1}, water) and 2.0 mL of sodium carbonate solution (7.5% m v^{-1}, water). The reaction was kept at 50 °C for 5 min, and the absorbance was recorded at 760 nm using a UV–Vis spectrometer (Thermo Fisher Scientific, Waltham, MA, USA, model Genesys 10S UV–Vis). The obtained results were expressed in mg gallic acid equivalent (GAE) per gram of extracts/tinctures using a calibration curve of gallic acid. All the analyses were carried out in triplicate.

2.4. Antioxidant Activity

2.4.1. DPPH Radical Assay

The DPPH procedure was performed according to the method described by Quaresma et al. [15]. In a conical glass test tube protected from light was added 0.2 mL of the extracts/tinctures prepared in different concentrations (HE—50 to 333.3 µg mL^{-1}, EE and DT—0.08 to 8.30 µg mL^{-1}, and BHT—0.14 to 1.70 µg mL^{-1}) and 2.8 mL of the 2,2-diphenyl-1-picrylhydrazyl (DPPH) reagent (140 mg mL^{-1}). The mixture was allowed to rest for 1 h at room temperature, and the absorbance was measured at 517 nm (UV–Vis spectrometer, Thermo Scientific model Genesys 10S UV–Vis). Each experiment was performed in triplicate. The CE$_{50}$ values were determined by the equation:

$$AA = DPPH_{sequestered}(\%) = \left(\frac{Abs_{control}(Abs_{sample} - Abs_{blank})}{Abs_{control}} \right) 100 \qquad (1)$$

where Abs$_{control}$ is the absorbance of the methanolic solution of the radical DPPH, Abs$_{sample}$ is the absorbance of the mixture (DPPH + sample), and Abs$_{blank}$ is the absorbance of the sample in methanol.

2.4.2. Ferric Reducing Antioxidant Power Assay (FRAP)

The FRAP analysis was adapted as described by Malta and Liu [16]. Solutions of 10 mM of the 2,4,6-tris(2-pyridyl)-s-triazine (TPTZ) dissolved in 40 mM HCl, 20 mM of ferric chloride dissolved in water, 0.3 M of the acetate buffer (pH 3.6), 1600 µM of Trolox in methanol (and standard curve between 25 and 480 µM), and 50 µM of extracts/tinctures in methanol were prepared separately. The FRAP reagent solution was prepared by mixing 100 mL of acetate buffer, 10 mL of TPTZ, and 10 mL of ferric chloride. The mixture was kept at 37 °C for 30 min. 0.15 mL of the Trolox/extracts/tinctures and 2.85 mL of the FRAP solution (previously prepared) were added in a conical glass test tube protected from light. The mixture was kept for 10 min, protected from light, and then the absorbance was measured at 593 nm (UV–vis spectrometer, Thermo Fisher Scientific model Genesys 10S UV–Vis). The results were expressed in µmol Trolox equivalent (TE) per gram of extracts/tinctures using a calibration curve of Trolox.

2.4.3. ABTS

The ABTS evaluation was adapted as described in Malta and Liu [16]. Solutions of 7.4 mM ABTS, 2.6 mM of potassium persulfate, 25 to 600 mM of Trolox, and 50 µM of extracts/tinctures were prepared. The reaction mixture was obtained by mixing equal volumes of ABTS and potassium sulfate solutions and then kept in a dark environment for 12 h at room temperature. Subsequently, 1 mL of the reaction mixture was added to 60 mL of methanol to achieve an absorbance of 1.10 ± 0.02 A.U. at 734 nm. Fresh solutions were produced for each analysis. In a conical glass test tube protected from light, 0.15 mL of extract (HE, EE, and tincture) and 2.85 mL of the reaction mixture were added. The mixture was kept for 2 h, protected from light, and then the absorbance was measured at 734 nm (UV–Vis spectrometer, Thermo Fischer Scientific model Genesys 10S UV–Vis). The results were expressed in µmol of Trolox per gram of extract.

2.5. Determination of the Minimum Inhibitory Concentration Using the Microdilution Method

The antimicrobial activity was determined using the broth microdilution method according to the Clinical and Laboratory Standards Institute for Bacteria [17,18]. The microorganisms evaluated were from the American Type Culture Collection (ATCC): *Escherichia coli* (ATCC 8739), *Staphylococcus aureus* (ATCC 6538), *Pseudomonas aeruginosas* (ATCC 9027), and *Candida albicans* (ATCC 10231). The extracts/tinctures concentrations evaluated were 0.39 to 8000 µg mL^{-1} in dimethylsulfoxide (DMSO; Sigma-Aldrich, St. Louis, MO, USA; 5%, v/v). The positive control antibiotics analyzed were gentamicin for strains of *E. coli*, *S. aureus*, and *P. aeruginosa* at concentrations of 0.0115 to 5.9 µg mL^{-1} and amphotericin B for strains of *C. albicans* at concentrations of 0.031 to 16 µg mL^{-1}. For bacteria, the inoculum in Mueller-Honton broth was adjusted in a spectrophotometer to give a cell concentration of 5×10^5 colony-forming units per mL (CFU/mL) in the 96-well microplates.

For yeast, inoculum suspension was prepared in RPMI 1640 broth at 37 °C for 24 h and diluted to 1.2×10^3 UFC mL^{-1}. In the 96-well microplates, one inoculated well was included to control broth adequacy for microorganism growth, and one non-inoculated well free of antimicrobial agents was also used to ensure medium sterility. The microplates were incubated at 37 °C for 24 h. After that, 30 µL of resazurin (Sigma-Aldrich) at a concentration of 0.02% was added to each well. Resazurin is a redox probe that allows immediate observation of microbial growth. Blue and red represent the absence and presence of microbial growth, respectively [19]. The MIC is the lowest concentration that inhibits the growth of microorganisms.

2.6. Cytotoxicity Assessment

According to Riss et al. [20], cytotoxicity assessment was performed using the resazurin colorimetric assay. The human non-tumor keratinocyte cell line (HaCat) was used in the cytotoxic analysis. Cells were cultured in Dulbecco's modified Eagle medium (DMEM, Sigma-Aldrich) supplemented with 10% fetal bovine serum (Cultilab, Campinas, São Paulo, Brazil), antibiotics (0.01 mg mL^{-1} streptomycin and 0.005 mg mL^{-1} penicillin; Sigma-Aldrich), and 2.38 mg mL^{-1} Hepes (Sigma-Aldrich), at 36.5 °C with 5% CO_2. The extracts/tinctures were dissolved in DMSO (Sigma-Aldrich; 1%) at concentrations ranging from 39 to 5000 µg mL^{-1}. Therefore, 1×10^4 HaCat cells were seeded in a 96-well plate. Negative (DMSO 1%) and positive (DMSO 25%) control cultures were included. After 24 h of treatment, the culture medium was removed, and the cells were washed with PBS (phosphate-buffered saline solution) to remove the treatments and exposed to 80 µL of Ham's Nutrient Mixture F10 culture medium (HAM-F10) without phenol red (Sigma-Aldrich). Then, 20 µL of resazurin (0.15 mg mL^{-1}) was added to each well. The 96-well plate was incubated at 36.5 °C for 4 h. The absorbance of the samples was determined using a multiplate reader (Biochrom, Cambridge, UK, model ELISA—Asys—UVM 340/MikroWin 2000) at a wavelength of 570 nm and a reference length of 600 nm. The experiments were performed in triplicate. A non-linear regression analysis was performed using the GraphPad Prism program to calculate the sample concentration that inhibits 50% of cell viability (IC_{50}).

2.7. Analysis of the Extracts by Liquid Chromatography Coupled with Mass Spectrometry (LC-ESI-MS/MS)

The analyses were performed on the Maxis Impact mass spectrometer (Bruker) with an ESI-Q-TOF configuration and coupled to Nexera high-performance liquid chromatography (Shimadzu, Kyoto, Japan). The internal calibration of ESI-Q-TOF was performed with a solution of 100 µM sodium formate in water/acetonitrile (1:1) and data-dependent acquisition mode (DDA/AutoMS) with isolation/fragmentation of three precursors per cycle. The flow rate of the ESI-Q-TOF nebulizer gas was 8.0 L min^{-1} and all the data were acquired in negative modes. The extract solution was prepared at 2 mg mL^{-1} in methanol. The LC separation was performed with an Ascentis® C18 column (150 mm × 1 mm × 3 µm, Ascentis, Supelco). The mobile phase A has 0.1% formic acid, and the mobile phase B has methanol. 10 µL of sample was injected at a flow rate of 0.075 mL min^{-1}, and the separation was performed using the following gradient: 5–75% B (1 to 5 min); 75–100% B (5–15 min); 100% B (15–24 min); 100–5% B (24–25.2 min); 5% B (25.2–32 min).

2.8. Preparation of Topical Moisturizing Cream and Evaluation of the Preservative Activity of the L. replicata

Four topical moisturizing creams were prepared according to the Brazilian Resolution [21] in collaboration with D'brisse® company (Minas Gerais, Brazil). The cream base contains cetearyl alcohol, polysorbate 60, glycerin, PEG 100 stearate, rosehip, and water. The extracts/tinctures were dissolved in glycerin following their incorporation into the base cream until homogenized. The cream composition of the four formulations contained 5% (mass/mass) of glycerin and 0.4% (mass/mass) of EE, dried tincture, commercial tincture, or methylparaben.

2.9. Statistical Analysis

The Analysis of Variance (ANOVA) method was used to evaluate the results obtained in the analyses to determine the total phenol content and antioxidant activity. Those results with a significance level of less than 5% ($p < 0.05$) were considered statistically different. Tukey's test was used to determine significant differences between the means. The analyses were carried out using SigmaPlot 11.0.

3. Results and Discussion

3.1. Determination of Total Phenolic Content and Antioxidant Activity

The total phenolic content (TP) was determined using an analytical curve relating the absorbance generated by the reaction with the Folin–Ciocalteu reagent to the concentration of gallic acid. A linear equation was obtained (y = 0.0147x + 0.0554) with a coefficient of determination (r^2) of 0.9904. The TP content was expressed in milligrams of gallic acid equivalent per gram of extract (mg EAG g^{-1}). If a high value of phenolic compounds is found in an extract, it may partly explain its antioxidant activity. The results obtained for TP content and the antioxidant activities (DPPH, FRAP, and ABTS methods) are shown in Table 1.

Table 1. Total phenol content and antioxidant activities by DPPH, FRAP, and ABTS methods of the extracts from leaves of the *L. replicata*.

Sample	Total Phenol Content (mg GAE g^{-1})	Antioxidant Activity		
		DPPH CE$_{50}$ (µg mL^{-1})	FRAP (µmol TE g^{-1})	ABTS (µmol TE g^{-1})
HE	43.4 ± 1.8 [a]	>200	84.67 ± 3.47 [a]	111.3 ± 2.7 [a]
EE	253.7 ± 2.6 [b]	4.24 ± 0.16 [a]	473.79 ± 6.82 [b]	473.79 ± 46.3 [b]
DT	230.2 ± 2.1 [c]	3.50 ± 0.06 [b]	681.16 ± 5.23 [c]	479.7 ± 12.0 [c]

Note: HE: hexane extract; EE: ethanolic extract; DT: dried tincture. Analyses with the same letter showed no significant difference between the means by Tukey's test at 5% for the same test. A *p*-value < 0.01 was obtained for all correlations with different means.

TP contents in EE and DT were 230 and 253.7 mg GAE g^{-1}. Table 1 shows the variation in the reducing power of the extracts using the FRAP methodology and the variation in radical scavenging (DPPH and ABTS). Both extracts showed high values obtained by both analyzed methods, which were associated with the high total phenol content.

The DPPH analysis revealed that these samples showed CE_{50} value very close to the BHT (2.58 ± 0.01 µg mL^{-1}), a synthetic antioxidant used as a positive control. The tincture and ethanolic extract indicate promising results for the L. replicata plant.

Extraction with hexane revealed a low TP content in its composition. This difference can be explained by the difference in polarity between hexane and the polar properties of the ethanol used in the other extraction process. The hexane extract's low number of phenolic compounds results in low antioxidant activity.

It has been reported that phenolic compounds are responsible for antioxidant activity through several potential pathways. The main one is probably through free radical scavenging, in which phenolic molecules can break down the free radical chain reaction [22]. Phenolic acids, flavonoids, coumarins, and tannins are compounds found in the species that may be directly related to this antioxidant activity [23].

3.2. Antimicrobial Activity and Cytotoxicity Assessment

The Brazilian resolution [24] establishes parameters for the microbiological control of personal hygiene products, cosmetics, and perfumes. The microorganisms S. aureus, P. aeruginosa and fecal coliforms were evaluated because they are likely found in cosmetics. The minimum inhibitory concentration of each extract was assessed separately for each microorganism, considering its possible use as a preservative in creams. Table 2 shows the results of the antimicrobial activity.

Table 2. Minimum inhibitory concentrations (MIC) of the extracts from leaves of the L. replicata.

Microorganisms	Minimum Inhibitory Concentration (MIC)—µg mL^{-1}				
	Extracts			Antibiotics	
	HE	EE	DT	Gentamicin	Anthofericin B
S. aureus	>800	50	25	0.36	-
E. coli	>800	>800	>800	0.36	-
P. aeruginosas	>800	100	50	0.36	-
C. albicans	400	3.12	3.12	-	0.5

Note: HE: hexane extract; EE: ethanolic extract; DT: dried tincture.

All the ethanolic extracts and tinctures of the plant inhibited the growth of the evaluated microorganisms, except for E. coli, with MIC values between 25 and 100 µg mL^{-1}. The tincture was the most active against S. aureus and P. aerugionosas, with MICs of 25 µg mL^{-1} and 50 µg mL^{-1}, respectively. For the fungus C. albicans, all the extracts except the hexane extract presented a MIC < 100 µg mL^{-1}; moreover, low values (3.12 µg mL^{-1}) were obtained, showing good antifungal activity.

These results are relevant and promising since the ethanolic extract and tincture showed MIC values lower than 100 µg mL^{-1} for three of the microorganisms evaluated [25], as well as showing better results when compared to the L. pacari species (the ethanolic extracts of the leaves and stem showed a MIC of 312.5 and 625 µg mL^{-1} against S. aureus, respectively) [26]. Another study showed that the hydroalcoholic extract of L. pacari leaves had a MIC of 250 µg mL^{-1} for P. aeruginosa [27].

Cytotoxicity was measured using the non-tumor human keratinocyte cell line (HaCat), with the results expressed as the concentration that inhibits 50% of cell viability (IC_{50}). Table 3 shows the results obtained and the selectivity index calculated.

Table 3. Cytotoxicity assessment of the extracts from leaves of the L. replicata.

Samples	IC$_{50}$ (µg mL^{-1})	Selectivity Index			
		S. aureus	E. coli	P. aeruginosas	C. albicans
EE	397.23 ± 1.70	0.90	−0.30	0.60	2.10
DT	396.87 ± 20.00	1.20	−0.30	0.90	2.10

Note: EE: ethanolic extract; DT: dried tincture; IC$_{50}$: the concentration that inhibits 50% of the viability of human keratinocytes.

The selectivity index (SI) was determined to obtain a relationship between the cytotoxic concentration and the antimicrobial activity, using Equation (2):

$$\text{SI} = \log\left(\frac{IC50}{MIC}\right) \qquad (2)$$

The SI of the extracts showed that for bacteria with a MIC of less than 50 µg mL^{-1}, the extracts had low toxicity (SI > 0).

3.3. Analysis of the Effectiveness of the Preservative System of the Creams—"The Challenge Test"

Motivated by the promising results of total phenol content and antimicrobial and antioxidant activity for the ethanolic extract and tincture of L. replicata, the next step was to evaluate its use as a preservative in cosmetics. The Brazilian resolution [28] establishes the substances permitted for personal care products, cosmetics, and perfumes. Based on this resolution, a moisturizing cream was made with 0.4% methylparaben, one of the most widely used preservatives in the cosmetics industry.

The challenge test is used during a product's development to determine the preservative's effectiveness and stability over time. It is carried out by inoculating a known quantity of microorganisms (bacteria and fungi) protected from light and incubating for 28 days. The results are shown in Table 4.

Results in Table 4 show that all the creams presented a progressive reduction in the microbial load over time. From the seventh day onwards, there was a reduction in viable bacteria from the initial count, followed by a continuous decrease until the end of the test. This profile also occurred for the fungus C. albicans. For the cream produced with methylparaben, by the seventh day, there were no longer any microorganisms in the product. For the creams made with the tincture and ethanolic extract of the leaves of L. replicata, microorganisms were reduced substantially after seven days, and after 21 days, no more microorganisms were found.

Two creams were made using the tincture. One used 2% commercial tincture (CT), and the other used 0.4% dried tincture (DT). For the cream using 2% CT, 21 days were needed for complete inhibition of the microorganisms. In comparison, for the cream using 0.4% DT, there was no more proliferation of bacteria and fungi after 14 days. This can be explained by the fact that 2% CT in 350 g of cream corresponds to a concentration of approximately 0.3% DT.

For the cream to which no preservative system was added, there was no reduction in the number of microorganisms in the cream. Although the tincture and ethanolic extract of the plant's leaves needed more days to inhibit microbial growth when compared to methylparaben, both are within the legal limit for use as preservatives (total inhibition of microbial growth for up to 28 days) [17].

Therefore, the extracts obtained from this species should be considered a promising natural source for other activities.

Table 4. Challenge test for the creams without any preservative (control) and with methylparaben and extracts from leaves of L. replicata to evaluate stability for 0, 7, 14, 21, and 28 days.

Preservatives	Microorganisms	0 Days log/CFU	7 Days log/CFU	14 Days log/CFU	21 Days log/CFU	28 Days log/CFU
-	A. brasiliensis	5.70	5.65	5.63	5.69	5.72
	C. albicans	6.91	6.74	6.74	6.88	6.84
	E. coli	7.20	7.11	7.31	7.20	7.12
	P. aeruginosa	6.18	6.14	6.14	6.21	6.26
	S. aureus	6.26	6.35	6.35	6.25	6.43
Methylparaben	A. brasiliensis	5.00	<1.00	<1.00	<1.00	<1.00
	C. albicans	6.08	<1.00	<1.00	<1.00	<1.00
	E. coli	6.14	<1.00	<1.00	<1.00	<1.00
	P. aeruginosa	6.15	<1.00	<1.00	<1.00	<1.00
	S. aureus	6.36	<1.00	<1.00	<1.00	<1.00
CT	A. brasiliensis	5.86	1.24	1.24	<1.00	<1.00
	C. albicans	6.43	1.70	1.70	<1.00	<1.00
	E. coli	6.60	3.21	2.30	<1.00	<1.00
	P. aeruginosa	6.49	3.18	3.18	<1.00	<1.00
	S. aureus	6.91	<1.00	<1.00	<1.00	<1.00
DT	A. brasiliensis	5.78	1.74	<1.00	<1.00	<1.00
	C. albicans	6.33	2.01	<1.00	<1.00	<1.00
	E. coli	6.46	2.12	<1.00	<1.00	<1.00
	P. aeruginosa	6.89	2.34	<1.00	<1.00	<1.00
	S. aureus	6.47	2.14	<1.00	<1.00	<1.00
EE	A. brasiliensis	5.63	<1.00	<1.00	<1.00	<1.00
	C. albicans	6.89	1.00	<1.00	<1.00	<1.00
	E. coli	6.89	2.31	<1.00	<1.00	<1.00
	P. aeruginosa	6.88	2.90	2.04	<1.00	<1.00
	S. aureus	6.71	1.30	<1.00	<1.00	<1.00

Note: CT: commercial tincture; DT: dried tincture; EE: ethanolic extract.

3.4. Chemical Composition

The ethanolic extract and tincture showed better antioxidant and antimicrobial activity when compared to the hexane extract. For this reason, high-performance liquid chromatography coupled with mass spectrometry (HPLC-MS) with an electrospray ionization (ESI) system was used to obtain the chemical composition of the ethanolic extract and tincture. The analyses were carried out in negative mode. By analyzing the TIC chromatograms in Figure 1 and comparing the values of the m/z peaks obtained in the mass spectrum with a database of m/z values, it was possible to identify the composition of the main substances, as shown in Table 5.

Table 5. Proposed identification of compounds in the extracts and tincture from leaves of the *L. replicata*.

	R_t (min)	Compound	Molecular Formula	m/z exp. [M − H]$^-$	m/z Calculated [M − H]$^-$	Error (ppm)	Fragments MS^2	Extracts	Refs.
1	1.967	Sorbitol	$C_6H_{13}O_6^-$	181.0724	181.0718	3.31	163	EE; DT	[29,30]
2	1.967	HHDP—glucose	$C_{20}H_{17}O_{14}^-$	481.0635	481.0622	2.70	421, 301, 275	EE; DT	[31,32]
3	2.094	Pedunculagin isomer I (di-HHDP—glucose)	$C_{34}H_{23}O_{22}^-$	783.0685	783.0686	−0.12	481, 301, 291, 275, 249, 145	EE; DT	[31,32]
4 *	2.097	Pedunculagin isomer I (di-HHDP—glucose)	$C_{34}H_{23}O_{22}^-$	391.0303	391.0307	−1.02	301, 291, 275, 145	DT	[31,32]
5	2.178	Sucrose + formic acid	$C_{12}H_{21}O_{11}^-$	387.1166	387.1144	5.68	341, 179, 161, 143	EE; DT	[30]
6	2.220	Punicalin	$C_{34}H_{21}O_{22}^-$	781.0530	781.0530	0.00	765, 721, 601, 575, 481, 393, 301, 299, 273	EE; DT	[31,32]
7	2.473	HHDP galloyl glucose	$C_{27}H_{21}O_{18}^-$	633.0733	633.0733	0.00	463, 275, 301, 249, 169	EE; DT	[31,32]
8 *	2.490	Punicalagin isomers	$C_{48}H_{27}O_{30}^-$	541.0269	541.0260	1.66	601, 531, 402, 301, 124	EE; DT	[32]
9	2.515	Galloyl glucose	$C_{13}H_{15}O_{10}^-$	331.0686	331.0671	4.53	304, 170, 169, 139, 125	EE; DT	[31]
10	2.541	Trisgaloyl—HHDP glucose	$C_{41}H_{27}O_{27}^-$	951.0784	951.0740	4.62	907, 605, 425, 341, 301, 275	EE	[31]
11	2.566	Galloyl punicalin	$C_{41}H_{25}O_{26}^-$	933.0627	933.0640	−1.39	631, 451, 425, 301	EE; DT	[31]
12	2.617	Gallic acid	$C_7H_5O_5^-$	169.0139	169.0142	−1.77	125	EE; DT	[31]
13	2.687	Flavogalonic acid	$C_{21}H_9O_{13}^-$	469.0051	469.0049	0.43	470, 425, 407, 299	DT	[31]
14	2.768	Terflavin A	$C_{48}H_{29}O_{30}^-$	1085.0754	1085.074	0.46	933, 783, 631, 601, 451, 301	EE; DT	[31]
15	2.937	Punicalagin isomers	$C_{48}H_{27}O_{30}^-$	1083.0578	1083.059	−1.38	781, 601, 451, 301	EE; DT	[31,32]
16	3.131	HHDP—galloyl glucose isomer II	$C_{27}H_{21}O_{18}^-$	633.0725	633.0733	−1.26	301, 275	EE; DT	[31,32]
17	4.251	HHDP—galloyl glucose isomer III	$C_{27}H_{21}O_{18}^-$	633.0686	633.0733	−7.42	301, 275	EE; DT	[31,32]
18	4.639	HHDP—galloyl glucose isomer IV	$C_{27}H_{21}O_{18}^-$	633.0736	633.0733	0.47	301, 275	EE; DT	[31,32]
19	9.385	Trigaloyl hexoside	$C_{27}H_{23}O_{18}^-$	635.0895	635.0890	0.79	483, 465, 313, 301, 169, 125	EE; DT	[31]

Table 5. Cont.

	R_t (min)	Compound	Molecular Formula	m/z exp. $[M-H]^-$	m/z Calculated $[M-H]^-$	Error (ppm)	Fragments MS^2	Extracts	Refs.
20	9.385	Pterocarinin C	$C_{41}H_{29}O_{26}^-$	937.0968	937.0953	1.60	785, 635, 465, 301, 275, 169, 125	DT	[33]
21	9.764	Tetragaloyl hexose	$C_{34}H_{27}O_{22}^-$	787.0977	787.0999	−2.79	635, 617, 465, 169	EE; DT	[34]
22	10.101	Ethyl gallate	$C_9H_9O_5^-$	197.0456	197.0455	0.51	169, 125	EE; DT	[31]
23	10.312	Pentagaloyl hexoside	$C_{41}H_{31}O_{26}^-$	939.1112	939.1109	0.32	787, 617, 465, 393, 241, 169	DT	[35]
24	10.481	Quercetin dihexoside	$C_{27}H_{29}O_{17}^-$	625.1412	625.1410	0.32	581, 579, 487, 463, 301, 300, 271, 169, 151	EE; DT	[36]
25	10.481	Isorhamnetin hexoside	$C_{22}H_{21}O_{12}^-$	477.1036	477.1038	−0.42	433, 314, 313, 301, 271, 169, 125	DT	[35,36]
26	10.565	Quercetin arabinoglycoside	$C_{26}H_{27}O_{16}^-$	595.1315	595.1305	1.68	300, 301, 271, 169	EE; DT	[37]
27	10.776	Quercetin galloyl hexoside	$C_{28}H_{23}O_{16}^-$	615.0982	615.0992	−1.62	463, 300, 301, 271	EE; DT	[38]
28	11.071	Quercetin hexoside	$C_{21}H_{19}O_{12}^-$	463.0882	463.0881	0.22	301, 300, 271	EE; DT	[38]
29	11.404	Galoyl quercetin	$C_{28}H_{23}O_{15}^-$	599.1040	599.1042	−0.33	463, 301, 285	EE; DT	[35,39]
30	11.362	Ellagic acid	$C_{14}H_5O_8^-$	300.9996	300.9984	3.99	284, 173, 145, 133	EE; DT	[32]
31	11.573	Hexoside kaempferol	$C_{21}H_{19}O_{11}^-$	447.0934	447.0933	0.22	285, 284, 255, 227	EE; DT	[32,35,40]
32	12.206	Quercetin	$C_{15}H_9O_7^-$	301.0370	301.0354	5.31	273, 169, 151, 134	EE; DT	[35]

Note: DT: dried tincture; EE: ethanolic extract/Rt: retention time; HHDP: hexahydroxydiphenol group, * $[M-2H]^{2-}$.

(a)

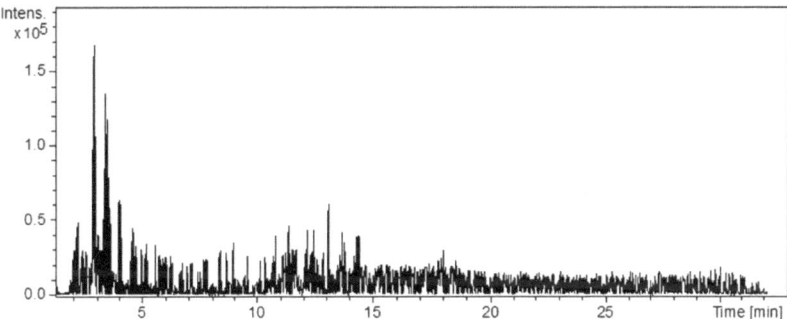

(b)

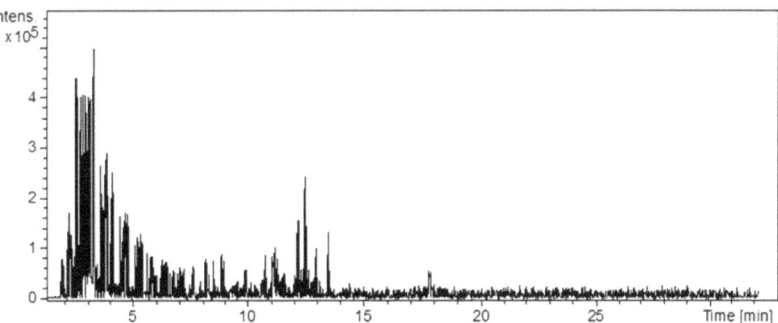

Figure 1. TIC chromatograms for (**a**) extracts and (**b**) tincture from leaves of the *L. replicata*.

EE and DT samples showed a predominance of tannins, flavonoids, and phenolic acids. Ellagitannins are hydrolyzable tannins that have attracted much attention because they have several beneficial properties for human health, including reducing the risk of diabetes, anticancer, and antioxidant activities [41].

Some studies have confirmed the biological activities of these compounds, such as antimicrobial, antioxidant, anti-inflammatory, and antimalarial activities [22,42]. Another compound belonging to the ellagitannin class that was also identified was pedunculagin (compound **3**), with a molecular ion at m/z 783, which produced a fragment ion at m/z 481, which corresponds to HHDP-glucose (compound **2**).

It is possible to note the many phenolic compounds identified by the Folin–Ciocalteu method, including acids, flavonoids, and tannins, which are possibly responsible for the extracts with high antioxidant and antimicrobial activity. Compounds **4, 13, 20, 23** and **25** were found only in the DT sample. This is probably due to the longer extraction time.

4. Conclusions

The ethanolic, hexane, and tincture extracts of *L. replicata* were evaluated for the first time regarding chemical composition and biological activities. It is possible to correlate the presence of these compounds with the high content of total phenols and the antioxidant and antimicrobial activities of the extracts. LC-ESI-MS/MS enabled the identification of tannins, flavonoids, and phenolic acids in the most active extracts.

By analyzing the content of total phenols, it was possible to verify that both the ethanolic extract and the tincture showed a large number of phenolic compounds, high antioxidant activity ($CE_{50} < 50$ µg mL^{-1}) for the DPPH method, values greater than 350 µmol ET g^{-1} for the FRAP and ABTS methods, and high antimicrobial activity (CIM < 50 µg mL^{-1}). On the other hand, the hexane extract showed a low PT content and, consequently, low antioxidant and antimicrobial activity.

When comparing the creams prepared with extracts from *L. replicata* leaves and the cream prepared with methylparaben, it was observed that the extracts have the same ability to be used as preservatives, as they were able to reduce viable bacteria and fungi by at least 99.9% of the initial count. This statement also correlates to the activities found in this study. One can then consider the extracts obtained from *L. replicata* as a promising natural source for the study of other activities as well as a possible ally for the production of cosmetics.

Supplementary Materials: The following supporting information can be downloaded at: https://www.mdpi.com/article/10.3390/plants13152011/s1, Figure S1: (−)-MS of compounds **1** and **2** (*m/z* 181.0724 e 481.0635 [M − H]$^-$) of Table 4. Figure S2: (−)-ESI-MSMS of compound **1** (*m/z* 181.0724 [M − H]$^-$) of Table 4. Figure S3: (−)-ESI-MSMS of compound **2** (*m/z* 481.0635 [M − H]$^-$) of Table 4. Figure S4: (−)-MS of compound **3** (mz 783.0685 [M − H]-) of Table 4. Figure S5: (−)-ESI-MSMS of compound **3** (*m/z* 783.0685 [M − H]$^-$) of Table 4. Figure S6: (−)-MS of compound **4** (*m/z* 391.0303 [M−2H]$^{2-}$) of Table 4. Figure S7: (−)-ESI-MSMS of compound **4** (*m/z* 391.0303 [M − 2H]$^{2-}$) of Table 4. Figure S8: (−)-MS of compound **5** (*m/z* 387.1166 [M − H]$^-$) of Table 4. Figure S9: (−)-ESI-MSMS of compound **5** (*m/z* 387.1166 [M − H]$^-$) of Table 4. Figure S10: (−)-MS of compound **6** (*m/z* 781.0530 [M − H]$^-$) of Table 4. Figure S11: (−)-ESI-MSMS of compound **6** (*m/z* 781.0530 [M − H]$^-$) of Table 4. Figure S12: (−)-MS of compound **7** (*m/z* 633.0733 [M − H]$^-$) of Table 4. Figure S13: (−)-ESI-MSMS of compound **7** (*m/z* 633.0733 [M − H]$^-$) of Table 4. Figure S14: (−)-MS of compound **8** (*m/z* 541.0249 [M − 2H]$^{2-}$) of Table 4. Figure S15: (−)-ESI-MSMS of compound **8** (*m/z* 541.0249 [M − 2H]$^{2-}$) of Table 4. Figure S16: (−)-MS of compound **9** (*m/z* 331.0686 [M − H]$^-$) of Table 4. Figure S17: (−)-ESI-MSMS of compound **9** (*m/z* 331.0686 [M − H]$^-$) of Table 4. Figure S18: (−)-MS of compound **10** (*m/z* 951.0784 [M − H]$^-$) of Table 4. Figure S19: (−)-ESI-MSMS of compound **10** (*m/z* 951.0784 [M − H]$^-$) of Table 4. Figure S20: (−)-MS of compound **11** (*m/z* 933.0627 [M − H]$^-$) of Table 4. Figure S21: (−)-ESI-MSMS of compound **11** (*m/z* 933.0627 [M − H]$^-$) of Table 4. Figure S22: (−)-MS of compound **12** (*m/z* 169.0139 [M − H]$^-$) of Table 4. Figure S23: (−)-ESI-MSMS of compound **12** (*m/z* 169.0139 [M − H]$^-$) of Table 4. Figure S24: (−)-MS of compound **13** (*m/z* 469.0051 [M − H]$^-$) of Table 4. Figure S25: (−)-ESI-MSMS of compound **13** (*m/z* 469.0051 [M − H]$^-$) of Table 4. Figure S26: (−)-MS of compound **14** (*m/z* 1085.0754 [M − H]$^-$) of Table 4. Figure S27: (−)-ESI-MSMS of compound **14** (*m/z* 1085,0754 [M − H]$^-$) of Table 4. Figure S28: (−)-MS of compound **15** (*m/z* 1083.0578 [M − H]$^-$) of Table 4. Figure S29: (−)-ESI-MSMS of compound **15** (*m/z* 1083.0578 [M − H]$^-$) of Table 4. Figure S30: (−)-MS of compound **16** (*m/z* 633.0725 [M − H]$^-$) of Table 4. Figure S31: (−)-ESI-MSMS of compound **16** (*m/z* 633.0725 [M − H]$^-$) of Table 4. Figure S32: (−)-MS of compound **17** (*m/z* 633.0686 [M − H]$^-$) of Table 4. Figure S33: (−)-ESI-MSMS of compound **17** (*m/z* 633.0686 [M − H]$^-$). Figure S34: (−)-MS of compound **18** (*m/z* 633.0736 [M − H]$^-$) of Table 4. Figure S35: (−)-ESI-MSMS of compound **18** (*m/z* 633.0736 [M − H]$^-$) of Table 4. Figure S36: (−)-MS of compounds **19** and **20** (*m/z* 635.0895 e *m/z* 937.0968 [M − H]$^-$) of Table 4. Figure S37: (−)-ESI-MSMS of compound **19** (*m/z* 635.0895 [M − H]$^-$) of Table 4. Figure S38: (−)-ESI-MSMS of compound **20** (*m/z* 937.0968 [M − H]$^-$) of Table 4. Figure S39: (−)-MS of compound **21** (*m/z* 787.0977 [M − H]$^-$) of Table 4. Figure S40: (−)-ESI-MSMS of compound **21** (*m/z* 787.0977 [M − H]$^-$) of Table 4. Figure S41: (−)-MS of compound **22** (*m/z* 197.0456 [M − H]$^-$) of Table 4. Figure S42: (−)-ESI-MSMS of compound **22** (*m/z* 197.0456 [M − H]$^-$) of Table 4. Figure S43: (−)-MS of compound **23** (*m/z* 939.1112 [M − H]$^-$) of Table 4. Figure S44: (−)-ESI-MSMS of compound **23** (*m/z* 939.1112 [M − H]$^-$) of Table 4. Figure S45: (−)-MS of compounds **24** and **25** (*m/z* 625.1412 e *m/z* 477.1036 [M − H]$^-$) of Table 4. Figure S46: (−)-ESI-MSMS of compound **24** (*m/z* 625.1412 [M − H]$^-$) of Table 4. Figure S47: (−)-ESI-MSMS of compound **25** (*m/z* 477.1036 [M − H]$^-$) of Table 4. Figure S48: (−)-MS of compound **26** (*m/z* 595.1315 [M − H]$^-$) of Table 4. Figure S49: (−)-ESI-MSMS of compound **26** (*m/z* 595.1315 [M − H]$^-$) of Table 4. Figure S50: (−)-MS of compound **27** (*m/z* 615.0982 [M − H]$^-$) of Table 4. Figure S51: (−)-ESI-MSMS of compound **27** (*m/z* 615.0982 [M − H]$^-$) of Table 4. Figure S52: (−)-MS of compound **28** (*m/z* 463.0881 [M − H]$^-$) of Table 4. Figure S53: (−)-ESI-MSMS of compound **28** (*m/z* 615.0982 [M − H]$^-$) of Table 4. Figure S54: (−)-MS of compound **29** (*m/z* 599.1040 [M − H]$^-$) of Table 4. Figure S55: (−)-ESI-MSMS of compound **29** (*m/z* 599.1040 [M − H]$^-$) of Table 4. Figure S56: (−)-MS of compound **30** (*m/z* 301.0010 [M − H]$^-$) of Table 4. Figure S57: (−)-ESI-MSMS of compound **30** (*m/z* 301.0010 [M − H]$^-$) of Table 4. Figure S58: (−)-MS of compound **31** (*m/z* 447.0934 [M − H]$^-$) of Table 4. Figure S59: (−)-ESI-MSMS of compound **31** (*m/z* 447.0934 [M − H]$^-$) of Table 4. Figure S60: (−)-MS of compound **32** (*m/z* 301.370 [M − H]$^-$) of Table 4. Figure S61: (−)-ESI-MSMS of compound **32** (*m/z* 301.0370 [M − H]$^-$) of Table 4.

Author Contributions: Conceptualization, R.M.F.S.; Methodology, D.M.d.L., R.M.F.S., A.L.O.S., M.R.S.d.M., D.C.T. and C.H.G.M.; Validation, D.M.d.L., A.L.O.S., M.R.S.d.M., D.C.T. and C.H.G.M.; Formal analysis, D.M.d.L., A.L.O.S. and M.R.S.d.M.; Investigation, D.M.d.L. and R.M.F.S.; Writing—original draft, D.M.d.L. and R.M.F.S.; Supervision, R.M.F.S., D.C.T. and C.H.G.M.; Project administration, R.M.F.S.; Funding acquisition, R.M.F.S. All authors have read and agreed to the published version of the manuscript.

Funding: This research was funded by the following Brazilian agencies: CNPq (312313/2022-5); CAPES (financial code 001); FAPEMIG (APQ-00919-22, APQ-02391-22, and CEX-APQ-04104-22).

Data Availability Statement: Data are contained within the article and Supplementary Materials.

Acknowledgments: The authors also thank the Biomolecule Mass Spectrometry Centre (CEMBIO) of the Federal University of Rio de Janeiro (UFRJ) for LC-ESI-MS/MS analysis.

Conflicts of Interest: The authors declare no conflict of interest.

References

1. Cosmetics Market Size, Share & Industry Analysis, by Category (Hair Care, Skin Care, Makeup, and Others), by Gender (Men and Women), by Distribution Channel (Specialty Stores, Hypermarkets/Supermarkets, Online Channels, and Others), and Regional Forecasts, 2024–2032. Available online: https://www.fortunebusinessinsights.com/cosmetics-market-102614 (accessed on 3 November 2022).
2. Lóreal. *Annual Growth of the Global Cosmetics Market from 2004 to 2021*; Statista Research Department: New York, NY, USA, 2022.
3. Couteau, C.; Girard, E.; Coiffard, L. Analysis of 275 DIY recipes for eye cosmetics and their possible safety issues. *Int. J. Cosmet. Sci.* **2022**, *44*, 403–413. [CrossRef] [PubMed]
4. Nowak, K.; Jabłońska, E.; Ratajczak-Wrona, W. *Controversy around Parabens: Alternative Strategies for Preservative Use in Cosmetics and Personal Care Products*; Elsevier Inc.: Amsterdam, The Netherlands, 2021; Volume 198, p. 110488.
5. Fransway, A.F.; Fransway, P.J.; Belsito, D.V.; Yiannias, J.A. Paraben Toxicology. *Dermatitis* **2019**, *30*, 32–45. [CrossRef] [PubMed]
6. Fowler, J.F., Jr.; Woolery-Lloyd, H.; Waldorf, H.; Saini, R. Innovations in natural ingredients and their use in skin care. *J. Drugs Dermatol.* **2010**, *9*, S72–S81; quiz s82-73. [PubMed]
7. Cerrado. Ministério do Meio Ambiente e Mudança Climática. Available online: https://www.gov.br/mma/pt-br/assuntos/ecossistemas-1/biomas/cerrado (accessed on 8 November 2022).
8. Solon, S.; Lopes, L.; Teixeira de Sousa, P., Jr.; Schmeda-Hirschmann, G. Free radical scavenging activity of Lafoensia pacari. *J. Ethnopharmacol.* **2000**, *72*, 173–178. [CrossRef]
9. Nunes, G.P.; Silva, M.F.d.; Resende, U.M.; Siqueira, J.M.d. Plantas medicinais comercializadas por raizeiros no Centro de Campo Grande, Mato Grosso do Sul. *Rev. Bras. Farmacogn.* **2003**, *13*, 83–92. [CrossRef]
10. Mundo, S.R.; Duarte, M.d.R. Morfoanatomia Foliar e Caulinar de Dedaleiro: *Lafoensia pacari* A. St.-Hil. (Lythraceae). *Lat. Am. J. Pharm.* **2007**, *26*, 522–529.
11. Sobrinho, F.C.B.; Almeida, A.L.S.; Monteiro, J.M. Estudo etnofarmacológico sobre *Lafoensia replicata* Pohl. no leste do Maranhão, Brasil: Uma promissora espécie para bioprospecção. *Desenvolv. Meio Ambiente* **2016**, *39*, 207–216. [CrossRef]
12. Vieira, I.d.S.; da Silva, S.d.S.F.; de Souza, J.S.N.; Monteiro, J.M. Relação entre parâmetros biométricos e teores de fenóis totais em *Lafoensia replicata* Pohl.—Um estudo de caso. *Sci. Plena* **2017**, *13*. [CrossRef]
13. Monteiro, J.; Souza, J.; Lins Neto, E.; Scopel, K.; Trindade, E. Does total tannin content explain the use value of spontaneous medicinal plants from the Brazilian semi-arid region? *Rev. Bras. Farmacogn.* **2014**, *24*, 116–123. [CrossRef]
14. Anvisa. *Formulário de Fitoterápicos da Farmacopéia Brasileira/Agência Nacional de Vigilância Sanitária*, 6th ed.; Anvisa: Brasilia, Brazil, 2019; Volume 1.
15. Quaresma, D.M.O.; Justino, A.B.; Sousa, R.M.F.; Munoz, R.A.A.; de Aquino, F.J.T.; Martins, M.M.; Goulart, L.R.; Pivatto, M.; Espindola, F.S.; de Oliveira, A. Antioxidant compounds from *Banisteriopsis argyrophylla* leaves as α-amylase, α-glucosidase, lipase, and glycation inhibitors. *Bioorg. Chem.* **2020**, *105*, 104335. [CrossRef]
16. Malta, L.G.; Liu, R.H. Analyses of Total Phenolics, Total Flavonoids, and Total Antioxidant Activities in Foods and Dietary Supplements. In *Encyclopedia of Agriculture and Food Systems*, 2nd ed.; Alfen, N.K.V., Ed.; Academic Press: Cambridge, UK, 2014; Volume 1, pp. 305–314.
17. CLSI. Methods for dilution antimicrobial susceptibility tests for bacteria that grow aerobically. In *CLSI Document M7-A9*, 9th ed.; Clinical and Laboratory Standards Institute: Wayne, PA, USA, 2012; Volume 32, p. 63.
18. CLSI. Methods for antimicrobial susceptibility testing of anaerobic bacteria. In *CLSI document M11-A8*, 8th ed.; Clinical and Laboratory Standards Institute: Wayne, PA, USA, 2012; Volume 32, p. 39.
19. Sarker, S.D.; Nahar, L.; Kumarasamy, Y. Microtitre plate-based antibacterial assay incorporating resazurin as an indicator of cell growth, and its application in the in vitro antibacterial screening of phytochemicals. *Methods* **2007**, *42*, 321–324. [CrossRef] [PubMed]

20. Riss, T.L.; Moravec, R.A.; Niles, A.L.; Duellman, S.; Benink, H.A.; Worzella, T.J.; Minor, L. Cell Viability Assays. In *Assay Guidance Manual [Internet]*; Markossian, S., Grossman, A., Brimacombe, K., Arkin, M., Auld, D., Austin, C., Baell, J., Chung, T.D.Y., Coussens, N.P., Dahlin, J.L., et al., Eds.; Eli Lilly & Company and the National Center for Advancing Translational Sciences: Bethesda, MD, USA, 2013. Available online: https://www.ncbi.nlm.nih.gov/books/NBK144065/ (accessed on 10 January 2024).
21. Anvisa. *Resolução da Diretoria Colegiada-Resolução RDC no. 14, de 14 de Março de 2013*; Agência Nacional de Vigilância Sanitária: Brasília, Brazil, 2013.
22. Lee, C.-J.; Chen, L.-G.; Liang, W.-L.; Wang, C.-C. Anti-inflammatory effects of *Punica granatum* Linne in vitro and in vivo. *Food Chem.* **2010**, *118*, 315–322. [CrossRef]
23. Recuenco, M.C.; Lacsamana, M.S.; Hurtada, W.A.; Sabularse, V.C. Total Phenolic and Total Flavonoid Contents of Selected Fruits in the Philippines. *Philipp. J. Sci.* **2016**, *145*, 275–281.
24. Anvisa. *Resolução da Diretoria Colegiada-Resolução RDC no. 630, de 10 de Março de 2022*; Agência Nacional de Vigilância Sanitária: Brasília, Brazil, 2022.
25. Ríos, J.L.; Recio, M.C. Medicinal plants and antimicrobial activity. *J. Ethnopharmacol.* **2005**, *100*, 80–84. [CrossRef] [PubMed]
26. Lima, M.R.F.d.; Ximenes, E.C.P.A.; Luna, J.S.; Sant'Ana, A.E.G. The antibiotic activity of some Brazilian medicinal plants. *Rev. Bras. Farmacogn.* **2006**, *16*, 300–306. [CrossRef]
27. Porfírio, Z.; Melo-Filho, G.C.; Alvino, V.; Lima, M.R.F.; Sant'Ana, A.E.G. Atividade antimicrobiana de extratos hidroalcoólicos de *Lafoensia pacari* A. St.-Hil., Lythraceae, frente a bactérias multirresistentes de origem hospitalar. *Rev. Bras. Farmacogn.* **2009**, *19*, 785–789. [CrossRef]
28. Anvisa. *Resolução da Diretoria Colegiada-Resolução RDC no. 528, de 4 de Agosto de 2021*; Agência Nacional de Vigilância Sanitária: Brasília, Brazil, 2021.
29. Albuquerque Nerys, L.L.D.; Jacob, I.T.T.; silva, P.A.; da Silva, A.R.; de Oliveira, A.M.; Rocha, W.R.V.D.; Pereira, D.T.M.; da Silva Abreu, A.; da Silva, R.M.F.; da Cruz Filho, I.J.; et al. Photoprotective, biological activities and chemical composition of the non-toxic hydroalcoholic extract of *Clarisia racemosa* with cosmetic and pharmaceutical applications. *Ind. Crops Prod.* **2022**, *180*, 114762. [CrossRef]
30. Jin, Y.; Lao, J.; Zhou, R.; He, W.; Qin, Y.; Zhong, C.; Xie, J.; Liu, H.; Wan, D.; Zhang, S.; et al. Simultaneous identification and dynamic analysis of saccharides during steam processing of rhizomes of *Polygonatum cyrtonema* by HPLC–QTOF–MS/MS. *Molecules* **2018**, *23*, 2855. [CrossRef]
31. Singh, A.; Bajpai, V.; Kumar, S.; Sharma, K.R.; Kumar, B. Profiling of gallic and ellagic acid derivatives in different plant parts of *Terminalia arjuna* by HPLC-ESI-QTOF-MS/MS. *Nat. Prod. Commun.* **2016**, *11*, 239–244. [CrossRef]
32. Mena, P.; Calani, L.; Dall'Asta, C.; Galaverna, G.; García-Viguera, C.; Bruni, R.; Crozier, A.; Del Rio, D. Rapid and comprehensive evaluation of (poly)phenolic compounds in pomegranate (*Punica granatum* L.) Juice by UHPLC-MSn. *Molecules* **2012**, *17*, 14821–14840. [CrossRef]
33. Tan, H.P.; Ling, S.K.; Chuah, C.H. Characterisation of galloylated cyanogenic glucosides and hydrolysable tannins from leaves of *Phyllagathis rotundifolia* by LC-ESI-MS/MS. *Phytochem. Anal.* **2011**, *22*, 516–525.
34. Abu-Reidah, I.M.; Ali-Shtayeh, M.S.; Jamous, R.M.; Arráez-Román, D.; Segura-Carretero, A. HPLC-DAD-ESI-MS/MS screening of bioactive components from *Rhus coriaria* L. (Sumac) fruits. *Food Chem.* **2015**, *166*, 179–191. [CrossRef] [PubMed]
35. Abu-Reidah, I.M.; Ali-Shtayeh, M.S.; Jamous, R.M.; Arráez-Román, D.; Segura-Carretero, A. Comprehensive metabolite profiling of *Arum palaestinum* (Araceae) leaves by using liquid chromatography-tandem mass spectrometry. *Food Res. Int.* **2015**, *70*, 74–86. [CrossRef]
36. Hamed, A.I.; Said, R.B.; Kontek, B.; Al-Ayed, A.S.; Kowalczyk, M.; Moldoch, J.; Stochmal, A.; Olas, B. LC-ESI-MS/MS profile of phenolic and glucosinolate compounds in samh flour (*Mesembryanthemum forsskalei* Hochst. ex Boiss) and the inhibition of oxidative stress by these compounds in human plasma. *Food Res. Int.* **2016**, *85*, 282–290. [CrossRef] [PubMed]
37. Correira, V.T.V.; D'Angelis, D.F.; Macedo, M.C.C.; Ramos, A.L.C.C.; Vieira, A.L.S.; Queiroz, V.A.V.; Augusti, R.; Ferreira, A.A.; Fante, C.A.; Melo, J.O.F. Chemical profile of extruded sorghum flour the genotype BRS 305 by paper spray. *Res. Soc. Dev.* **2021**, *10*, e40710111414.
38. Pereira, V.V.; da Fonseca, F.A.; Bento, C.S.O.; Oliveira, P.M.; Rocha, L.L.; Augusti, R.; Mendonça Filho, C.V.; Silva, R.R. Electrospray ionization mass spectrometry fingerprint of the Byrsonima species. *Rev. Virtual Quim.* **2015**, *7*, 2539–2548. [CrossRef]
39. Erşan, S.; Güçlü Üstündağ, Ö.; Carle, R.; Schweiggert, R.M. Identification of phenolic compounds in red and green pistachio (*Pistacia vera* L.) hulls (exo- and mesocarp) by HPLC-DAD-ESI-(HR)-MSn. *J. Agric. Food Chem.* **2016**, *64*, 5334–5344. [CrossRef] [PubMed]
40. Costa Silva, T.d.; Justino, A.B.; Prado, D.G.; Koch, G.A.; Martins, M.M.; Santos, P.d.S.; Morais, S.A.L.d.; Goulart, L.R.; Cunha, L.C.S.; Sousa, R.M.F.d.; et al. Chemical composition, antioxidant activity and inhibitory capacity of α-amylase, α-glucosidase, lipase and non-enzymatic glycation, in vitro, of the leaves of *Cassia bakeriana* Craib. *Ind. Crops Prod.* **2019**, *140*, 111641. [CrossRef]
41. Vivas, N.; Nonier, M.-F.; de Gaulejac, N.V.; de Boissel, I.P. Occurrence and partial characterization of polymeric ellagitannins in *Quercus petraea* Liebl. and *Q. robur* L. wood. *Comptes Rendus Chim.* **2004**, *7*, 945–954. [CrossRef]
42. Reddy, M.; Kasimsetty, S.; Jacob, M.; Khan, S.; Ferreira, D. Antioxidant, Antimalarial and Antimicrobial Activities of Tannin-Rich Fractions, Ellagitannins and Phenolic Acids from *Punica granatum* L. *Planta Med.* **2007**, *73*, 461–467. [CrossRef]

Disclaimer/Publisher's Note: The statements, opinions and data contained in all publications are solely those of the individual author(s) and contributor(s) and not of MDPI and/or the editor(s). MDPI and/or the editor(s) disclaim responsibility for any injury to people or property resulting from any ideas, methods, instructions or products referred to in the content.

Article

Chemical Variation of Leaves and Pseudobulbs in *Prosthechea karwinskii* (Orchidaceae) in Oaxaca, Mexico

Gabriela Soledad Barragán-Zarate, Beatriz Adriana Pérez-López, Manuel Cuéllar-Martínez, Rodolfo Solano * and Luicita Lagunez-Rivera *

Laboratorio de Extracción y Análisis de Productos Naturales Vegetales, Centro Interdisciplinario de Investigación para el Desarrollo Integral Regional Unidad Oaxaca, Instituto Politécnico Nacional, Hornos 1003, Santa Cruz Xoxocotlán C.P. 71230 Oaxaca, Mexico; gbarraganz@ipn.mx (G.S.B.-Z.); estudioorquideasdeoaxaca@gmail.com (B.A.P.L.); mcuellarm@ipn.mx (M.C.-M.)
* Correspondence: asolanog@ipn.mx (R.S.); llagunez@ipn.mx (L.L.-R.)

Academic Editors: Luis Ricardo Hernández, Eugenio Sánchez-Arreola and Edgar R. López-Mena

Received: 11 January 2025
Revised: 12 February 2025
Accepted: 13 February 2025
Published: 18 February 2025

Citation: Barragán-Zarate, G.S.; Pérez-López, B.A.; Cuéllar-Martínez, M.; Solano, R.; Lagunez-Rivera, L. Chemical Variation of Leaves and Pseudobulbs in *Prosthechea karwinskii* (Orchidaceae) in Oaxaca, Mexico. *Plants* 2025, 14, 608. https://doi.org/10.3390/plants14040608

Copyright: © 2025 by the authors. Licensee MDPI, Basel, Switzerland. This article is an open access article distributed under the terms and conditions of the Creative Commons Attribution (CC BY) license (https://creativecommons.org/licenses/by/4.0/).

Abstract: *Prosthechea karwinskii* is an endemic orchid of Mexico with significant value for its traditional uses: ornamental, ceremonial, and medicinal. The pharmacological activity of this plant has been studied using specimens recovered from religious use during Holy Week in Oaxaca, Mexico, sourced from various localities within this state. Geographical variability can influence the chemical composition of plants, as environmental factors affect the production of their secondary metabolites, which impact their biological properties. This research evaluated the variability in the chemical composition of leaves and pseudobulbs of *P. karwinskii* obtained from different localities in Oaxaca, comprising 95–790 g and 376–3900 g of fresh material for leaves and pseudobulbs, respectively, per locality. Compounds were identified using UHPLC-ESI-qTOF-MS/MS following ultrasound-assisted hydroethanolic extraction. Twenty-one compounds were identified in leaves and twenty in pseudobulb. The findings revealed differences in chemical composition across localities and between leaves and pseudobulbs of the species. The Roaguia locality exhibited the highest extraction yield and pharmacological potential in leaves. For pseudobulbs, Cieneguilla specimens showed the highest yield, and El Lazo had the lowest yield but the highest pharmacological potential. This study represents the first comprehensive analysis of the variation in the chemical composition of a native Mexican orchid. In all localities, leaves and pseudobulbs contained compounds with known biological activity, validating the use of the species in traditional medicine and highlighting its potential for medical and biological applications.

Keywords: bioactive compounds; medicinal plants; extraction yield; UHPLC-ESI-qTOF-MS/MS; geographical variation

1. Introduction

The orchid *Prosthechea karwinskii* (Mart.) J.M.H. Shaw is endemic to the mountains of southern Mexico and is considered one of the most striking species of Mexican flora due to its ornamental and cultural value. In Oaxaca, this orchid is traditionally used for religious purposes during Holy Week celebrations and has applications in traditional medicine. The leaves are used to reduce glucose levels in individuals with diabetes; the pseudobulbs are used for healing wounds and burns, as well as for treating coughs and diabetes; and the flowers are used to prevent miscarriage, help in labor, and relieve coughs [1].

Previous studies have identified compounds in the leaves, pseudobulbs, and flowers of *P. karwinskii* that are associated with its medicinal properties. The leaves are rich in phenols and flavonoids, which contribute to the orchid's antioxidant activity [2]. Extracts from leaves, pseudobulbs, and flowers have been shown to reduce body fat, glucose, total cholesterol, and triglyceride levels in Wistar rats with induced metabolic syndrome [3]. Furthermore, leaf extracts demonstrate antioxidant and anti-inflammatory properties, along with a gastroprotective effect against damage caused by non-steroidal anti-inflammatory drugs (NSAIDs) in Wistar rats [4]. These extracts also mitigate obesity, insulin resistance, inflammation, and cardiovascular risk in Wistar rats with induced metabolic syndrome [5]. Additionally, the extracts reduce oxidative stress by regulating the activity of antioxidant enzymes such as superoxide dismutase and catalase [6], offer cardiovascular protection [7], and have the potential for treating atherothrombosis [8].

The biological activity of *P. karwinskii* reported in these studies has been evaluated using extracts from specimens recovered after religious use in Zaachila, Oaxaca. These specimens were originally collected from their natural habitats for such purposes [9]. However, the potential effects of geographic variability on the chemical composition of this species were not considered. The content of secondary metabolites in plants is influenced by environmental factors that induce various types of stress. These stress conditions can be classified as abiotic or biotic. Abiotic stress factors include chemical or physical imbalance in the environment, temperature fluctuations, drought, salinity, light exposure, flooding, nutrient availability, altitude, and phytotoxins. In contrast, biotic stress factors include viroids, fungi, viruses, bacteria, nematodes, oomycetes, protists, mycoplasmas, invertebrates, and competing plants [10–12]. Plants respond to stress at morphological, anatomical, biochemical, and molecular levels [10], which influences the synthesis of secondary metabolites. These changes can, in turn, impact the quality and properties of plants used for medicinal purposes [10–12]. Additionally, different plant parts may produce distinct secondary metabolites [11].

Regarding orchids, environmental factors influence the chemical composition of *Epidendrum ciliare* L. floral fragrances, causing variation among localities [13]. Other studies have described variations in the colors, scents, nectar, and essential oils produced by orchid flowers [14–17]. However, in these plants, studies analyzing chemical composition variation in structures other than their flowers are scarce. A recent study demonstrated geographic variation in the floral morphology of *P. karwinskii* among different localities in Oaxaca, identifying traits that are useful for distinguishing local forms [18]. While populations of this species inhabit different regions with similar vegetation types, variations in elevation may influence environmental conditions. As previously mentioned, such variability can affect the chemical composition of plants. Therefore, in this study, we expect geographical differences (among localities) in the composition of the biocompounds present in the leaves and pseudobulbs of *P. karwinskii*. To explore this further, the objective of this research was to evaluate the variation in the chemical composition of the leaves and pseudobulbs of *P. karwinskii* sampled from six different localities representing its geographical distribution in Oaxaca, Mexico.

2. Results

2.1. Extraction Yields

Table 1 presents the extraction yields of leaves and pseudobulbs of *P. karwinskii* across the different localities. Significant differences in yield were observed both between plant parts and between localities, with leaves generally producing higher percentage yields. Among the leaf extracts, Roaguia showed the highest yield, while San Sebastian de las Grutas displayed the lowest. For pseudobulbs, the highest yield was recorded in Cieneguilla, and the lowest in El Lazo.

Table 1. Extraction yields of leaves and pseudobulbs of *Prosthechea karwinskii* collected in six localities in the state of Oaxaca, Mexico. Yield values are reported as the mean ± standard deviation of three replicates.

Locality, Municipality	Voucher	Altitude	Habitat	Plant Part	Yield (%)
Cieneguilla, Santo Domingo Yanhuitlan	Solano 4246 OAX	2409	Oak-Pine Forest	Leaves	24.3 ± 5.0
				pseudobulbs	23.3 ± 3.0
El Molino, San Pedro y San Pablo Teposcolula	Solano s.n. OAX	2411	Oak-Pine Forest	Leaves	25.0 ± 3.6
				pseudobulbs	14.0 ± 1.0
Roaguia, San Lorenzo Albarradas	Solano 4441 OAX	2450	Oak-Pine Forest	Leaves	30.0 ± 6.0
				pseudobulbs	22.0 ± 4.3
San Sebastian de las Grutas, San Miguel Sola de Vega	Solano 4334 OAX	1720	Oak Forest	Leaves	18.0 ± 6.4
				pseudobulbs	15.0 ± 1.0
El Lazo, San Miguel Sola de Vega	Solano s.n. OAX	1840	Oak Forest	Leaves	21.6 ± 1.5
				pseudobulbs	12.0 ± 1.0
Amialtepec, Santa Catarina Juquila	Solano 1806 OAX	2105	Oak Forest	Leaves	22.3 ± 7.3
				pseudobulbs	19.3 ± 9.0

2.2. Compounds Detected by UHPLC-ESI-qTOF-MS/MS Analysis

Tables 2 and 3 provide detailed information on the retention time (RT) of each compound, its *m/z* value, error, the fragments formed, tentative identification with the corresponding formula, and relative intensity in the extracts of leaves and pseudobulbs of *P. karwinskii* from the six evaluated localities.

Table 2. Summary of the values for the compounds detected by UHPLC-ESI-qTOF-MS/MS in the leaf extracts of *Prosthechea karwinskii* from six localities in Oaxaca, Mexico. PN: Peak number, RT: Retention Time. CIE: Cieneguilla, Santo Domingo Yanhuitlan. EMO: El Molino, San Pedro y San Pablo Teposcolula. ROA: Roaguia, San Lorenzo Albarradas. ELA: El Lazo, San Miguel Sola de Vega. SSG: San Sebastian de las Grutas, San Miguel Sola de Vega. AMI: Amialtepec, Santa Catarina Juquila.

PN	RT	m/z (M-H)	Error (ppm)	Fragments	Compound (Chemical Formula) *	CIE	EMO	ROA	ELA	SSG	AMI
1	0.6	179.0555	2.2	89.0225, 101.0233	D-tagatose [a,b] ($C_6H_{12}O_6$)	15,932	19,041	35,079	21,016	21,294	12,347
2	0.7	191.0557	1.3	85.0293, 87.0078, 111.0443, 127.6945	Quinic Acid [b], [19,20] ($C_7H_{12}O_6$)	273,254	198,656	269,103	258,808	206,762	192,818
3	0.8	133.0140	0.6	115.0032	Malic acid [b], [20–22] ($C_4H_6O_5$)	214,624	148,194	185,426	109,841	165,901	102,223
4	1.1	191.0189	1.3	85.0290	Isocitric acid [b] ($C_6H_8O_7$)	95,658	122,188	107,847	96,552	70,179	65,244
5	1.2	117.0191	1.1	73.0290, 99.0072	Succinic acid [b], [23] ($C_4H_6O_4$)	16,916	9855	15,878	23,167	8367	11,657
6	1.4	243.0609	2.4	110.0247	Uridine [b] ($C_9H_{12}N_2O_6$)	4501	5508	10,765	6992	8776	7766
7	2.3	164.0712	1.6	72.0072, 103.0539, 147.0442	L-(-)-Phenylalanine [a,b] ($C_9H_{11}NO_2$)	10,384	6797	9201	11,228	6829	6067
8	2.7	282.0833	5.7	108.5347, 133.0157, 150.0429	Guanosine [a,b], [23] ($C_{10}H_{13}N_5O_5$)	5940	5697	8539	8199	6524	6254
9	6.0	353.0867	3.4	173.0430, 179.0365, 191.0556	Neochlorogenic acid [b], [19,23] ($C_{16}H_{18}O_9$)	32,839	39,103	73,155	57,463	58,317	36,874
10	6.3	353.0866	3.3	173.0452, 179.0365, 191.0556	Chlorogenic acid [b], [24] ($C_{16}H_{18}O_9$)	82,543	82,525	184,816	144,387	145,486	92,587
11	6.5	609.1438	2.4	300.0266, 301.0335	Rutin [a,b], [19,24–26] ($C_{27}H_{30}O_{16}$)	184,308	162,646	266,057	201,362	96,138	164,073
12	6.6	593.1489	3.1	284.0314, 285.0393	Kaempferol-3-O-rutinoside [a,b] ($C_{27}H_{30}O_{15}$)	244,814	282,177	345,170	356,051	286,492	236,781
13	6.8	187.0970	2.2	97.0653, 125.0963, 169.0889	Azelaic acid [b], [27] ($C_9H_{16}O_4$)	25,721	13,536	14,839	28,839	19,554	23,313

Table 2. Cont.

PN	RT	m/z (M-H)	Error (ppm)	Fragments	Compound (Chemical Formula) *	CIE	EMO	Relative Intensity ROA	ELA	SSG	AMI
14	7.2	201.1127	3.7	139.1128, 183.1021	Sebacic acid [b], [19] ($C_{10}H_{18}O_4$)	20,350	18,634	19,024	27,785	22,903	22,182
15	7.8	327.2157	4.5	171.1023	9,12,13-Trihydroxy-10(E),15(Z)-octadecadienoic acid [19] ($C_{18}H_{32}O_5$)	4569	3034	6130	14,026	9583	6106
16	8.2	329.2321	2.5	171.1023, 229.1436	Pinellic acid [27,28] ($C_{18}H_{34}O_5$)	9574	5869	7558	19,591	13,093	15,067
17	9.2	329.1383	3.9	179.0717	Gibberellin A7 [b] ($C_{19}H_{22}O_5$)	11,912	26,421	19,198	16,877	14,109	27,180
18	11.0	309.2055		291.1951 197.1187	Eicosenoic acid [b], [29] ($C_{20}H_{38}O_2$)	5351	1621	4374	13,745	11,528	8779
19	11.3	311.2215	4.1	293.2102 171.1012 185.1190 197.1171	9-hydroperoxy-10E,12Z-octadecadienoic acid [b]	6665	5344	7237	22,682	19,019	12,549
20	11.5	293.2112	5.6	223.1685, 235.1680, 275.2013	Embelin [27] ($C_{17}H_{26}O_4$)	123,583	26,104	19,527	171,609	128,598	105,675
21	11.8	295.2270	3.8	277.2161 278.2182 183.1384 195.1375	12,13-epoxy-9Z-octadecenoic acid [b] ($C_{18}H_{32}O_3$)	147,178	30,182	24,249	156,918	129,353	115,606

* Superscripts correspond to the sources included in the References section used in the identification of the compounds, except for [a], which was obtained from MetaboBase, and [b], which was obtained from MassBank.

Table 3. Summary of the values for the compounds detected by UHPLC-ESI-qTOF-MS/MS in the pseudobulb extracts of Prosthechea karwinskii from six localities in Oaxaca, Mexico. PN: Peak number, RT: Retention Time. CIE: Cieneguilla, Santo Domingo Yanhuitlan. EMO: El Molino, San Pedro y San Pablo Tepoxcolula. ROA: Roaguia, San Lorenzo Albarradas. ELA: El Lazo, San Miguel Sola de Vega. SSG: San Sebastian de las Grutas, San Miguel Sola de Vega. AMI: Amialtepec, Santa Catarina Juquila.

PN	RT	m/z (M-H)	Error (ppm)	Fragments	Compound (Chemical Formula) *	CIE	EMO	Relative Intensity ROA	ELA	SSG	AMI
1	0.6	179.0554	5.4	89.0225, 101.0233	D-tagatose [a,b,c] ($C_6H_{11}O_6$)	101,469	80,866	76,466	69,493	50,654	49,727
2	0.7	191.0552	4.9	85.0293, 127.0396	Quinic acid [b], [19,20] ($C_7H_{11}O_6$)	67,473	59,443	65,657	62,586	41,841	67,841
3	0.8	133.0142	4.7	115.0032	Malic acid [b], [20–22] ($C_4H_5O_5$)	243,423	331,052	221,701	251,164	148,232	225,769
4	1.1	191.0189	1.3	85.0290	Isocitric acid [b] ($C_6H_8O_7$)	15,627	21,045	11,672	18,533	5750	9828
5	1.2	117.0188	4.3	73.0290, 99.0072	Succinic acid [b], [23] ($C_4H_6O_4$)	9209	7648	12,577	6499	2721	2200
6	6.1	145.0499	3.6	83.0491, 101.0606	3-Methylglutaric acid [a] ($C_6H_{10}O_4$)	9759	12,854	13,009	11,827	12,204	10,093
7	6.1	359.0967	8.9	153.0550 197.0452 198.0479	Deoxyloganic acid [b] ($C_{16}H_{24}O_9$)	5272	17,344	8526	17,829	15,064	6639
8	6.3	353.0861	3.4	173.0430, 179.0365, 191.0556	Neoclorogénic acid [b], [19,23] ($C_{16}H_{18}O_9$)	6892	12,576	13,998	15,489	7185	10,367
9	6.4	353.0854	5.6	173.0452, 179.0365, 191.0556	Clorogénic acid [b], [24] ($C_{16}H_{18}O_9$)	4149	11,438	12,090	13,169	6001	9235
10	6.5	609.1433	5.6	300.0266, 301.0335	Rutin [a,b], [19,24–26] ($C_{27}H_{30}O_{16}$)	15,281	32,907	32,777	45,907	16,529	19,530
11	6.6	607.1272		299.0148 300.0248 301.0338 607.1246 608.1348 609.1458	Diosmin [b], [30] ($C_{28}H_{32}O_{15}$)	9326	27,964	22,464	16,871	7442	7482
12	6.8	187.0967	3.5	97.0653, 125.0963, 169.0889	Azelaic acid [b], [27] ($C_9H_{16}O_4$)	42,386	56,461	40,943	165,652	90,303	122,443
13	7.1	312.1231	2.2	148.0526 178.0502 297.0991	Feruloyltyramine [b] ($C_{18}H_{19}NO_4$)	18,859	7459	952	1344	3131	7354

200

Table 3. Cont.

PN	RT	m/z (M-H)	Error (ppm)	Fragments	Compound (Chemical Formula) *	CIE	EMO	ROA	ELA	SSG	AMI
14	7.2	201.1125	2.8	139.1128, 183.1021	Sebacic acid [b], [19] ($C_{10}H_{18}O_4$)	29,726	40,137	28,181	97,495	55,532	55,222
15	8.2	329.2315	4.7	171.1023, 229.1436	Pinellic acid [27,28] ($C_{18}H_{34}O_5$)	32,062	39,424	20,701	101,363	66,913	92,206
16	8.4	273.1119	3.6	121.0297, 98.0359, 137.0624	Gigantol [31] ($C_{16}H_{18}O_4$)	13,688	10,737	16,754	37,417	36,557	9725
17	11.3	311.2215	4.1	293.2102, 171.1012, 185.1190, 197.1171	9-hydroperoxy-10E,12Z-octadecadienoic acid [b] ($C_{18}H_{32}O_4$)	23,380	65,305	31,562	100,234	76,017	120,398
18	11.5	293.2112	5.6	223.1685, 235.1680, 275.2013	Embelin [27] ($C_{17}H_{26}O_4$)	48,675	97,200	50,540	76,229	54,746	70,383
19	11.8	295.2270	3.8	277.2161, 278.2182, 183.1384, 195.1375	12,13-epoxy-9Z-octadecenoic acid [b] ($C_{18}H_{32}O_3$)	355,025	564,410	245,784	382,412	343,826	405,753
20	11.9	311.2212	4.1	293.2102, 113.0966, 195.1376	13-Hydroperoxyoctadeca-9,11-dienoic acid [b] ($C_{18}H_{32}O_4$)	47,923	327,127	353,295	780,595	593,111	894,679

* Superscripts correspond to the sources included in the References section used in the identification of the compounds, except for [a], which was obtained from MetaboBase, [b], which was obtained from MassBank, and [c], which was obtained from Plant metabolites.

Figure 1 provides comparative chromatograms of the hydroethanolic extracts from leaf (Figure 1A) and pseudobulb (Figure 1B) of *P. karwinskii*, respectively, collected from different localities in Oaxaca. While the same compounds were present in each plant part across all localities, variations in the relative intensity of the peaks were observed. This indicates that the concentration of secondary metabolites differs depending on the geographical origin of the plant material.

2.3. Heatmap Analysis of the Variation in Compounds in Leaves and Pseudobulbs of P. karwinskii

Figure 2A presents a heatmap showing the relative intensity of the compounds identified in the leaf extracts of *P. karwinskii* from six localities in Oaxaca, Mexico. The extract from Roaguia exhibited higher intensities of rutin, chlorogenic acid, uridine, guanosine, neochlorogenic acid, and D-tagatose. El Lazo's extract displayed elevated levels of kaempferol-3-O-rutinoside, azelaic acid, sebacic acid, succinic acid, L-(-)-phenylalanine, embelin, 9,12,13-trihydroxy-10(E),15(Z)-octadecadienoic acid, pinelic acid, eicosenoic acid, 9-hydroperoxy-10E,12Z-octadecadienoic acid, and 12,13-epoxy-9Z-octadecadienoic acid. Cieneguilla extracts showed the highest intensities for quinic acid and malic acid. Meanwhile, the extracts from El Molino and Amialtepec exhibited the highest intensities for isocitric acid and gibberellin A7, respectively. According to the Heatmap colorations, the compound found at a higher intensity in *P. karwinskii* leaves is kaempferol-3-O-rutinoside.

Figure 2B presents a heatmap showing the relative intensity of compounds identified in the extracts of *P. karwinskii* pseudobulbs from six different localities in Oaxaca. The extract from Cieneguilla exhibited higher intensities of D-tagatose and feruloyltyramine. In contrast, the extract from El Molino showed greater intensities of malic acid, isocitric acid, diosmin, embelin, and 12,13-epoxy-9Z-octadecenoic acid. The extract from Roaguia displayed the highest intensities of succinic acid and 3-methylglutaric acid. El Lazo's extract demonstrated the highest intensities for deoxyloganic acid, neochlorogenic acid, chlorogenic acid, rutin, azelaic acid, sebacic acid, pinelic acid, and gigantol. Meanwhile, the extract from Amialtepec had the highest intensities of quinic acid, 9-hydroperoxy-10E,12Z-octadecadienoic acid, and 13-hydroperoxyoctadeca-9,11-dienoic acid. According to the Heatmap staining, the compound found at a higher intensity in the pseudobulbs of *P. karwinskii* is 13-hydroperoxyoctadeca-9,11-dienoic acid.

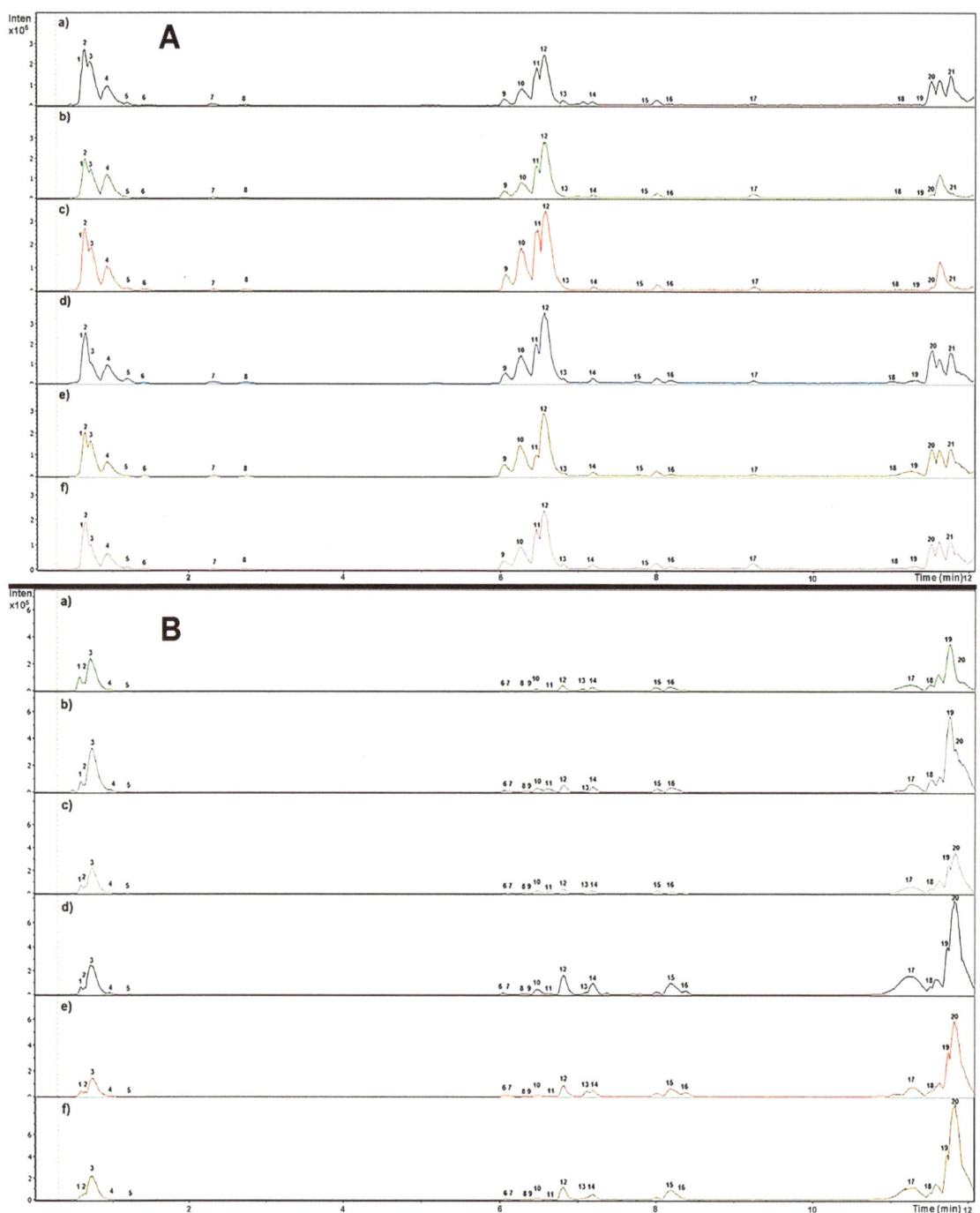

Figure 1. Chromatograms obtained with UHPLC-ESI-qTOF-MS/MS from the leaf (**A**) and pseudobulb (**B**) extracts of *Prosthechea karwinskii* from six localities of Oaxaca, Mexico. (**a**) Cieneguilla, Santo Domingo Yanhuitlan, (**b**) El Molino, San Pedro y San Pablo Teposcolula, (**c**) Roaguia, San Lorenzo Albarradas; (**d**) El Lazo, San Miguel Sola de Vega, (**e**) San Sebastian de las Grutas, San Miguel Sola de Vega, (**f**) Amialtepec, Santa Catarina Juquila.

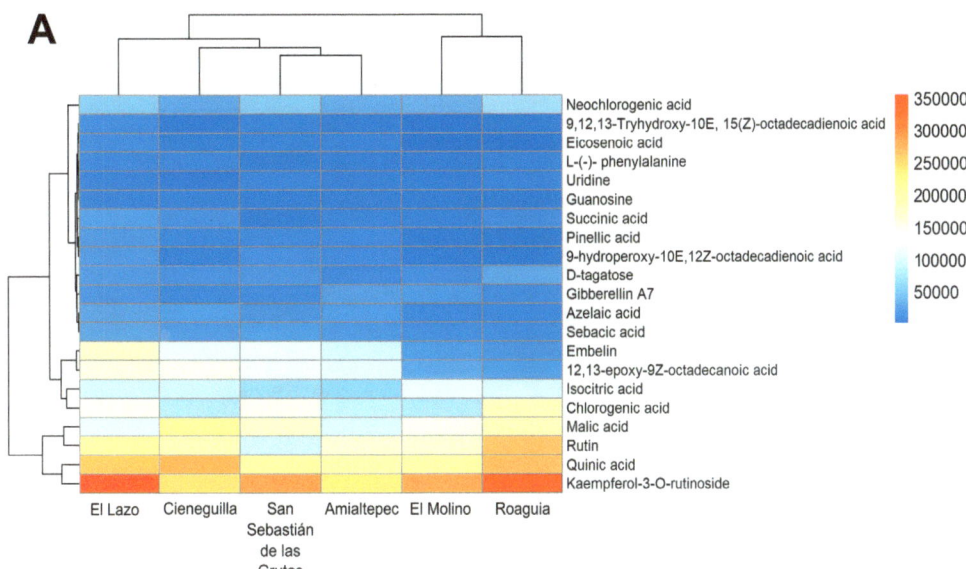

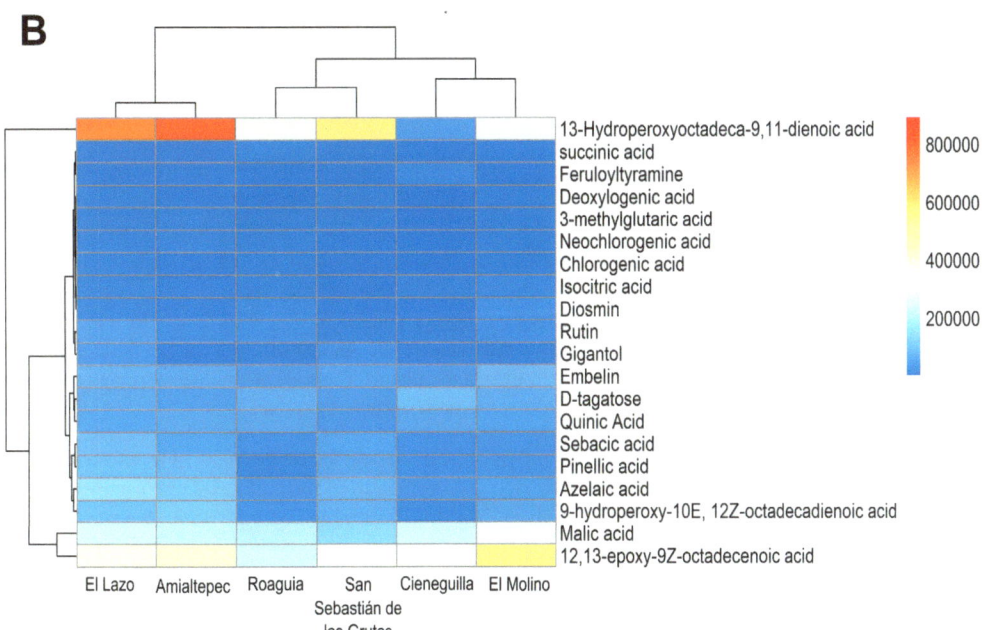

Figure 2. Heatmap showing the relative intensity of compounds identified in the leaf (**A**) and pseudobulb (**B**) extracts of *Prosthechea karwinskii* from six localities of Oaxaca, Mexico. The relative intensity scale on the right ranges from red, indicating the highest values, to blue, indicating the lowest.

Figure 3 presents a comparative heatmap showing the relative intensity of compounds identified in the leaf and pseudobulb extracts of *P. karwinskii* across six localities. In the leaf extracts, 21 compounds were detected, seven of which were unique to this part of the plant: uridine, L-(-)-phenylalanine, guanosine, kaempferol-3-O-rutinoside, 9,12,13-trihydroxy-10(E),15(Z)-octadecadienoic acid, eicosenoic acid, and 9-hydroperoxy-10E,12Zoctadecadienoic acid. Conversely, 20 compounds were identified in the pseudobulb extracts, six of which were unique to this part of the plant: 3-methylglutaric acid, deoxyloganic acid, diosmin, feruloyltyramine, gigantol, and 13-hydroperoxyoctadeca-9,11-dienoic acid. Fifteen compounds were common to both plant parts, though their intensities varied depending on both the plant part and the locality.

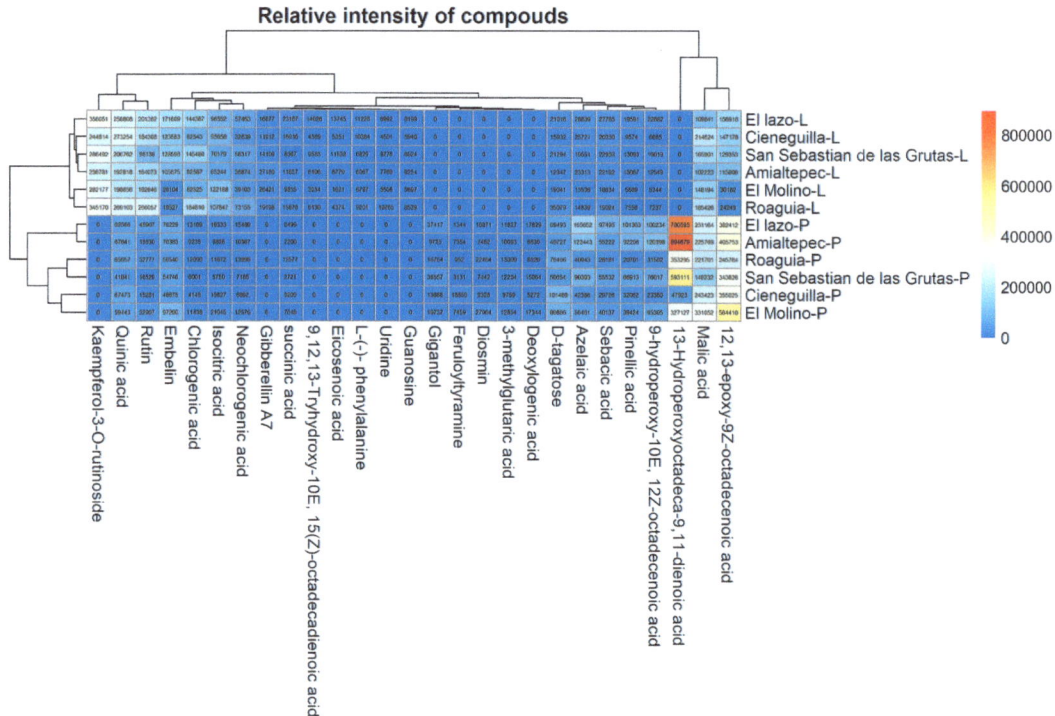

Figure 3. Heatmap showing the relative intensity of compounds identified in the leaf (L) and pseudobulb (P) extracts of *Prosthechea karwinskii* from six localities in Oaxaca, Mexico. Each cell value represents the relative intensity of a compound. The scale on the right indicates relative intensity, with red denoting the highest values and blue the lowest.

Figure 4 shows a dendrogram illustrating the clustering of detected compounds in the leaf and pseudobulb extracts of *P. karwinskii* from six localities. The clustering is determined by the plant part from which the extracts were obtained: pseudobulb extracts are grouped separately from leaf extracts.

Within the pseudobulb group, two subgroups are observed. One subgroup includes the localities in southern Oaxaca (El Lazo and Amialtepec), while the other comprises Roaguia (Valles Centrales), San Sebastian de las Grutas (southern Oaxaca), and the Mixteca localities (El Molino and Cieneguilla). The grouping of leaf extracts reveals a different pattern. Cieneguilla forms a subgroup with El Lazo, San Sebastian de las Grutas, and Amialtepec, while another subgroup consists solely of El Molino and Roaguia.

Table 4 highlights the compounds found in the leaves and pseudobulbs of *P. karwinskii* that are associated with biological activities relevant to the medicinal use of this species, including their ability to improve glucose metabolism, anti-inflammatory properties, and antioxidant activity. The table also indicates the sampled locality where each compound is found in the highest intensity.

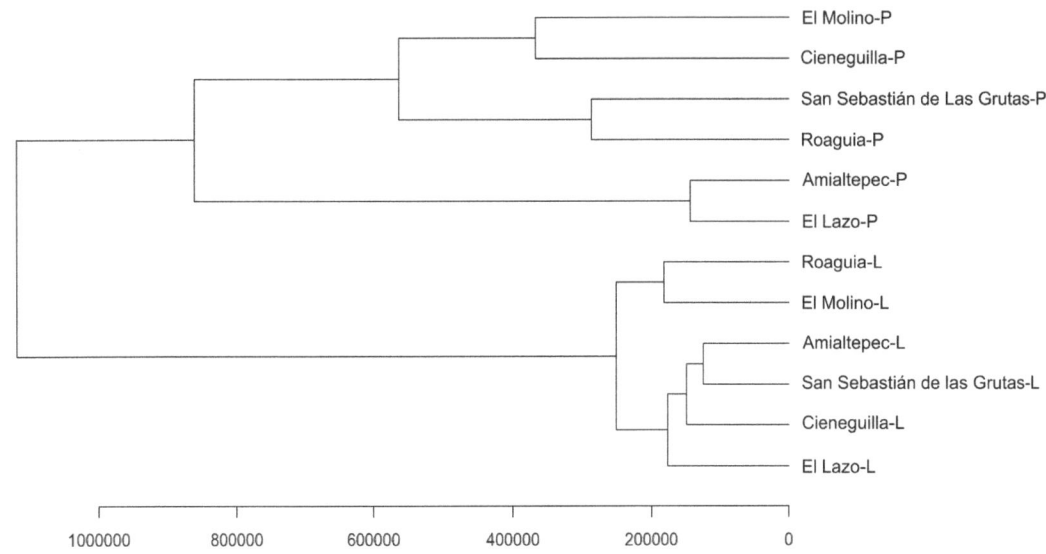

Figure 4. Dendrogram generated using a hierarchical clustering algorithm and clustering index, illustrating the grouping of localities based on the detected chemical compounds in the leaf (L) and pseudobulb (P) extracts of *Prosthechea karwinskii*.

Table 4. Compounds with biological activities related to the medicinal use of leaves and pseudobulbs of *Prosthechea karwinskii* collected from different localities in Oaxaca, Mexico. (-) means that the compound is not present in that part of the plant.

Compound	Type of Compound	Locality with Higher Relative Intensity		Biological Activity (Reference)
		Leaves	Pseudobulbs	
Quinic Acid	Carboxylic acid, cyclitol	Cieneguilla	Amialtepec	Anti-inflammatory [32]
Uridine	Pyrimidine nucleoside	Roaguia	-	Anti-inflammatory [33]
Guanosine	Purine nucleoside	Roaguia	-	Anti-inflammatory [33] Antioxidant [34]
Neochlorogenic acid	Phenolic acid	Roaguia	El Lazo	Anti-inflammatory [35,36] Anti-inflammatory and Antioxidant [37] Improves glucose metabolism [38,39]
Chlorogenic acid	Phenolic acid	Roaguia	El Lazo	Anti-inflammatory [35,36,40] Antioxidant [41,42] Improves glucose metabolism [43]
Rutin	Flavonoid glycoside	Roaguia	El Lazo	Antioxidant and Anti-inflammatory [44,45] Antioxidant [46]
Kaempferol-3-O-rutinoside	Flavonoid glycoside	El Lazo	-	Antioxidant [47] Anti-inflammatory [48]
Azelaic acid	Dicarboxylic acid	El Lazo	El Lazo	Antioxidant [49,50] Anti-inflammatory [51]
Pinellic acid	Trihydroxyoctadecenoic acid	El Lazo	El Lazo	Anti-inflammatory [52] Improves glucose metabolism [53]
Embelin	Benzoquinone	El Lazo	El Molino	Anti-inflammatory [54] Antioxidant and Anti-inflammatory [55] Improves glucose metabolism and Anti-inflammatory [56]
Diosmin	Flavonoid glycoside	-	El Molino	Antioxidant and Anti-inflammatory [57,58] Improves glucose metabolism, Anti-inflammatory and Antioxidant [59]
Gigantol	Bibenzyl	-	El Lazo	Anti-inflammatory [60,61]

3. Discussion

According to Cruz-García et al. [1], both the leaves and pseudobulbs of *P. karwinskii* are used in traditional medicine for treating diabetes. Additionally, pseudobulbs are utilized to treat coughs, as well as wounds and burns. These medicinal uses are linked to inflammatory processes. For example, respiratory issues are associated with bronchial inflammation [33]; wound and burn healing involves various types of inflammatory cells [62]; and both inflammation and oxidative stress are implicated in the pathophysiology of diabetes [63]. Furthermore, both plant parts exhibit a similar ability to inhibit reactive oxygen species (ROS) [2], which could be related to the similarity in their chemical profiles. Elevated glucose levels in diabetes are known to increase ROS production [64], and excessive ROS can lead to oxidative stress. Inflammation also contributes to oxidative stress, and in turn, oxidative stress exacerbates inflammation [41]. Thus, the plant's effects on ROS and oxidative stress may be linked to its traditional medicinal uses.

Since the traditional medicinal uses of *P. karwinskii* leaves and pseudobulbs are related to inflammation, oxidative stress, and glucose metabolism disorders [1,2], and both plant parts contain compounds with these biological activities (Table 4), it can be inferred that extracts from localities where these compounds are more abundant may have greater pharmacological potential.

Environmental conditions play a crucial role in determining the concentration of secondary metabolites in medicinal plants, as plants produce specific types and amounts of metabolites necessary to counteract the environmental stress they experience [11]. According to Sampaio et al. [65], drought stress can reduce photosynthetic activity, increasing ROS production and stimulating the production of phenolic compounds as a defense mechanism. Heat stress during drought can affect metabolic regulation, water permeability, and CO_2 levels, thereby enhancing antioxidant properties and increasing carbohydrate availability. Additionally, UV-B radiation stress can lead to the production of phenolic compounds that absorb and/or dissipate solar energy, helping to prevent the formation of free radicals and other oxidative species. Phenolic compounds, such as flavonoids, accumulate in plants in response to abiotic stress and are considered a fundamental mechanism for the elimination of ROS [10].

The review by Punetha et al. [10], analyzing the effect of abiotic stress on secondary metabolite production in medicinal plants, indicates that phenolic acids and flavonoids are enhanced by increased UV radiation, drought, high temperatures, and salinity. In addition, the review by Pant et al. [11] found that these compounds are promoted by high temperatures, elevated concentrations of carbon dioxide and ozone, and UV radiation. Additionally, rutin and chlorogenic acid, compounds identified in *P. karwinskii*, are favored by higher light intensity.

Sun et al. [66] compared the metabolites of Lushan Yunwu tea leaves from different geographical regions, linking the observed variations to altitude. At higher altitudes, diffuse light promotes nitrogen metabolism, amino acid production, and the synthesis of nitrogen compounds, while increasing the content of organic acids and flavonoids, though total phenol content decreases. Our results did not show a similar pattern for *P. karwinskii* extracts. However, chlorogenic acid was found in higher concentrations in the leaf extract from the highest altitude locality (Cieneguilla), but not in the pseudobulb extract. Conversely, rutin content was higher in the pseudobulb extract from the lower-altitude locality (San Sebastian de las Grutas), but not in the leaf extract.

In contrast to the previous analysis [2] of Soxhlet extracts of leaves and pseudobulbs of *P. karwinskii* collected from churches in Zaachila Oaxaca, in the present study, more compounds were identified for each part of the plant. This variation could be due to the method used for the extraction of the compounds, as well as to the loss or degradation of compounds due to the time they remained on the altars of the churches because of their ceremonial use and the manipulation of the plants. In the previous study, the origin of the plant was not considered, as they had been collected from different parts of the state for ceremonial purposes.

This study represents the first analysis of chemical composition variations in the orchid *P. karwinskii*, as well as other Mexican orchid species from different localities. The findings highlight the need for continued research into how these chemical variations correlate with physical and environmental factors of the localities where the orchid is found. The results presented here, along with those obtained in previous studies on *P. karwinskii*, lead to questions for future research. The habitat in each of the sampling localities for *P. karwinskii* is very similar, with this epiphytic orchid growing on the same host trees, *Quercus* sp. Could the variation in chemical composition detected among the species' localities be related to the environmental factors that characterize them? Several compounds identified in the leaves and pseudobulbs of *P. karwinskii* are known for their anti-inflammatory, antioxidant, and glucose metabolism-regulating activities. When extracting compounds from these localities, could any of them show better performance in evaluating these activities using an *in vivo* model?

It is also interesting to compare these results with those of Santos-Escamilla et al. [18], who documented variations in the floral morphology of *P. karwinskii* across different populations in the state of Oaxaca, where Albarradas presented the most differentiated population of *P. karwinskii*. Notably, the leaf extract from the locality of Roaguia, in the municipality of San Lorenzo Albarradas, contained a higher intensity of compounds with reported biological activity associated with the plant's traditional uses (Table 4). This suggests a possible link between floral morphological variation and chemical composition in this species. Could the compounds identified here serve as chemical markers to recognize intraspecific variation in *P. karwinskii*?

4. Materials and Methods

4.1. Plant Material

Plant material was collected from six localities representing the distribution of *P. karwinskii* in Oaxaca, Mexico. For this, a scientific collector's permit was obtained from the Mexican Ministry of Environment and Natural Resources (00851/06 and 02228/17). Table 1 provides information on these localities, while the map in Figure 5 indicates its geographic location. Sampling was conducted from January to February 2018. From each adult plant (>15 pseudobulbs), only divisions containing 2–3 pseudobulbs with their leaves were collected, comprising 95–790 g and 376–3900 g of fresh material for leaves and pseudobulbs, respectively, per locality, leaving the remaining parts of the plant on its host tree to ensure its survival. Additionally, a voucher specimen was collected from each locality, herborized, and deposited in the OAX Herbarium of Instituto Politécnico Nacional (see Table 1). The taxonomic identity of the species was validated by one of the authors (R.S.), following Villasenor's checklist in the assigned scientific name [67].

4.2. Conditioning of Plant Material and Obtaining Extracts

The plant material from each locality was separated into leaves and pseudobulbs. Each part was individually weighed, washed with water, and dried at 50 °C. After drying, the material was ground and sieved (Figure 6). Ultrasound-assisted extraction was performed on the pulverized leaf and pseudobulb samples from each locality, following the method described by Barragán-Zárate et al. [4]. The extraction was carried out using an ultrasonic processor (VCX 750, Cientifica Senna, Mexico City, Mexico) set at a frequency of 20 kHz and an amplitude of 30%. A 50% (v/v) ethanol–water mixture was used as the solvent, with a sample-to-solvent ratio of 1 g:18 mL. The extraction was performed at 40 °C for 20 min, with cycles of 10 s of operation followed by 5 s of rest.

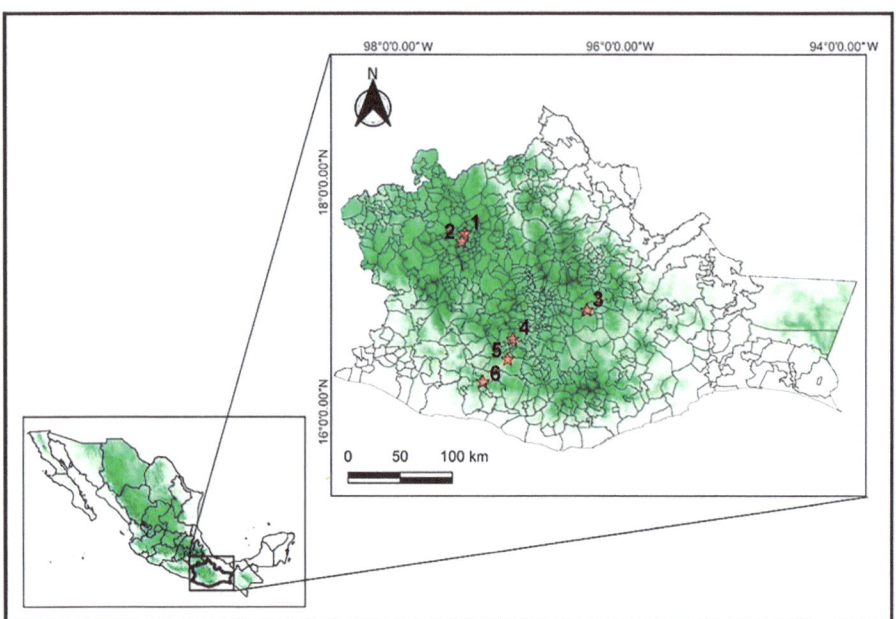

Figure 5. Map of the six localities of *Prosthechea karwinskii* in Oaxaca, Mexico sampled in this study. 1: Cieneguilla, Santo Domingo Yanhuitlan, 2: El Molino, San Pedro y San Pablo Teposcolula, 3: Roaguia, San Lorenzo Albarradas, 4: El Lazo, San Miguel Sola de Vega, 5: San Sebastian de las Grutas, San Miguel Sola de Vega, 6: Amialtepec, Santa Catarina Juquila.

Figure 6. Process scheme for the extraction of compounds from leaves and pseudobulbs of *Prosthechea karwinskii*.

4.3. Compound Identification by UHPLC-ESI-qTOF-MS/MS

To determine the compound profile of the leaves and pseudobulbs from each locality, the method described by Barragán-Zárate et al. [2] was followed. One mg of each extract was analyzed using an ultra-high-performance liquid chromatography (UHPLC) system (Thermo Scientific Ultimate 3000, Waltham, MA, USA) coupled with an Impact II mass spectrometer (Bruker, Billerica, MA, USA). The mass spectrometer operated with electrospray ionization (ESI) and quadrupole time-of-flight (qTOF) detection. The analysis was performed using a Thermo Scientific Acclaim 120 C18 column (2.2 µm, 120 Å, 50 × 2.1 mm). The mobile phase consisted of: (A) 0.1% formic acid in water and (B) acetonitrile. The gradient elution program was as follows: 0% B (0–2 min), 1% B (2–3 min), 3% B (3–4 min),

32% B (4–5 min), 36% B (5–6 min), 40% B (6–8 min), 45% B (8–9 min), 80% B (9–11 min), and 0% B (12–14 min). The injection temperature was 25 °C, and the flow rate was set at 0.35 mL/min. The analysis was conducted in negative electrospray ionization mode at 0.4 bar (5.8 psi) with autoMSMS, in the mass range 50–700 m/z. The ionization of the capillary voltage (Vcap) was 2700 V. Data obtained from the UHPLC-ESI-qTOF-MS/MS analysis were processed using DataAnalysis software 3.1 (Bruker) and imported into MetaboScape 3.0 (Bruker) for further analysis, including peak extraction. For each compound, retention time (RT), m/z, fragmentation patterns, and peak intensities were recorded. Compounds were identified by comparing their exact mass and MS/MS spectra with those in the MetaboBase 3.0 (Bruker) and MassBank libraries, as well as by consulting relevant scientific literature.

4.4. Statistical Analysis

The heatmap is a visualization that simultaneously reveals row and column hierarchical cluster structure in a data matrix [68]. Each heatmap was generated based on the relative intensity values of the peaks from the analyses [69,70] to visualize differences in chemical compounds between leaves and pseudobulbs, as well as across localities. The heatmaps were created with the Pheatmap 1.0.12 package [71] with complete linkage as the agglomeration method. Additionally, to explore the chemical composition dataset and assess whether it could be summarized into several clusters that were similar within themselves and distinct from others, a cluster analysis was conducted.

The hierarchical cluster analysis was performed with the hclust function of the stats package, with the complete linkage as the agglomeration method, where the distance or similarity between clusters occurs by the maximum distance or minimal similarity among its components [72]. Both analyses were carried out using RStudio 2023.03.0 [73] and using Euclidian distance.

5. Conclusions

Prosthechea karwinskii exhibits variation in the chemical composition of extracts derived from its leaves and pseudobulbs, as well as differences based on the locality of origin of the plant material. The chemical composition is more consistent among localities for the same plant part than between leaves and pseudobulbs from the same locality. Leaf extracts from El Lazo and Roaguia showed higher intensities of compounds previously reported to have biological activities associated with the medicinal use of the species. Similarly, pseudobulb extracts from El Lazo also exhibited a higher intensity of these bioactive compounds. Extracts from these localities demonstrate greater potential for addressing health issues related to glucose metabolism, inflammation, and oxidative stress. In terms of extraction yield, Roaguia produced the highest yields for both leaves and pseudobulbs, while El Lazo had the lowest yield for pseudobulbs. The findings of this study raise questions for future research on *P. karwinskii*. Despite growing in similar habitats, variations in chemical composition among localities may be influenced by environmental factors. Additionally, some identified compounds have bioactive properties, prompting the question of whether extracts from certain localities may perform better within *in vivo* models. Finally, these compounds could serve as chemical markers for intraspecific variation in *P. karwinskii*.

Author Contributions: Conceptualization, R.S. and L.L.-R.; Methodology, B.A.P.L.; Data Curation, G.S.B.-Z. and M.C.-M.; Writing—original draft preparation, G.S.B.-Z.; Writing—review and editing, G.S.B.-Z., R.S., L.L.-R., B.A.P.L. and M.C.-M.; Visualization, R.S. and L.L.-R.; Supervision, R.S. and L.L.-R.; funding acquisition, R.S. and L.L.-R. All authors have read and agreed to the published version of the manuscript.

Funding: This research was funded by CONACYT (project 270428, postdoctoral grants 392827 and 612625), and Instituto Politécnico Nacional (projects SIP-20161270, SIP-2016RE/50, SIP-20220649, SIP-20220893, SIP-20231610, SIP-20240944, SIP-20241893).

Data Availability Statement: Data are contained within the article.

Acknowledgments: We thank the authorities of the different communities who allowed the collection of the plant material. The comments of two anonymous reviewers helped improve the manuscript.

Conflicts of Interest: The authors declare no conflicts of interest.

Abbreviations

NSAIDs: non-steroidal anti-inflammatory drugs, UHPLC-ESI-qTOF-MS/MS: ultra-high performance liquid chromatography, combined with a mass spectrometer with electrospray ionization and quadrupole time-of-flight, analysis masses, RT: Retention time, PN: Peak number, ROS: reactive oxygen species.

References

1. Cruz-Garcia, G.; Solano-Gomez, R.; Lagunez-Rivera, L. Documentation of the medicinal knowledge of *Prosthechea karwinskii* in a Mixtec community in Mexico. *Rev. Bras. Farmacogn.* **2014**, *24*, 153–158. [CrossRef]
2. Barragán-Zarate, G.S.; Lagunez-Rivera, L.; Solano, R.; Carranza-Álvarez, C.; Hernández-Benavides, D.M.; Vilarem, G. Validation of the Traditional Medicinal Use of a Mexican Endemic Orchid (*Prosthechea Karwinskii*) through UPLC-ESI-QTOF-MS/MS Characterization of Its Bioactive Compounds. *Heliyon* **2022**, *8*, e09867. [CrossRef] [PubMed]
3. Rojas-Olivos, A.; Solano-Gómez, R.; Alexander-Aguilera, A.; Jiménez-Estrada, M.; Zilli-Hernández, S.; Lagunez-Rivera, L. Effect of *Prosthechea karwinskii* (Orchidaceae) on obesity and dyslipidemia in Wistar rats. *Alex. J. Med.* **2017**, *53*, 311–315. [CrossRef]
4. Barragán-Zarate, G.S.; Lagunez-Rivera, L.; Solano, R.; Pineda-Peña, E.A.; Landa-Juárez, A.Y.; Chávez-Piña, A.E.; Carranza-Álvarez, C.; Hernández-Benavides, D.M. *Prosthechea karwinskii*, an orchid used as traditional medicine, exerts anti-inflammatory activity and inhibits ROS. *J. Ethnopharmacol.* **2020**, *253*, 112632. [CrossRef]
5. Barragán-Zarate, G.S.; Alexander-Aguilera, A.; Lagunez-Rivera, L.; Solano, R.; Soto-Rodríguez, I. Bioactive compounds from *Prosthechea karwinskii* decrease obesity, insulin resistance, pro-inflammatory status, and cardiovascular risk in Wistar rats with metabolic syndrome. *J. Ethnopharmacol.* **2021**, *279*, 114376. [CrossRef]
6. Barragán-Zarate, G.S.; Lagunez-Rivera, L.; Alexander-Aguilera, A.; Solano, R.; Vilarem, G. Extraction, Characterization, and Nutraceutical Potential of *Prosthechea karwinskii* Orchid for Insulin Resistance and Oxidative Stress in Wistar Rats. *Foods* **2024**, *13*, 2432. [CrossRef] [PubMed]
7. Lagunez-Rivera, L.; Barragan-Zarate, G.S.; Solano, R.; Alexander-Aguilera, A.; Chavez-Piña, A.E. Mexican Orchid (*Prosthechea karwinskii*) and Use in Cardiovascular Protection: Cellular and Physiological Aspects. In *Ancient and Traditional Foods, Plants, Herbs and Spices Used in Cardiovascular Health and Disease*; Taylor and Francis: Abingdon, UK, 2023. [CrossRef]
8. Barragán-Zarate, G.S.; Lagunez-Rivera, L.; Solano, R.; Hernández Juárez, J.; López Pérez, A.; Carranza Álvarez, C. *Prosthechea karwinskii* leaves extract with potential for the treatment of atherothrombosis. *Biotecnia* **2024**, *26*, e2378. [CrossRef]
9. Solano, R.; Cruz-Lustre, G.; Martinez-Feria, A.; Lagunez-Rivera, L. Plantas utilizadas en la celebración de la semana santa en Zaachila, Oaxaca, México. *Polibotánica* **2010**, *29*, 263–279.
10. Punetha, A.; Kumar, D.; Suryavanshi, P.; Padalia, R.C.; Katanapalya-Thimmaiah, V. Environmental abiotic stress and secondary metabolites production in medicinal plants: A review. *Tarim Bilim. Derg.* **2022**, *28*, 351–362. [CrossRef]
11. Pant, P.; Pandey, S.; Dall'Acqua, S. The Influence of environmental conditions on secondary metabolites in medicinal plants: A literature review. *Chem. Biodivers.* **2021**, *18*, e2100345. [CrossRef] [PubMed]
12. Shahrajabian, M.H.; Kuang, Y.; Cui, H.; Fu, L.; Sun, W. Metabolic Changes of Active Components of Important Medicinal Plants on the Basis of Traditional Chinese Medicine under Different Environmental Stresses. *Curr. Org. Chem.* **2023**, *27*, 782–806. [CrossRef]
13. Moya, S.; Ackerman, J.D. Variation in the floral fragrance of *Epidendrum ciliare* (Orchidaceae). *Nord. J. Bot.* **1993**, *13*, 41–47. [CrossRef]
14. Brzosko, E.; Mirski, P. Floral nectar chemistry in orchids: A short review and meta-analysis. *Plants* **2021**, *10*, 2315. [CrossRef] [PubMed]
15. Dormont, L.; Joffard, N.; Schatz, B. Intraspecific variation in floral color and odor in orchids. *Int. J. Plant Sci.* **2019**, *180*, 1036–1058. [CrossRef]

16. Perkins, J.; Hayasi, T.; Peakal, R.; Flematti, G.R.; Bohman, B. The volatile chemistry of orchid pollination. *Nat. Prod. Rep.* **2023**, *40*, 819–839. [CrossRef]
17. Robustelli della Cuna, F.S.; Cortis, P.; Esposito, F.; De Agosti, A.; Sottani, C.; Sanna, C. Chemical composition of essential oil from four sympatric orchids in NW-Italy. *Plants* **2022**, *11*, 826. [CrossRef] [PubMed]
18. Santos-Escamilla, M.H.; Cruz-Lustre, G.; Cuéllar-Martínez, M.; Lagunez-Rivera, L.; Solano, R. Variation in the Floral Morphology of *Prosthechea karwinskii* (Orchidaceae), a Mexican Endemic Orchid at Risk. *Plants* **2024**, *13*, 1984. [CrossRef]
19. Ma, T.; Lin, J.; Gan, A.; Sun, Y.; Sun, Y.; Wang, M.; Wan, M.; Yan, T.; Jia, Y. Qualitative and quantitative analysis of the components in flowers of Hemerocallis citrina Baroni by UHPLC–Q-TOF-MS/MS and UHPLC–QQQ-MS/MS and evaluation of their antioxidant activities. *J. Food Compos. Anal.* **2023**, *120*, 105329. [CrossRef]
20. Ghareeb, M.A.; Sobeh, M.; Rezq, S.; El-Shazly, A.M.; Mahmoud, M.F.; Wink, M. HPLC-ESI-MS/MS profiling of polyphenolics of a leaf extract from *Alpinia zerumbet* (Zingiberaceae) and its anti-inflammatory, anti-nociceptive, and antipyretic activities in vivo. *Molecules* **2018**, *23*, 3238. [CrossRef]
21. El-Hawary, S.S.; Sobeh, M.; Badr, W.K.; Abdelfattah, M.A.O.; Ali, Z.Y.; El-Tantawy, M.E.; Rabeh, M.A.; Wink, M. HPLC-PDA-MS/MS profiling of secondary metabolites from *Opuntia ficus-indica* cladode, peel and fruit pulp extracts and their antioxidant, neuroprotective effect in rats with aluminum chloride induced neurotoxicity. *Saudi J. Biol. Sci.* **2020**, *27*, 2829–2838. [CrossRef]
22. Ragheb, A.Y.; Masoud, M.A.; El Shabrawy, M.O.; Farid, M.M.; Hegazi, N.M.; Mohammed, R.S.; Marzouk, M.M.; Aboutabl, M.E. MS/MS-based molecular networking for mapping the chemical diversity of the pulp and peel extracts from Citrus japonica Thunb.; in vivo evaluation of their anti-inflammatory and anti-ulcer potential. *Sci. Afr.* **2023**, *20*, e01672. [CrossRef]
23. Zhang, Q.Q.; Dong, X.; Liu, X.G.; Gao, W.; Li, P.; Yang, H. Rapid separation and identification of multiple constituents in Danhong Injection by ultra-high performance liquid chromatography coupled to electrospray ionization quadrupole time-of-flight tandem mass spectrometry. *Chin. J. Nat. Med.* **2016**, *14*, 147–160. [CrossRef] [PubMed]
24. de Souza, M.P.; Bataglion, G.A.; da Silva, F.M.A.; de Almeida, R.A.; Paz, W.H.P.; Nobre, T.A.; Marinho, J.V.N.; Salvador, M.J.; Fidelis, C.H.V.; Acho, L.D.R.; et al. Phenolic and aroma compositions of pitomba fruit (*Talisia esculenta* Radlk.) assessed by LC-MS/MS and HS-SPME/GC-MS. *Food Res. Int.* **2016**, *83*, 87–94. [CrossRef]
25. Zhao, M.; Linghu, K.G.; Xiao, L.; Hua, T.; Zhao, G.; Chen, Q.; Xiong, S.; Shen, L.; Yu, J.; Hou, X.; et al. Anti-inflammatory/antioxidant properties and the UPLC-QTOF/MS-based metabolomics discrimination of three yellow camellia species. *Food Res. Int.* **2022**, *160*, 111628. [CrossRef]
26. Sun, X.; Xue, S.; Cui, Y.; Li, M.; Chen, S.; Yue, J.; Gao, Z. Characterization and identification of chemical constituents in Corni Fructus and effect of storage using UHPLC-LTQ-Orbitrap-MS. *Food Res. Int.* **2023**, *164*, 112330. [CrossRef]
27. Dahibhate, N.L.; Dwivedi, P.; Kumar, K. GC–MS and UHPLC-HRMS based metabolite profiling of *Bruguiera gymnorhiza* reveals key bioactive compounds. *S. Afr. J. Bot.* **2022**, *149*, 1044–1048. [CrossRef]
28. Cong, W.; Schwartz, E.; Tello, E.; Simons, C.T.; Peterson, D.G. Identification of non-volatile compounds that negatively impact whole wheat bread flavor liking. *Food Chem.* **2021**, *364*, 130362. [CrossRef] [PubMed]
29. Cebo, M.; Schlotterbeck, J.; Gawaz, M.; Chatterjee, M.; Lämmerhofer, M. Simultaneous targeted and untargeted UHPLC-ESI-MS/MS method with data-independent acquisition for quantification and profiling of (oxidized) fatty acids released upon platelet activation by thrombin. *Anal. Chim. Acta* **2020**, *1094*, 57–69. [CrossRef]
30. Taamalli, A.; Arráez-Román, D.; Abaza, L.; Iswaldi, I.; Fernández-Gutiérrez, A.; Zarrouk, M.; Segura-Carretero, A. LC-MS-based metabolite profiling of methanolic extracts from the medicinal and aromatic species *Mentha pulegium* and *Origanum majorana*. *Phytochem. Anal.* **2015**, *26*, 320–330. [CrossRef] [PubMed]
31. Singh, D.; Kumar, S.; Pandey, R.; Hasanain, M.; Sarkar, J.; Kumar, B. Bioguided chemical characterization of the antiproliferative fraction of edible pseudo bulbs of *Malaxis acuminata* D. Don by HPLC-ESI-QTOF-MS. *Med. Chem. Res.* **2017**, *26*, 3307–3314. [CrossRef]
32. Jang, S.A.; Park, D.W.; Kwon, J.E.; Song, H.S.; Park, B.; Jeon, H.; Sohn, E.H.; Koo, H.J.; Kang, S.C. Quinic acid inhibits vascular inflammation in TNF-α-stimulated vascular smooth muscle cells. *Biomed. Pharmacother.* **2017**, *96*, 563–571. [CrossRef]
33. Luo, Y.; Chen, H.; Huang, R.; Wu, Q.; Li, Y.; He, Y. Guanosine and uridine alleviate airway inflammation via inhibition of the MAPK and NF-κB signals in OVA-induced asthmatic mice. *Pulm. Pharmacol. Ther.* **2021**, *69*, 102049. [CrossRef]
34. Courtes, A.A.; de Carvalho, N.R.; Gonçalves, D.F.; Hartmann, D.D.; da Rosa, P.C.; Dobrachinski, F.; Franco, J.L.; de Souza, D.O.G.; Soares, F.A.A. Guanosine protects against Ca^{2+}-induced mitochondrial dysfunction in rats. *Biomed. Pharmacother.* **2019**, *111*, 1438–1446. [CrossRef] [PubMed]
35. Wu, T.Y.; Chen, C.C.; Lin, J.Y. Anti-inflammatory in vitro activities of eleven selected caffeic acid derivatives based on a combination of pro−/anti-inflammatory cytokine secretions and principal component analysis—A comprehensive evaluation. *Food Chem.* **2024**, *458*, 140201. [CrossRef] [PubMed]
36. Freitas, M.; Ribeiro, D.; Janela, J.S.; Varela, C.L.; Costa, S.C.; da Silva, E.T.; Fernandes, E.; Roleira, F.M.F. Plant-derived and dietary phenolic cinnamic acid derivatives: Anti-inflammatory properties. *Food Chem.* **2024**, *459*, 140080. [CrossRef]

37. Park, S.Y.; Jin, M.L.; Yi, E.H.; Kim, Y.; Park, G. Neochlorogenic acid inhibits against LPS-activated inflammatory responses through up-regulation of Nrf2/HO-1 and involving AMPK pathway. *Environ. Toxicol. Pharmacol.* **2018**, *62*, 1–10. [CrossRef]
38. Ma, Y.; Gao, M.; Liu, D. Chlorogenic acid improves high fat diet-induced hepatic steatosis and insulin resistance in mice. *Pharm. Res.* **2015**, *32*, 1200–1209. [CrossRef] [PubMed]
39. Singh, A.K.; Rana, H.K.; Singh, V.; Chand Yadav, T.; Varadwaj, P.; Pandey, A.K. Evaluation of antidiabetic activity of dietary phenolic compound chlorogenic acid in streptozotocin induced diabetic rats: Molecular docking, molecular dynamics, in silico toxicity, in vitro and *in vivo* studies. *Comput. Biol. Med.* **2021**, *134*, 104462. [CrossRef]
40. Xu, P.; Chen, S.; Fu, Q.; Zhu, S.; Wang, Z.; Li, J. Amelioration effects of chlorogenic acid on mice colitis: Anti-inflammatory and regulation of gut flora. *Food Biosci.* **2024**, *61*, 104942. [CrossRef]
41. Song, D.; Zhang, S.; Chen, A.; Song, Z.; Shi, S. Comparison of the effects of chlorogenic acid isomers and their compounds on alleviating oxidative stress injury in broilers. *Poult. Sci.* **2024**, *103*, 103649. [CrossRef] [PubMed]
42. Elbasan, F.; Arikan, B.; Ozfidan-Konakci, C.; Tofan, A.; Yildiztugay, E. Hesperidin and chlorogenic acid mitigate arsenic-induced oxidative stress via redox regulation, photosystems-related gene expression, and antioxidant efficiency in the chloroplasts of *Zea mays*. *Plant Physiol. Biochem.* **2024**, *208*, 108445. [CrossRef] [PubMed]
43. Ghorbani, A. Biomedicine & Pharmacotherapy Mechanisms of antidiabetic effects of flavonoid rutin. *Biomed. Pharmacother. J.* **2017**, *96*, 305–312. [CrossRef]
44. Akash, S.R.; Tabassum, A.; Aditee, L.M.; Rahman, A.; Hossain, M.I.; Hannan, M.A.; Uddin, M.J. Pharmacological insight of rutin as a potential candidate against peptic ulcer. *Biomed. Pharmacother.* **2024**, *177*, 116961. [CrossRef]
45. Li, F.; Zhang, L.; Zhang, X.; Fang, Q.; Xu, Y.; Wang, H. Rutin alleviates Pb-induced oxidative stress, inflammation and cell death via activating Nrf2/ARE system in SH-SY5Y cells. *NeuroToxicology* **2024**, *104*, 1–10. [CrossRef] [PubMed]
46. Londero, É.P.; Bressan, C.A.; Pês, T.S.; Saccol, E.M.H.; Baldisserotto, B.; Finamor, I.A.; Pavanato, M.A. Rutin-added diet protects silver catfish liver against oxytetracycline-induced oxidative stress and apoptosis. *Comp. Biochem. Physiol. Part-C Toxicol. Pharmacol.* **2021**, *239*, 108848. [CrossRef]
47. Wang, Y.; Tang, C.; Zhang, H. Hepatoprotective effects of kaempferol 3-O-rutinoside and kaempferol 3-O-glucoside from *Carthamus tinctorius* L. on CCl4-induced oxidative liver injury in mice. *J. Food Drug Anal.* **2015**, *23*, 310–317. [CrossRef] [PubMed]
48. Hu, W.H.; Dai, D.K.; Zheng, B.Z.Y.; Duan, R.; Chan, G.K.L.; Dong, T.T.X.; Qin, Q.W.; Tsim, K.W.K. The binding of kaempferol-3-O-rutinoside to vascular endothelial growth factor potentiates anti-inflammatory efficiencies in lipopolysaccharide-treated mouse macrophage RAW264.7 cells. *Phytomedicine* **2021**, *80*, 153400. [CrossRef]
49. Akamatsu, H.; Komura, J.; Asada, Y.; Miyachi, Y.; Niwa, Y. Inhibitory effect of azelaic acid on neutrophil functions: A possible cause for its efficacy in treating pathogenetically unrelated diseases. *Arch. Dermatol. Res.* **1991**, *283*, 162–166. [CrossRef]
50. Jiang, Y.; Qin, Y.; Chandrapala, J.; Majzoobi, M.; Brennan, C.; Sun, J.; Zeng, X.A.; Sun, B. Investigation of interactions between Jiuzao glutelin with resveratrol, quercetin, curcumin, and azelaic and potential improvement on physicochemical properties and antioxidant activities. *Food Chem. X* **2024**, *22*, 101378. [CrossRef]
51. Sieber, M.A.; Hegel, J.K.E. Azelaic acid: Properties and mode of action. *Ski. Pharmacol. Physiol.* **2013**, *27*, 9–17. [CrossRef]
52. Choi, H.G.; Park, Y.M.; Lu, Y.; Chang, H.W.; Na, M.; Lee, S.H. Inhibition of prostaglandin D2 production by trihydroxy fatty acids isolated from *Ulmus davidiana* var. *japonica*. *Phytother. Res.* **2013**, *27*, 1376–1380. [CrossRef] [PubMed]
53. Durg, S.; Veerapur, V.P.; Neelima, S.; Dhadde, S.B. Antidiabetic activity of *Embelia ribes*, embelin and its derivatives: A systematic review and meta-analysis. *Biomed. Pharmacother.* **2017**, *86*, 195–204. [CrossRef] [PubMed]
54. Bai, X.; Wang, J.; Ding, S.; Yang, S.; Pei, B.; Yao, M.; Zhu, X.; Jiang, M.; Zhang, M.; Mu, W.; et al. Embelin protects against apoptosis and inflammation by regulating PI3K/Akt signaling in IL-1β-stimulated human nucleus pulposus cells. *Tissue Cell* **2023**, *82*, 102089. [CrossRef] [PubMed]
55. Goal, A.; Raj, K.; Singh, S.; Arora, R. Protective effects of Embelin in Benzo[α]pyrene induced cognitive and memory impairment in experimental model of mice. *Curr. Res. Neurobiol.* **2024**, *6*, 100122. [CrossRef]
56. Rajasekar, M.; Baskaran, P.; Mary, J.; Sivakumar, M.; Selvam, M. Revisiting diosmin for their potential biological properties and applications. *Carbohydr. Polym. Technol. Appl.* **2024**, *7*, 100419. [CrossRef]
57. Mohtadi, S.; Shariati, S.; Mansouri, E.; Khodayar, M.J. Nephroprotective effect of diosmin against sodium arsenite-induced renal toxicity is mediated via attenuation of oxidative stress and inflammation in mice. *Pestic. Biochem. Physiol.* **2023**, *197*, 105652. [CrossRef]
58. Zhao, J.; Zhang, M.; Zhang, H.; Wang, Y.; Chen, B.; Shao, J. Diosmin ameliorates LPS-induced depression-like behaviors in mice: Inhibition of inflammation and oxidative stress in the prefrontal cortex. *Brain Res. Bull.* **2024**, *206*, 110843. [CrossRef]
59. Mustafa, S.; Akbar, M.; Khan, M.A.; Sunita, K.; Parveen, S.; Pawar, J.S.; Massey, S.; Agarwal, N.R.; Husain, S.A. Plant metabolite diosmin as the therapeutic agent in human diseases. *Curr. Res. Pharmacol. Drug Discov.* **2022**, *3*, 100122. [CrossRef] [PubMed]
60. Fang, H.; Hu, X.; Wang, M.; Wan, W.; Yang, Q.; Sun, X.; Gu, Q.; Gao, X.; Wang, Z.; Gu, L.; et al. Anti-osmotic and antioxidant activities of gigantol from *Dendrobium aurantiacum* var. *denneanum* against cataractogenesis in galactosemic rats. *J. Ethnopharmacol.* **2015**, *172*, 238–246. [CrossRef] [PubMed]

61. Chowdhury, R.; Bhuia, S.; Rakib, A.I.; Al Hasan, S.; Shill, M.C.; El-Nashar, H.A.S.; El-Shazly, M.; Islam, M.T. Gigantol, a promising natural drug for inflammation: A literature review and computational based study. *Nat. Prod. Res.* **2024**, 1–17. [CrossRef]
62. Wang, P.; Huang, S.; Hu, Z.; Yang, W.; Lan, Y.; Zhu, J. In situ formed anti-inflammatory hydrogel loading plasmid DNA encoding VEGF for burn wound healing. *Acta Biomater.* **2019**, *100*, 191–201. [CrossRef]
63. Halim, M.; Halim, A. The effects of inflammation, aging and oxidative stress on the pathogenesis of diabetes mellitus (type 2 diabetes). *Diabetes Metab. Syndr. Clin. Res. Rev.* **2019**, *13*, 1165–1172. [CrossRef] [PubMed]
64. de Lima Júnior, J.P.; Franco, R.R.; Saraiva, A.L.; Moraes, I.B.; Espindola, F.S. *Anacardium humile* St. Hil as a novel source of antioxidant, antiglycation and α-amylase inhibitors molecules with potential for management of oxidative stress and diabetes. *J. Ethnopharmacol.* **2021**, *268*, 113667. [CrossRef]
65. Sampaio, B.L.; Edrada-Ebel, R.; Da Costa, F.B. Effect of the environment on the secondary metabolic profile of *Tithonia diversifolia*: A model for environmental metabolomics of plants. *Sci. Rep.* **2016**, *6*, 29265. [CrossRef] [PubMed]
66. Sun, Q.; Wu, F.; Wu, W.; Yu, W.; Zhang, G.; Huang, X.; Hao, Y.; Luo, L. Identification and quality evaluation of Lushan Yunwu tea from different geographical origins based on metabolomics. *Food Res. Int.* **2024**, *186*, 114379. [CrossRef]
67. Villaseñor, J.L. Checklist of the native vascular plants of Mexico. *Rev. Mex. De Biodivers.* **2016**, *87*, 559–902. [CrossRef]
68. Wilkonson, L.; Friendly, M. The history of the cluster heat map. *Am. Stat.* **2008**, *63*, 179–184. [CrossRef]
69. Gerry, C.J.; Hua, B.K.; Wawer, M.J.; Knowles, J.P.; Nelson, S.D., Jr.; Verho, O.; Dandapani, S.; Wagner, B.K.; Clemons, P.A.; Booker-Milburn, K.I.; et al. Real-time biological annotation of synthetic compounds. *J. Am. Chem. Soc.* **2016**, *138*, 8920–8927. [CrossRef]
70. Auman, J.T.; Boorman, G.A.; Wilson, R.E.; Travlos, G.S.; Paules, R.S. Heat Map Visualization of high-density clinical chemistry data. *Physiol. Genom.* **2007**, *31*, 352–356. [CrossRef]
71. Kolde, R. *Pheatmap: Pretty Heatmaps*, version 1.0.12; R. package; R Foundation for Statistical Computing: Vienna, Austria, 2019.
72. Everitt, B. *Cluster Analysis*; Heinemann Educ. Books: London, UK, 1974.
73. R Core Team. *R: A Language and Environment for Statistical Computing*; R Foundation for Statistical Computing: Vienna, Austria, 2023.

Disclaimer/Publisher's Note: The statements, opinions and data contained in all publications are solely those of the individual author(s) and contributor(s) and not of MDPI and/or the editor(s). MDPI and/or the editor(s) disclaim responsibility for any injury to people or property resulting from any ideas, methods, instructions or products referred to in the content.

MDPI AG
Grosspeteranlage 5
4052 Basel
Switzerland
Tel.: +41 61 683 77 34

Plants Editorial Office
E-mail: plants@mdpi.com
www.mdpi.com/journal/plants

Disclaimer/Publisher's Note: The title and front matter of this reprint are at the discretion of the Guest Editors. The publisher is not responsible for their content or any associated concerns. The statements, opinions and data contained in all individual articles are solely those of the individual Editors and contributors and not of MDPI. MDPI disclaims responsibility for any injury to people or property resulting from any ideas, methods, instructions or products referred to in the content.

www.ingramcontent.com/pod-product-compliance
Lightning Source LLC
LaVergne TN
LVHW072335090526
838202LV00019B/2425